CLEAN WATER: FACTORS THAT INFLUENCE ITS AVAILABILITY, QUALITY AND ITS USE

CLEAN WATER: FACTORS THAT INFLUENCE ITS AVAILABILITY, QUALITY AND ITS USE

INTERNATIONAL CLEAN WATER CONFERENCE
held in La Jolla, California, 28–30 November 1995

Edited by

WINSTON CHOW
ROBERT W. BROCKSEN
JOE WISNIEWSKI

with editorial assistance of

Billy M. McCormac
Editor-in-Chief, WASP

Reprinted from *Water, Air, and Soil Pollution*
Volume 90, Nos. 1–2, July 1996

KLUWER ACADEMIC PUBLISHERS
DORDRECHT / LONDON / BOSTON

A C.I.P. Catalogue record for this book is available from the Library of Congress

ISBN-13: 978-94-010-6619-8   e-ISBN-13: 978-94-009-0299-2
DOI: 10.1007/978-94-009-0299-2

Published by Kluwer Academic Publishers,
P.O. Box 17, 3300 AA Dordrecht, The Netherlands.

Kluwer Academic Publishers incorporates
the publishing programmes of
D. Reidel, Martinus Nijhoff, Dr W. Junk and MTP Press.

Sold and distributed in the U.S.A. and Canada
by Kluwer Academic Publishers,
101 Philip Drive, Norwell, MA 02061, U.S.A.

In all other countries, sold and distributed
by Kluwer Academic Publishers Group,
P.O. Box 322, 3300 AH Dordrecht, The Netherlands.

Printed on acid-free paper

PREFACE   VII
ACKNOWLEDGEMENTS   IX

PART I
CONFERENCE SUMMARY STATEMENT

R.W. BROCKSEN, W. CHOW, E.D. DAUGHERTY, Y.G. MUSSALLI, J. WISNIEWSKI
  and A.L. WOODIS / Clean Water: Factors that Influence its Availability, Quality
  and its Use: Summary of the International Water Conference   3–7

PART II
WATER RESOURCE OVERVIEWS

S. PECK / Managing and Protecting Our Water Resources   11–20
R. BROCKSEN, W. CHOW and K. CONNOR / Addressing Electric Utility Surface Water
  Challenges   21–29
C. LOHSE-HANSON / Lake Superior Binational Program: The Role of Electric Utilities   31–40
J.A. VEIL and D.O. MOSES / Consequences of Proposed Changes to Clean Water Act
  Thermal Discharges   41–52

PART III
ECOLOGICAL / HEALTH RISKS

c. SEIGNEUR, E. CONSTANTINOU and L. LEVIN / Multipathway Health Risk
  Assessment of Power Plant Water Discharges   55–64
C.W. CHEN, J. HERR, R.A. GOLDSTEIN, F.J. SAGONA, K.E. RYLANT and G.E.
  HAUSER / Watershed Risk Analysis Model for TVA's Holston River Basin   65–70
S. FERSON, L.R. GINZBURG and R.A. GOLDSTEIN / Inferring Ecological Risk from
  Toxicity Bioassays   71–82
C. ARQUIETT, M. GERKE and I. DATSKOU / Evaluation of Contaminated Groundwater
  Cleanup Objectives   83–92
G.L. BOWIE, J.G. SANDERS, G.F. RIEDEL, C.C. GILMOUR, D.L. BREITBURG, G.A.
  CUTTER and D.B. PORCELLA / Assessing Selenium Cycling and
  Accumulation in Aquatic Ecosystems   93–104
D.W. RODGERS, J. SCHRÖDER and L. VEREECKEN SHEEHAN / Comparison of
  Daphnia Magna, Rainbow Trout and Bacterial-based Toxicity Tests of Ontario
  Hydro Aquatic Effluents   105–112
G.P. BEHRENS, D.A. ORR, R.G. WETHEROLD and B.T. O'NEIL / Use of the Pisces
  Database: Power Plant Aqueous Stream Compositions   113–122
C.S. LEW, W.B. MILLS, K.J. WILKINSON and S.A. GHERINI / RIVRISK: A Model to
  Assess Potential Human Health and Ecological Risks from Chemical and
  Thermal Releases into Rivers   123–132
I. DATSKOU and K. NORTH / Risks Due to Groundwater Contamination at a Plutonium
  Processing Facility   133–141
K. LINDQUIST, M. MCGEE and L. COLE / TVA-EPRI River Resource Aid (Terra)
  Reservoir and Power Operations Decision Support System   143–150

PART IV
CONTROL STRATEGIES

M. ZIMMERMAN and C. MURPHY / Best Management Practices for Storm Water at
  Industrial Facilities   153–162
R.F. MADDALONE / Coal Sorbent System for the Extraction and Disposal of Heavy
  Metals and Organic Compounds   163–171

B. NOTT, K.A. SELBY, T. BRICE, D. GOLDSTROHM, D. MORRIS, S.R. PATE, D. SMAY and T. SPAULDING / The EPRI State-of-the-Art Cooling Water Treatment Research Project: A Tailored Collaboration Program   173–181

K.A. SELBY, P.R. PUCKORIUS and K.R. HELM / The Use of Reclaimed Water in Electric Power Stations and Other Industrial Facilities   183–193

R.E. BROSHEARS / Reactive Solute Transport in Acidic Streams   195–204

J.W. GOODRICH-MAHONEY / Constructed Wetland Treatment Systems Applied Research Program at the Electric Power Research Institute   205–217

D.W. RODGERS, D.W. EVANS and L. VEREECKEN SHEEHAN / Toxicity Reduction of Ontario Hydro Radioactive Liquid Waste   219–229

I. MOCH, F. DEPENBROCK and Y. MUSSALLI / Electricity and Water Desalination: Separate Sites Offer Value   231–241

PART V
MEASURING QUALITY

R.F. MADDALONE, J.W. SCOTT, J.K. RICE and B.R. NOTT / Impact on Discharge Monitoring of Recent EPA Initiatives in Water Quality Measurement   245–255

D.K. NORDSTROM / Trace Metal Speciation in Natural Waters: Computational vs. Analytical   257–267

M.H. MACH, B. NOTT, J.W. SCOTT, R.F. MADDALONE and N.T. WHIDDON / Metal Speciation: Survey of Environmental Methods of Analysis   269–279

A.J. HOROWITZ, K.R. LUM, J.R. GARBARINO, G.E.M. HALL, C. LEMIEUX and C.R. DEMAS / The Effect of Membrane Filtration on Dissolved Trace Element Concentrations   281–294

N.V. EKKAD and C.O. HUBER / Residual Sulfite After Dechlorination of Water   295–300

PART VI
WATER RESOURCES

G.H. LEAVESLEY, S.L. MARKSTROM, M.S. BREWER and R.J. VIGER / The Modular Modeling System (MMS) – The Physical Process Modeling Component of a Database-Centered Decision Support System for Water and Power Management   303–311

J.H. COPELAND / Validation and Sensitivity of a Convective Precipitation Model for Mountainous Areas   313–320

J.M. NELSON / Predictive Techniques for River Channel Evolution and Maintenance   321–333

K.L. BRUBAKER and A. RANGO / Response of Snowmelt Hydrology to Climate Change   335–343

Conference Participants   345–350

Author Index   351

Key Word Index   353–354

PREFACE

Approximately 95% of total water use in the United States satisfies three primary requirements: domestic/municipal, industrial and irrigation. The electric utility industry is among the prominent users of water for power plant needs and hydroelectric power generation, therefore, of water quality is important to the utility industry and the obligation to protect the integrity of water resources is of paramount importance.

Over the past few years, the focus of water quality management has shifted from point source discharge control of pollutants towards managing outcomes and impacts to the environment. The debate continues regarding the use of risk and cost-benefit assessments to balance human and ecological health against sustaining industrial growth and, for the electric utility industry, energy development. Toxics have surfaced to become one of this decade/s key issues. However, an informed clean water debate demands a better understanding of our environmental compartments, e.g., water sources, the industrial/power generation process systems, water impoundments, and basins, watersheds, wetlands, streams, estuaries and oceans.

The 1995 International Water Conference was held in La Jolla, California during 28–30 November 1995 with approximately 140 individuals attending from different affiliations: 64 attendees representing 43 USA and foreign utilities; 19 attendees representing various federal and state government organizations; and 61 attendees representing various worldwide universities, environmental organizations, associations and private sector entitities.

The Conference provided an opportunity for international information exchange on new research applications and timely policy issues affecting the availability, use and quality of water for power generation by the electric utility industry. Emphasis was placed on the future of water-related research and compliance with the U.S. Clean Water Act, related statutes and current and emerging international requirements. The conference/s five key objectives were as follows:

- To share current research and regulatory findings germane to utilities;
- To meet and maintain contact with key individuals in industry, the research community and regulatory agencies dealing with water issues;
- To address the diversity of technical and policy issues associated with water use by the utility industry;
- To define data gaps and R&D needs on water issues;
- To disseminate the results of this conference to all interested parties;
- To increase awareness of water use issues on an international scale.

Overall, we feel that the objectives have been met. Some 48 presentations were made at the Conference and 32 peer-reviewed papers are being published in this Special Issue of the Journal of *Water, Air and Soil Pollution* .

ACKNOWLEDGEMENTS

We acknowledge the contribution of all authors who made presentations and completed papers for the conference, as well as all individuals who served as technical reviewers for the papers. We further acknowledge the efforts of all of the session chairpersons who kept the discussions active in a very timely manner.

We acknowledge the main sponsor, the Electric Power Research Institute (EPRI); additional support was also received from: the California State Water Resources Control Board, OLPA; the Edison Electric Institute; the Empire State Electric Energy Research Corporation (ESEERCo); the International Energy Agency - Coal Research, U.K.; Power Gen, U.K.; and the U.S. DOE - PETC.

We specifically acknowledge the following individuals for their vital contributions: Christine Lillie of EPRI for overall administrative and logistical coordination of the Conference; Yusef Musalli and Val Chepeleff of Stone and Webster Engineering for assisting in soliciting the paper presenters and authors and participating in the initial screening of submitted papers; E. David Daugherty of Global Management Services and Amy Woodis of Woodis Associates for their overall assistance with the conference logistics; Steve Ragone of Ragone Associates for his solicitation of keynote speakers and technical papers for this conference; and Dr. Billy McCormac and Dee McCormac of the Journal of *Water, Air and Soil Pollution* for both the technical and copy-editing of the papers. Finally, we acknowledge Joe Wisniewski for overall editorial coordination and Josephine Gurnee-Hoffman for the overall administrative coordination and clerical production of this publication.

Winston Chow
Robert Brocksen

PART I

CONFERENCE SUMMARY STATEMENT

CLEAN WATER: FACTORS THAT INFLUENCE ITS AVAILABILITY, QUALITY AND ITS USE: SUMMARY OF THE INTERNATIONAL WATER CONFERENCE

R.W. BROCKSEN[1], W.CHOW[1], E D. DAUGHERTY[2], Y.G. MUSSALLI[3], J. WISNIEWSKI[4], A.L. WOODIS[5]

[1]Electric Power Research Institute, 3412 Hillview Avenue, Palo Alto, California 94304-1395, USA, [2]Global Management Services, 4779 Highway 58, Suite 7, Chattanooga, Tennessee 37416, USA, [3]Y. Mussalli, Stone & Webster Engineering Corporation, 245 Summit Street, Boston, Massachusetts 02210, USA, [4]J. Wisniewski, 6862 McLean Province Circle, Falls Church, Virginia 22043, USA, [5]A.L.Woodis, Woodis Associates, 811 Emerald Bay Lane, Foster City, California 94404, USA.

Abstract. The International Water Conference held in La Jolla, California during 28-30 November 1995, had an objective to provide an opportunity for international information exchange on new research applications and timely policy issues effecting the availability, use and quality of water for power generation. This Conference included the topical areas, the use of water and the management of chemical quality in process wastewaters; physical and biological transformation due to discharges in open surface waters, watersheds and wetlands; the role of risk and cost-benefit assessments; and discussion of the Clean Water Act containments. This Conference Summary contains an overview of the present state of the knowledge regarding the use of water by the electric utility industry and potential future issues as discussed in papers presented at the Conference and a summary of the most important findings and conclusions.

Key words. Clean water, fresh water, water resources, ecological risks, health risks, control strategies, measurement quality, modeling, electric utilities.

1. Introduction

The 1995 International Clean Water Conference, sponsored by the Electric Power Research Institute (EPRI), was held in La Jolla, California, during 28-30 November 1995. The focus of the conference was on issues relating to water quality, quantity and use by the electric utility industry. More than 70% of the earth's surface is covered by water; however, less than 1% of all the earth's fresh water is available for use by humankind. While it is generally agreed that enough fresh water is available to meet the world's present needs, the uneven distribution of fresh water will become a greater challenge in the future primarily due to population growth and economic development.

Ninety-five percent of total water use in the United States satisfies three primary requirements: domestic/municipal, industrial and agricultural irrigation. The electric utility industry is responsible for a large percentage of overall daily water withdrawals. Within the United States, regional disparities in fresh water supplies, most notably between the East and West, affect availability and apportionment among all users.

Water, Air and Soil Pollution **90**: 3-7, 1996.
© 1996 *Kluwer Academic Publishers.*

2. Regulatory Background

The passage of the Federal Water Pollution Control Act and the Clean Water Act of 1972 (CWA) introduced federally mandated control over water pollution in the United States. Effluent limits for point source discharges and water quality standards are among the most important issues affecting the electric utility industry. More stringent regulations and increasing public concerns regarding a broad range of surface water management issues have the potential to substantially impact electric utility operations and costs.

Reauthorizations of the CWA and the Safe Drinking Water Act and recent implementation of the Great Lakes Water Quality Initiative all have the potential to impact the electric utility industry. For example, § 316(b) of the CWA requires the "best technology available" for minimizing adverse fish impacts for water intake structures. The current study phase by the US Environmental Protection Agency (EPA) is scheduled for completion in 1997, with final Agency action scheduled for 2001. CWA § 316(a) currently allows a variance for thermal discharges into the nation's rivers and lakes. The 102nd Congress (1991-92) Senate debated S.1081 that, if passed, would have limited mixing zones and eliminated the § 316(a) variance. Estimated costs to the electric utility industry to retrofit for cooling towers and diffusers at existing once-through power plants if variances were disallowed range from $21.4 - $24.4 billion. Owners of the retrofitted facilities would also spend an estimated $10 - $18.4 billion in additional fuel costs and from $1.2 - $5.3 billion for additional capacity to compensate for energy losses. The §316(a) variance program has resulted in considerable cost savings to ratepayers and, to date, there appears to be little scientific justification to remove the variance.

3. Ecological and Health Risks

Guidance promulgated in 1995 by the EPA prescribes methods for ecological risk evaluation that are slightly modified from human health risk analysis methodologies. Although the goal of ecological risk evaluation is to protect populations, simply adopting methods from human health risk analysis, which focus on individuals, by EPA seems to be inappropriate for assessing ecological risk. Unlike human health risk assessment, where a single death is a cause for concern, ecological risk assessment normally does not elicit this level of concern over the same event.

Decision-making processes for water quality management are changing from the "command and control" approach of the federal government to consensus building at the local level. Instead of rigid technology-based requirements, stakeholders at the local level are becoming more involved in setting goals and strategies to address discharges on a watershed scale. Efforts have begun to develop a conceptual risk analysis framework for watershed management of point and non-point source pollution. Model development is being conducted which calculates water hydrology, waste load, quality and suitability for fisheries habitat. The goal is to produce models which can be used to provide cost-effective methods for improving water quality, including market-based pollution trading.

Policy makers have increasingly asked for integrated large-scale approaches to environmental management that allow for methodologies that are both environmentally responsible and economically viable. One such approach is the multi-attribute model used at the Oak Ridge National Laboratory's Center for Risk Management. The model is an attempt to effectively prioritize environmental projects for both cost and benefits associated with various cleanup tasks. Within this framework, multiple receptor and various pathways of exposure to specific contaminants are explored. Different levels of cleanup, of course, result in varying levels of risk and cost.

4. Control Strategies

Environmental regulations regarding the discharge of waste water from industrial facilities are remarkably similar throughout the world, since most of them are patterned after those used in the US and Europe. Major global lending institutions have added such environmental requirements and guidelines to facilities funded, regardless of location. Discharge regulations generally take into account the usage as well as the quality of receiving waters. A growing trend is the use of risk analyses to assess the potential impacts of waste water discharges and to define priorities for control strategies. Climate, fresh water availability, other users and any sensitive environmental receptors combine to influence the selection of the waste water treatment employed, in addition to facility economics, regional socioeconomic and other community concerns.

The addition of air pollution controls can sometimes raise concerns of consequent land and water degradation resulting from the release or disposal of control technology waste by-products. Contaminated storm water has also been identified as a significant source of degradation to waterways. In the US, the CWA was modified in 1987 to address storm water discharges. The EPA has finalized Phase II storm water regulations to cover "small facilities." Large facilities, such as electric utilities, are subject to the original Phase I regulations, where permits for storm water discharges are covered under the National Pollution Discharge Elimination System (NPDES) program.

The Comprehensive Environmental Response, Compensation and Liability Act (CERCLA) mandates that the EPA select remedies that utilize permanent solutions, maximize resource recovery, and/or significantly reduce the volume, toxicity or mobility of hazardous substances. Methods are being developed that combine treatment processes at electric utility facilities that meet these criteria. For example, one process combines in-situ flushing with an innovative iron coprecipitation process to remove arsenic from soil and ground water.

There has also been recent progress in the use of reclaimed water (treated municipal sewage plant effluent) for use in the cooling systems of electric utility plants and other industrial facilities. By using reclaimed water, facilities can reduce their need for water from higher quality sources, which consequently can be reserved for other purposes, such as drinking water. The use of reclaimed water may, however, have possible negative effects on equipment related to corrosion, deposition and biological fouling.

Recent programs to develop constructed wetland treatment systems as a cost-effective technology to treat metal-bearing electric utility aqueous discharges have begun. Data collected from such projects as well as existing data can be used to develop design criteria for the effective use of this technology to ensure that electric utilities will continue to meet imposed effluent discharge limits.

5. Measuring Quality

The 1987 CWA established the framework for water quality-based discharge permitting of toxic substances. Before 1987, NPDES permit limits for toxic substances were seldom based on in-stream water quality. Since 1987, however, states have established in-stream water quality criteria for the protection of aquatic life. These criteria are used in conjunction with background concentrations, receiving water body flow and discharge flow to develop actual discharge limits. As water quality-based effluent limitations become more established, electric utilities will face significant technological and regulatory challenges. For example, the determination of background concentrations can have significant implications in establishing standards at a particular site. Receiving water bodies may also be impacted by other discharges. As such, electric utilities will have to work closely with regulators to increase the chance that imposed regulations are scientifically defensible and technically appropriate.

In addition, human health and aquatic life requirements can result in the establishment of effluent limitations on power plant aqueous discharges that are below the quantification limits of many currently available sampling and analytical procedures. To address this situation for compliance monitoring, the EPA has developed analytical methods that lower detection and quantification levels by one or two orders of magnitude. To date, the EPA has not supported these methods with data to indicate that the stated detection limits can be met. Future work will begin to focus on speciating chemical constituents such as trace metals. Chemical speciation will facilitate more realistic assessments of health and ecological risks and better define the performance of water treatment technologies.

6. Water Resource Modeling

The interdisciplinary nature and complexity of water resource problems require the use of modeling approaches that incorporate knowledge from a broad range of scientific disciplines. EPRI and a number of government agencies are supporting the development of computer-based, integrated modeling systems to address these issues.

For example, in 1992 the Tennessee Valley Authority (TVA), in conjunction with EPRI, initiated the development of the Power and Reservoir System Scheduling Model (PRSYM) system. A major use of the model will be real-time scheduling for water allocation and hydropower dispatching that will ultimately lead to better management of the TVA water control system.

Another example is the Modular Modeling System, an integrated system of computer software that is being developed to provide the research and operational framework needed to support development testing and evaluation of physical-process algorithms. This system is being coupled with PRSYM to provide a decision-support system for making complex operational decisions on multipurpose reservoir systems and watersheds in Colorado, New Mexico, Arizona and Utah.

Other models designed to address global and regional climate issues are also under development since accurate analysis of snow cover data in mountainous regions is important for predicting the timing and quantity of water supply in these areas.

7. Summary

The spectrum of problems and solutions to water use by the electric utility industry was presented in the papers presented at the conference, with many of these peer-reviewed papers appearing in this Special Issue of The Journal of *Water, Air and Soil Pollution* . We believe this to be an exciting beginning to a continuing forum regarding the use of our water resources in an environmentally sound and productive manner.

Acknowledgements

We acknowledge the contribution of all authors who made presentations and completed papers for the conference, as well as all individuals who served as technical reviewers for the papers. We further acknowledge the efforts of all of the session chairpersons who kept the discussions active in a very timely manner.

We acknowledge the main sponsor, the Electric Power Research Institute (EPRI); additional support was also received from: the California State Water Resources Control Board, OLPA; the Edison Electric Institute; the Empire State Electric Energy Research Corporation (ESEERCo); the International Energy Agency - Coal Research, U.K.; Power Gen, U.K.; and the U.S. DOE - PETC.

PART II

WATER RESOURCE OVERVIEWS

MANAGING & PROTECTING OUR WATER RESOURCES

S. PECK

*Environment Group, Electric Power Research Institute, 3412 Hillview Ave.,
Palo Alto, CA 94303, U.S.A.*

Abstract. A number of factors have an impact on the quantity and quality of freshwater, particularly global population growth and economic development. In the U.S., the electric utility industry is a large consumer of water resources. Understanding the factors which affect water resource management is important for sustainable development as well as ecosystem management. The effects of past and anticipated legislation in the U.S. addressing clean water issues important to the electric utility industry are discussed.

Keywords: water resources; sustainable development; ecosystem management; legislation

1. Water - A Resource in Short Supply

The focus of this conference is on clean water and the factors that impact its availability and quality. While many of us treat water as a commodity, plentiful and relatively inexpensive, from a global perspective, usable, clean water can be considered a "scarce" resource. For example, even though more than 70% of the earth's surface is covered by water, 97% is contained in the oceans. Of the remaining 3% that is freshwater, approximately 76.5% of that is frozen in the polar ice caps or in glaciers. In addition, 23% of all freshwater is groundwater. Therefore, less than 1% is available for Earth's biota, including humankind, as shown in Figure 1. This remaining freshwater is stored in lakes, soils, the atmosphere and rivers. However, it is generally agreed that the total amount of freshwater available for use is adequate for the world's needs, at least for now and the immediate future. A major problem, however, is uneven distribution of this resource and this will become an even greater challenge as the worldwide human population continues to increase (Research & Exploration, 1993).

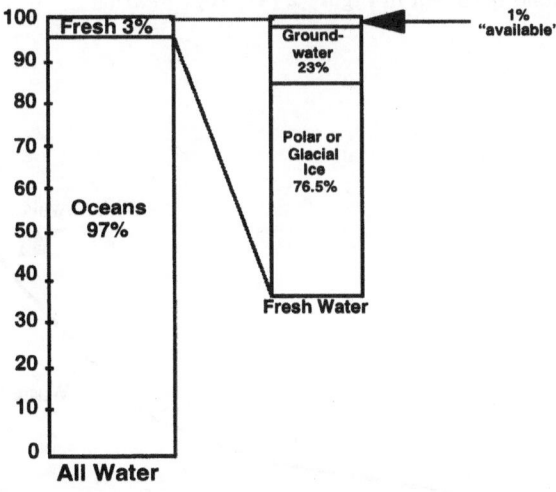

Figure 1. Global Water Resources

Water, Air and Soil Pollution **90**: 11-20, 1996.
© 1996 *Kluwer Academic Publishers.*

2. Factors Affecting Water Availability

Numerous factors affect freshwater availability. The most important of these include:

- Uneven worldwide and regional distribution of water resources;

- Variations in regional precipitation;

- Variations in regional availability, e.g., monsoons followed by dry seasons;

- Political concerns affecting shared water resources between nations;

- Generation of wastewater and water pollution;

- Selective absence of pricing mechanisms;

- Population growth; and

- Economic development.

Of these factors, population growth and economic development, which always increase the demand for water, will probably exert the greatest influence in the future.

3. Worldwide Water Uses

Withdrawals of freshwater have increased over time and are expected to increase dramatically in the future. This trend is shown in Figure 2 (Veltrop, 1993).

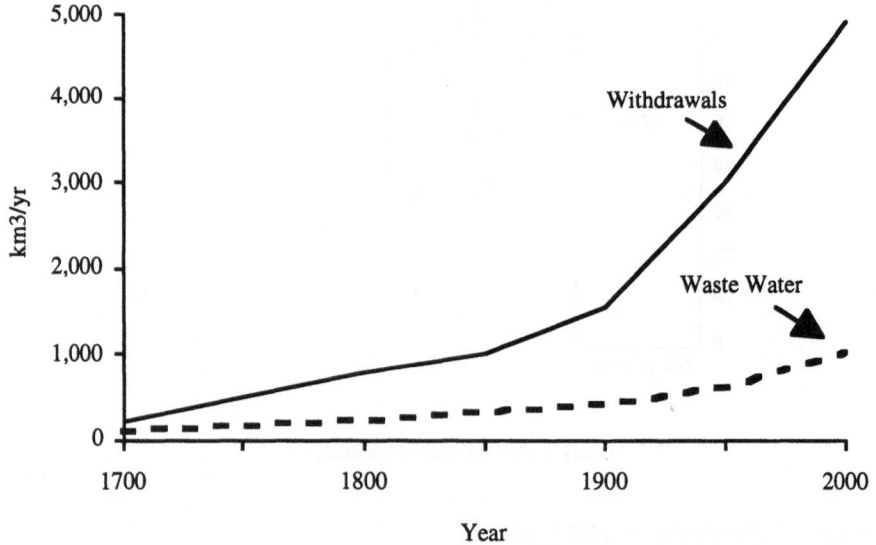

Figure 2. Projected Worldwide Water Withdrawals

Today, there are variations in water use on a continental basis, although 95% of total water use satisfies three primary requirements: domestic/municipal, industrial and irrigation.

4. United States Water Uses

As a developed nation, the United States uses fresh water in a variety of ways. Although agriculture is a major use, the electric utility industry also uses a large amount of water in its operations, as seen in Figure 3. These values reflect withdrawal usage, not consumptive use (van de Leeden, 1975).

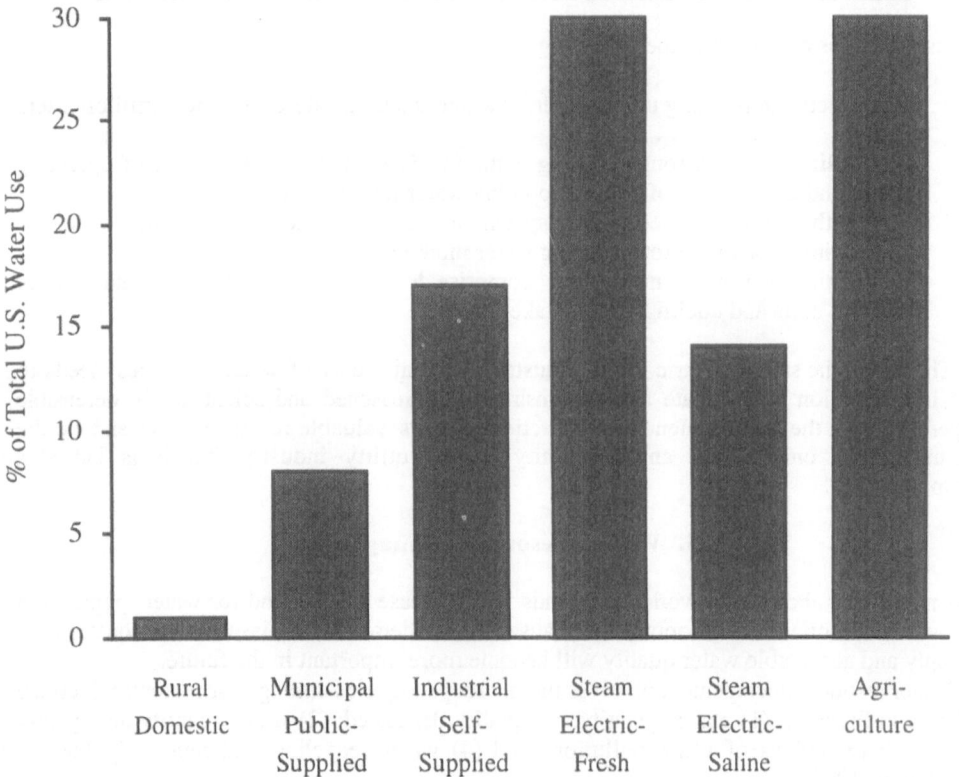

Figure 3. United States Water Uses

Within the electric utility industry, the primary use of water is for condenser cooling. Other water uses include: steam production, transportation of pollutants to treatment facilities, pollutant treatment and equipment cleaning (U.S. EPA, 1993). Table I summarizes cooling water use at U.S. steam electric power facilities. In addition, the United States uses and replaces approximately 10.6 trillion l/day for hydroelectric power generation (van der Leeden, p.363).

TABLE I

Cooling Water Use Within the U.S. Electric Utility Industry

Cooling System Type	Installed Generating Units		Cooling Water Intake Flow
	Number of Units	Percentage of Total	(million l/day)
Once-Through	1,211	59	680,000
Closed-Cycle	740	36	228,000
Combination	89	4.3	49,000
Mixed Mode	13	<1	7,500
Total	2,053	100	964,500

These statistics demonstrate the following:

• Steam electric generating units withdraw water at a total rate of 964,500 million liters per day;
• Units utilizing once-through cooling comprise 59% of the total number of operating units, withdrawing 71% of the total cooling water intake flow;
• Units with a closed-cycle cooling system are 36% of the total number of units, withdrawing 23% of the total cooling water intake flow; and
• Combination and mixed mode units comprise less than 6% of the total number of operating units and cooling water intake flow.

Therefore, the steam electric utility industry, as a major user of water resources, needs to be in a position to evaluate and demonstrate well reasoned and scientifically defensible approaches to the management and protection of this valuable resource, now and in the future. Not only is this an opportunity for the utility industry, but it is indeed a responsibility.

5. Water Resource Management

As mentioned above, two worldwide trends will increase the demand for water: population growth and economic development. Because of these demands, the issues of adequate water supply and acceptable water quality will become more important in the future.

Water resource management involves the interdependencies among four factors, which are shown in Figure 4: (1) water quantity and quality demanded; (2) water quantity and quality supplied; (3) effects of water pollution; and (4) water recycling and reuse (Spulber & Sabbaghi, 1990).

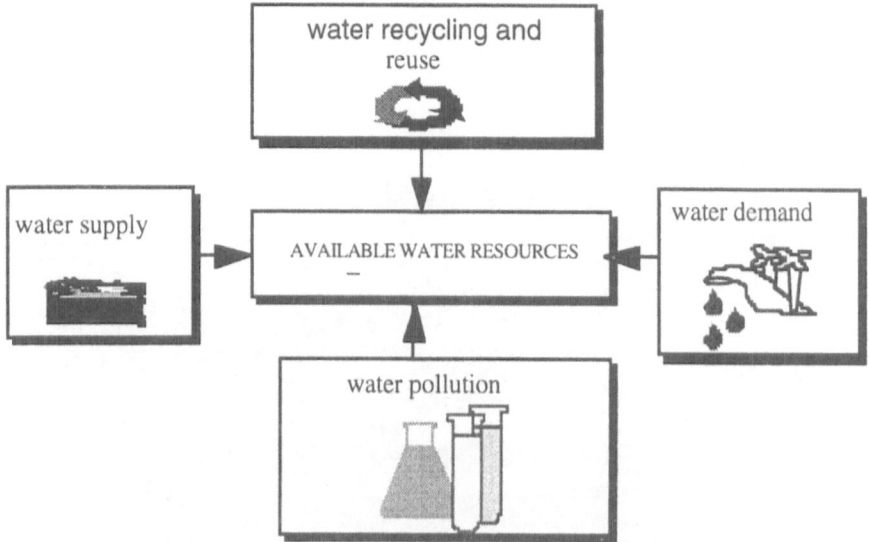

Figure 4. Factors Influencing Water Resource Management

Regionally, the significance of each of these factors can be determined, although they may vary over time. Understanding the importance of these factors is vital to making effective decisions regarding industrial, agricultural and residential development.

For example, within the United States, there is a great disparity in water supplies between the East and the West, as shown in Figure 5. In the Eastern portion of the country, an ample water supply is not usually a political or technical issue of great magnitude. Therefore, the focus in this region is on the quality and apportionment of the resource.

In contrast, in the Western United States, the availability of water supplies has often become a limiting factor in economic development. For example, irrigated agricultural interests often compete with municipal and industrial users for adequate, dependable water supplies. As a result, expansion of water development projects alone, has not solved the need for adequate water supply. It also has become necessary to incorporate improved management techniques to allocate the available water resources among competing users (*Id.*, pps 7-8).

6. Sustainable Development

The concept of sustainable development has been refined within the last decade to incorporate the constraints and conflicts inherent in the interdependencies among resource use (including water), economic development and environmental concerns. Current discussions of sustainable development generally focus on understanding the social, economic and political processes that will ensure that human needs can be met without depleting resources or significantly degrading the environment for future generations (Sadler, 1993).

Distribution of Water in the Contiguous United States

When considering policy options for water resource management, sustainable development should be amongst the foremost underpinning principles. For example, in the past, there has been a great reliance on technical and engineering solutions to water supply and quality problems. Sometimes this has led to major, unanticipated ecosystem problems. In the future, decision-makers will necessarily need to respond in a more "holistic" and less reactive manner in addressing critical water management issues (*Id.*, p. 377).

7. Ecosystem Management

Fortunately, new principles are emerging to address water resource management issues. These principles approach water management from a watershed (i.e., a drainage basin) perspective, rather than from a "per water project" view. An ecosystem management approach takes into consideration the following factors:

(1)   the need to maintain healthy ecosystems;

(2)   the recognition that human activities and impacts are an
        integral part of many ecosystems; and

(3)   the development of political and public policies that reflect
        these concerns (Moody, 1993).

The move toward a more holistic approach to managing water resources recognizes the interconnections of various components of ecosystems. An example of this trend is the move toward managing for biological diversity rather than for single species. Such an approach is more conducive to ensuring ecosystem integrity. This approach to water resource management will become more widely practiced in the future. There currently exists an inter-agency panel to plan and recommend a national ecosystem management implementation program.

8. The U.S. Experience: Clean Water Legislation

Widely supported environmental legislation was first enacted in the United States in the 1970's. With respect to water resources, Congress had enacted the Federal Water Pollution Control Act (FWPCA) of 1948, which delegated most pollution-control authority to individual states. The passage of the FWPCA amendments and the Clean Water Act of 1977 (CWA) introduced federally-mandated controls on water pollution, which reflected public concern over the degradation of America's surface water resources (Somach, 1989).

Of interest to the utility industry are the water quality controls that the current Clean Water Act addresses:

(1) the achievement of effluent limitations on point sources; and

(2) the achievement of water quality standards.

Under the provisions of the Clean Water Act, the discharge of pollutants into surface waters from point sources is limited and requires a National Pollutant Discharge Elimination System (NPDES) permit. Thus, point source discharges must meet specific effluent limitations intended to protect surface water resources.

With respect to the electric utility industry, besides complying with discharge standards under the Clean Water Act, other environmental laws are applicable, including:

- The National Environmental Policy Act of 1969;

- The Federal Endangered Species Act;

- Clean Water Act Section 401 (water quality certification);

- Clean Water Act Section 404 (dredge or fill permitting);

- The Federal Power Act (for hydroelectric power development); and

- Individual state environmental laws, e.g., The California Environmental Quality Act.

The U.S. House of Representatives passed H.R. 961, "The Clean Water Act Amendments of 1995" in May of this year. Numerous provisions are of interest to the electric utility industry. Some key provisions and a cursory summary of implications to the electric utility industry include (EEI, 1995):

- Risk Assessments: The U.S. Environmental Protection Agency (EPA) would be required to develop a risk assessment before issuing standards or limitations required under the CWA (permits excluded);

- Intake Credits: Dischargers could not generally be held responsible for pollutants already present in intake water;

- Waste Treatment Systems: These would be excluded from regulation;

- Transmission Lines in Wetlands: These would be exempt from CWA Section 404;

- Water Quality Standards: Increased flexibility would be introduced into the establishment and implementation of standards; and

- Mixing Zones: EPA could not prohibit or discontinue mixing zones established by individual states.

In addition, the development of regulations by the EPA are of great importance to the electric utility industry. Some examples of regulations being considered by the E.P.A. include (Bozek, 1995):

- CWA § 316(b): The development of regulations affecting cooling water intake structures under this section addresses such issues as (1) what constitutes an "adverse impact", (2) the appropriate data analysis techniques for the evaluation of impacts and (3) the impact of capital costs in assessing "best" technologies;

- Analytical Methods Assessment: Issues involving the detection, measurement and analysis of constituents in water are vital to the industry in the context of compliance and enforcement with CWA control requirements. Development of technically valid, sensitive and reliable analytical methods are critical to the effective evaluation of water quality;

- <u>Great Lakes Water Quality Initiative</u>: The establishment of water quality standards and antidegradation policies for the Great Lakes will set precedents for national and other regional water quality initiatives. Stringent measurement of pollutants, development of bioaccumulation factors and methodologies for minimizing pollutants present at levels below detection limits are issues affecting the industry; and

- <u>Bioavailability Guidance</u>: EPA is developing national guidance on how states should take into account the bioavailability of metals and other pollutants, using chemical or biological techniques for applying water quality standards. The development of valid, user-friendly techniques are essential to apply water quality standards.

9. Conclusion

The initial intent of the various rules and regulations governing the use and protection of water resources in the United States i.e., the prevention of wanton waste and acute pollution, I believe, has been met. We are now embarking on a new era of water resource management that should be designed to maximize the wise use of water while preventing chronic degradation. Many developing nations can learn from the intensive experience gained in the United States. As water is indeed a global resource, upon which we depend for our lives, the obligation exists for the most prudent approach to the use and management of this precious commodity.

The papers to be presented at this conference include the important topics of water protection, human and ecological effects of water pollution, strategies for control of water pollution, monitoring of water quality to ensure the integrity of our water resources, and tools for the day-to-day management of water use based upon availability. This conference is important and not only to the electric utility industry. We hope that it will also contribute to a national strategy for responsible management of an essential global resource.

References

Bozek, R.: 1995, *Water Use by the Electric Utility Industry*, Edison Electric Institute, (memorandum).

Edison Electric Institute (EEI): 1995, *Summary of H.R. 961, Issues of Concern to the Electric Utility Industry*.

Environmental Protection Agency (EPA):1993, *Background Paper Number 2: Cooling Water Use for Selected U.S. Industries and Summary of Selected EPA Regional and State Section 316(b) Activities*.

Moody, D.:1993, *Res. & Exploration*, Water Issue **9**, 81-85.

Research & Exploration, Water Issue: **9**, 1993, 1.

Sadler, B.: 1993, *Nat. Res. J.*, **33**, 375-396.

Somach, S.: 1989, *Pacific Law J.* **20**, 337-366.

Spulber, N. and Sabbaghi, A.: 1990, "Economics of Water Resources: From Regulation to Privatization", *Natural Resource Management and Policy*, edited by A. Dinar and D. Zilberman, Kluwer Publishers.

van der Leeden, F.: 1975,*Water Resources of the World*, Port Washington, NY: Water Information, Inc.

Veltrop, J.A.: 1993, "Importance of Dams for Water Supply and Hydropower", *Water for Sustainable Development in the Twenty-first Century*, edited by A.K. Biswas *et al.*, Oxford University Press, Delhi.

ADDRESSING ELECTRIC UTILITY SURFACE WATER CHALLENGES

R. Brocksen[1], W. Chow[1] and K. Connor[2]

[1]*Electric Power Research Institute, 3412 Hillview Avenue, Palo Alto, California 94303,* [2] *Decision Focus Incorporated, 650 Castro Street, Suite 300, Mountain View, California 94041*

Abstract. Growing regulatory pressures and increasing public concerns regarding a broad range of utility surface water management issues have the potential to substantially impact electric utility operations and costs. In order to pro-actively contribute and respond to the increasing number and scope of regulatory proposals, the electric utility industry must develop a focused and integrated plan for addressing key information and analytical needs. Over the past year, EPRI has conducted an extensive review of current and planned surface water-related research with the goal of targeting future research activity towards the highest priority industry needs. As a result, new or expanded research efforts in the areas of toxics sampling and analysis, field measurements of aqueous process and discharge streams, cost-effective metals control technologies, bioaccumulation and bioavailability of toxics, and watershed-wide risk assessment and management, are planned. The new initiatives will build on and complement current surface water research programs, as well as leverage recent efforts addressing utility stack emissions of trace substances (or air toxics). A risk management framework will be used to integrate the diverse research inputs and apply them to address key policy and utility management issues. This framework, and the tools and data developed through current and subsequent research efforts, will provide the comprehensive, multi-media data and analysis tools needed to respond to current and future utility environmental management challenges efficiently and cost effectively.

Keywords: Surface Water, R&D Planning, Risk Management

1. Introduction

Over the past several years and looking forward into the next decade, increasing regulatory and public attention has been focused on utility water management issues. These issues include potentially toxic discharges, water conservation, wetlands preservation, fish protection, and watershed management. Key regulations potentially affecting electric utility surface water management include reauthorization of the Clean Water Act (CWA) and Safe Drinking Water Act (SDWA), the "Great Waters" provisions of the Clean Air Act Amendments, the Endangered Species Act, hydro-relicensing requirements by the Federal Energy Regulatory Commission (FERC), surface water issues related to hazardous waste disposal (RCRA), toxics management (TSCA), and community right-to-know (EPCRA) requirements. Regional initiatives focused on specific water resources, such as the Great Lakes Water Quality Initiative (GLI), also have impacted the national debate. As a result of potential regulatory requirements, costs associated with electric utility surface water use and discharge could increase substantially. Even without new regulations or changes through reauthorization, expanded requirements and more stringent enforcement of existing National Pollutant

Water, Air and Soil Pollution **90**: 21-29, 1996.

Discharge Elimination System (NPDES) permits may lead to higher costs and greater uncertainty surrounding permit renewals.

A central tenet of many of these new regulations is the reduction or elimination of potentially toxic discharges and, in general, a broader, more holistic approach to environmental management and control. For example, key provisions of H.R. 961, the House Clean Water Act reauthorization bill, allow pollution prevention measures to be incorporated as part of a permit limitation if the measures are more effective than other measures in achieving *overall* pollutant reductions (including discharges to air and solid waste, as well as water) (U.S. House of Representatives, 1995). The bill also permits limited trading among sources within a watershed if greater pollutant reductions are achieved through this mechanism. Other regulations, such as the Endangered Species Act, Great Lakes Water Quality Initiative, and Great Waters Study, address the cycling of toxics through the environment and define ecosystem-wide impacts, including biodiversity and sustainable development, more broadly than previous statutes. Watershed-based management plans, currently under consideration in several states, may place additional constraints on the availability and allocation of water resources, as well as substantially increase sampling and monitoring requirements for electric utilities.

Given the growing number of initiatives addressing surface water issues, and the expanded scope of many of these initiatives, the potential costs to electric utilities are substantial. Compliance with increasingly stringent regulations could cost the industry up to $70 billion over the next decade. In light of these developments, it is imperative that the industry develop methods, information and tools necessary to identify, evaluate, and manage potential adverse impacts associated with utility water use and discharge. Because of the comprehensive nature of many new regulations, it also is imperative that utilities approach water use with an integrated perspective of their operations and impacts. Using such an integrated view, the industry has the opportunity to leverage previous investments in toxics management and pollution prevention focused on other media or operations.

2. An Integrated Approach To Surface Water R&D Planning

In response to the above existing, pending and proposed regulations, the Electric Power Research Institute (EPRI) initiated a comprehensive planning effort to define electric utility surface water research needs, assess the ability of current research plans to meet those needs, and define and prioritize new research initiatives. Because of the wide-ranging scope of the regulatory initiatives and the general shift from technology-based to water quality-based standards, the activities and research relevant to the water "life cycle" at a power plant were identified and evaluated. As shown in Figure 1, the life cycle perspective considers the interactions and feedbacks between all aspects of utility surface water management, including water supply, plant water use and re-use, water discharge and transport in the environment, and evaluation of the potential for health and environmental impacts. In order to address all components of the water life cycle, EPRI undertook a broad multi-disciplinary, multi-business unit review of current and planned research activities. The review considered research efforts in the areas of sampling and analysis methods, field measurements, treatment technology, facility design, pollution prevention, compliance guidance, ecological science, risk assessment, and risk management. While the immediate focus was on specific needs with respect to surface water management, this effort, combined with EPRI's recent multi-disciplinary air toxics research program, forms an integral part of EPRI's longer term mission to provide the information, guidance and tools necessary for integrated multi-media risk assessment and management for the electric power industry.

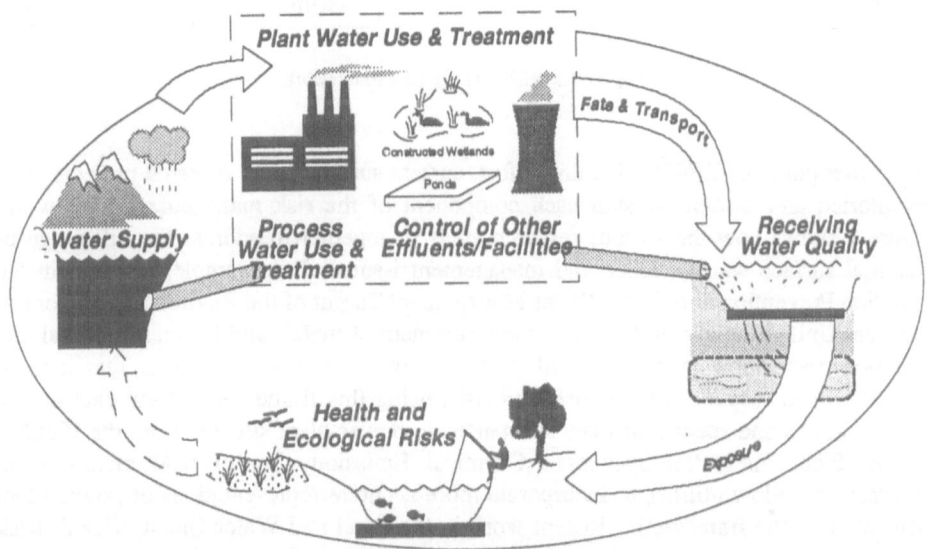

Fig. 1. The Water "Life Cycle"

The risk management framework in Figure 2 has been applied successfully in organizing and integrating EPRI's comprehensive air toxics research program (EPRI, 1994). This framework provides a useful paradigm for bringing a broad range of research results to bear on potential utility management and policy-related questions. As shown in Figure 2, the framework incorporates information on the source(s) of discharges, such as measurements and operating data; traces the transport and dispersion of effluents throughout the environment; relates the environmental concentrations of pollutants to exposures and doses in ecosystems and human populations; estimates potential health and environmental impacts based on these exposures; and ultimately can be used to evaluate potential risk and cost tradeoffs of various management or policy decisions. The risk management paradigm is particularly useful in responding to and communicating with regulators and the public regarding potential new regulations incorporating risk assessment and cost/benefit concepts.

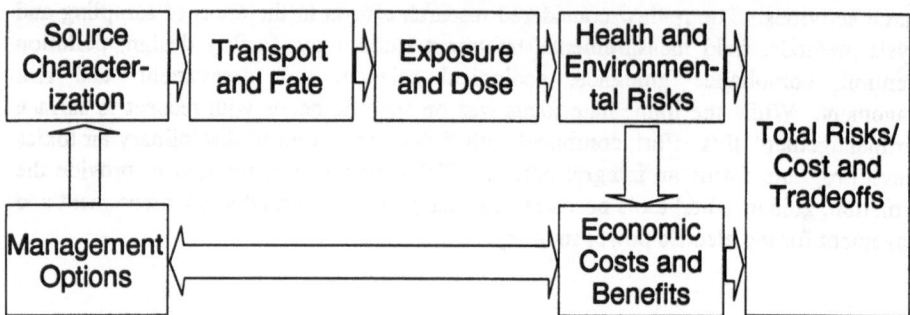

Fig. 2. The Risk Management Framework

In the first phase of EPRI's planning effort, current surface water research projects were inventoried and associated with each component of the risk management framework. Figure 3 shows how the various research areas fit together to address the full range of potential surface water science and management issues. For example, research in the Pollution Prevention and Waste/Water Management Target of the Environmental Control Business Unit on new technologies for measurement of metals and for their removal and pollution prevention guidelines will provide new management options that can be evaluated with respect to both costs and risks using this framework. New data on the concentrations and species of trace substances in power plant streams from the PISCES (Power Plant Integrated Systems: Chemical Emissions Study) field measurement program will allow utilities to incorporate more accurate representations of power plant effluents into the framework. Recent work in the Land and Water Quality/Health Risk Assessment Target of the Environment and Health Science Business Unit addressing the fate of metals in receiving waters and their chemical and biological availability will allow utilities to refine their estimates of the impact of utility discharges on receiving water quality. Additional research in this Business Unit investigating the availability and impacts of specific substances on aquatic organisms will help to provide a better measure of the actual exposures and doses received by humans and ecosystems. Finally, a broad range of integrating projects in both of EPRI's strategic Targets are designed to model the pathways of power plant chemicals and predict health, ecosystem, and economic impacts will provide a means of consistently evaluating and comparing utility surface water management options and the appropriate framework for communicating the results.

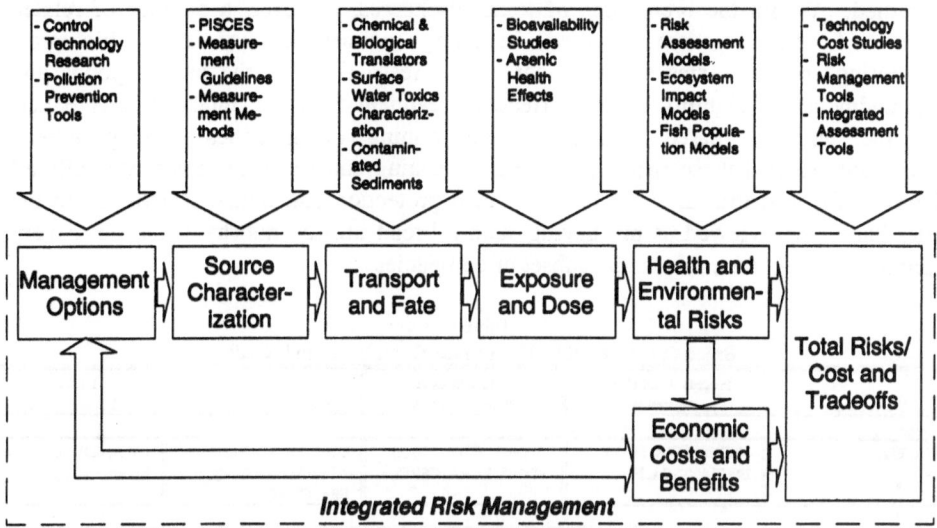

Fig. 3. EPRI's R&D Contributions

3. Utility Industry Surface Water Concerns And Priorities

After identifying and reviewing the current spectrum of surface water research projects, EPRI initiated a series of meetings with a "Water Focus Group" consisting of representatives from member utilities and other industry groups, including the Utility Water Advisory Group (UWAG) and the Edison Electric Institute (EEI). These meetings were extremely valuable in providing an "outside" review of EPRI's current plans, identifying potential new and shifting regulatory concerns that may need to be addressed in the future, and defining relative priorities among various research areas. Input from the Focus Group meetings, as well as EPRI's Pollution Prevention and Waste/Water Management and Land and Water Quality/Health Risk Assessment Research Advisory Committees, was incorporated into the planning process. This was done to insure that the resulting plans for both research areas would provide an integrated and focused strategy for addressing utility surface water research needs.

Table I lists a number of industry issues and concerns, in all phases of the water life cycle, that were discussed in the Focus Group meetings. For each phase of the life cycle (i.e., water supply, process water use and treatment, control of other effluents, etc.), Table I indicates potential regulatory drivers, key issues or topics of concern to utilities, and potential industry research needs to address these concerns. After reviewing and discussing the vast number of potential concerns and needs, two broad issue areas were identified as having both potentially large industry impacts and the need for a focused research effort to help develop sound scientific and technical data and analyze alternative utility management strategies. As shown in Table II, Toxics Discharge Standards and Management and Fish Protection/Intake Structures were the two key issue areas identified. Within the broad area of toxic discharges, however, a number of related issues and concerns were specified, including sediment quality issues, biological monitoring criteria, toxics measurement issues, restrictions on the use of ponds and impoundments, availability of cost effective treatment methods, intake credits issues, potential watershed-based regulation, and trading opportunities, among others. Participants viewed these issues as quite far-reaching and likely to impact virtually all electric utilities within the next decade. Fish protection and intake structures, on the other hand, were viewed as not having such a broad impact industry-wide, but a very significant impact on a substantial subset of companies.

Table I

Broad Categories of Utility Surface Water Issues and Needs

Water Supply	Process Water Use & Treatment	Control of Other Effluents/ Facilities	Receiving Water Quality	Health and Ecological Risks
Drivers:				
CWA, FPA	NPDES, CWA, 316(b), EPCRA, TSCA, FPA	Wetlands, CWA, 316 (a), NPDES, RCRA, CERCLA	CWA, 316(a) & (b), "Great Waters" Study, FPA, GLWQI	CWA, SDWA, Wetlands. Endangered Species, CAA
Topics/Issues:				
Intake water quality/biology Availability/Allocation Streamflow Impacts	Discharge standards Pollution prevention Control technologies Zero-discharge design Intake water treatment Fish entrainment/ impingement	Constructed wetlands Cooling water discharge Surface impoundments/ ponds Non-point sources/ runoff	Surface water quality Flow Thermal Mixing zones Sediment quality	Multi-media risks Watershed/ regional effects Human health Aquatic life/ Wildlife Biodiversity/ ecosystem effects
Industry Needs:				
Watershed resource management tools Fish compensation studies Stream flow impact studies Hydro plant resource management	Multi-media discharge characterizations & models Toxics reduction/ pollution prevention tools Sampling & analysis methods Treatment technology performance data Alternatives to chlorine for biofouling control Improved intake structure design New technology development for metals removal Water conservation/ management tools Boiler water treatment	Wetlands impacts/ compensation models Characterization of cooling water disinfection by-products Characterization of discharge from ponds/ impoundments Treatment of effluents from ponds/ impoundments Evaluation/ development of liner systems Characterization of non-point sources	Sampling & analysis methods for metals species Development of chemical/ biological translators Thermal impact analysis Sediment sampling methods/ results Water quality modeling, mixing zone studies Biological monitoring methods/ design Fate & transport models	Multi-media risk management tools Watershed/regional-level impact models Bioavailability/toxicity studies for metals Fish pop. dynamics/ compensation models Human health risk models Ecosystem impact models Wildlife impact models Economic analytical tools Integrated planning methods

Table II
High Priority Surface Water Issues

I.	**Toxics Discharge Standards and Management**
	– Sediment Quality
	– Biological Monitoring
	– Water Quality Criteria
	– Biodiversity/Ecosystem Effects
	– Risk Assessment
	– Multi-media Impacts
	– Intake Credits
	– Toxics Measurement
	– Chlorine Use/Discharge Restrictions
	– Restriction on Use of Ponds/Impoundments
	– Treatment/Management Options
	– Integrated Assessment/Economic Analysis
	– Watershed Management
II.	**Fish Protection/Intake Structure**

Based on this feedback, current and planned EPRI research projects addressing toxics and fish protection issues were re-evaluated, potential gaps or research needs were identified, and potential new initiatives were defined to address the gaps. Table III summarizes potential new research projects addressing high priority industry needs in both the Pollution Prevention and Waste/Water Management (PPWM) and Land and Water Quality/Health Risk Assessment (L&WQHRA) research. Among the highest-ranking surface water research priorities are improvements in toxics measurement methods and the acquisition of improved field data using these methods. In addition, the development and integration of economic data for new control technologies into utility management tools; better understanding of bioaccumulation of toxics and the implications with respect to water quality criteria; and the development and application of models and risk management tools for watershed-wide analyses are considered high-priority activities. These initiatives are now being evaluated in the context of EPRI's current research plans and, depending on final industry input, may replace or augment current research efforts.

Table III
Potential New EPRI Research Initiatives Addressing High Priority Needs

	PPWM	L&WQHRA
Toxics Measurement	• Expand pollutant detection work • Support development of trace Certified Reference Materials (CRM) • Expand PISCES field data	
Treatment & Management	• Economic assessment of metals control technologies • Source reduction studies	
Water Quality Criteria		• Bioaccumulation studies • Biological monitoring protocols • Review derivation of water quality criteria
Watershed Management		• Evaluate assimilative capacity models • Watershed risk management models • Expanded COMPMECH
Integrated Assessment	• Cost data for management options • Link plant and risk tools	• Cost-Benefit in risk models • Link plant and risk tools

4. EPRI Surface Water Activities

As indicated above, EPRI already has undertaken or initiated a number of projects designed to help electric utilities address new challenges with respect to surface water management. In addition, many of these projects are integrally linked with products and tools addressing other media (stack emissions, solid wastes), and will ultimately provide a multimedia perspective on utility operations and environmental management. This perspective is not only responsive to growing regulatory pressures for multimedia permitting, but is also consistent with economic pressures to improve efficiency and reduce waste across all utility operations.

In the area of toxics management and control, many of the methods and models developed to respond to air toxics issues in the 1990 Clean Air Act Amendments can be expanded upon to encompass surface water issues. For example, the pioneering sampling and analysis work for accurately measuring mercury and other metal species in stack emissions can be directly applied, or applied with some modifications, to field measurement programs for aqueous discharges. The power plant model and emissions database developed to help predict air emissions as part of EPRI's PISCES project will provide the initial structure for organizing and integrating information on plant water use and effluent characteristics, as well. Finally, data and tools developed to help predict human and ecosystem exposures to, and impacts from, airborne toxics, such as the Mercury Cycling Model (MCM) and the Total Risk of Utility Emissions (TRUE) model, provide comprehensive, multimedia frameworks for evaluating health and environmental risks that can be applied to aqueous discharges, as well as air emissions (Gillespie, et al., 1995). By building on this solid foundation of research addressing the nature and management of utility trace substance emissions, EPRI can provide solutions to electric utility surface water management challenges in a cost effective and timely manner.

While much has been accomplished with regard to surface water management in the past twenty years, there still remain critical information needs to reduce uncertainties regarding utility surface water uses and discharges, their impacts, and cost effective management.

5. Summary

New and expanded regulations addressing surface water issues are likely to lead to substantial changes in the way electric utilities manage this resource in the future. The emphasis of many new initiatives, including Clean Water Act reauthorization bills and the Great Lakes Water Quality Initiative, on the reduction or elimination of toxic discharges, as well as the broader multimedia (air, water, and land) approach taken in these initiatives will provide both challenges and opportunities for the utility industry in the next decade. The challenge is to develop the methods, information, tools and technologies needed to understand and cost effectively manage the potential adverse impacts associated with utility water use and discharge. The opportunity is to apply these tools and technologies in an integrated way that will provide both environmental and operating benefits.

EPRI's Environmental Control (EC) and Environment and Health Science (E&HS) Business Units have been working together to define and integrate the research components needed to help utilities address current and future surface water management challenges. New research in the areas of toxics sampling and analysis methods, field measurements, metals removal, and the bioaccumulation and bioavailability of toxics in ecosystems will build on and complement previous and ongoing surface water research programs, as well as leverage recent efforts and experiences in the area of air toxics. Ultimately, this collaborative effort will enable EPRI to provide member utilities with the data, technologies, and analysis tools needed to respond to the broad spectrum of business, engineering, environmental, and regulatory questions related to multimedia environmental management.

References

Electric Power Research Institute. Electric Utility Trace Substances Synthesis Report. EPRI TR-104614-V1 through V4, Palo Alto, CA:November 1994.

P.A. Gillespie, E. Constantinou, C. Seigneur, P. Pai, L. Levin. A Regionalized Approach to the Assessment of Health Risks Associated with Mercury Power Plant Emissions. Presented at the EPRI/DOE International Conference on Managing Hazardous and Particulate Pollutants, Session 2: Atmospheric Fate and Health and Risk Studies. Toronto, Canada. August 1995.

U.S. House of Representatives, Transportation and Infrastructure Committee, H.R. 961, Clean Water Act reauthorization bill, April 6, 1995.

LAKE SUPERIOR BINATIONAL PROGRAM:
THE ROLE OF ELECTRIC UTILITIES

C. Lohse-Hanson

Minnesota Pollution Control Agency, Water Quality Division, 520 Lafayette Road,
St. Paul, Minnesota 55155

Abstract. Lake Superior is called the greatest of the Great Lakes for good reason. It is the largest of the Great Lakes and also the cleanest. Although Lake Superior fish contain enough PCBs, mercury, and toxaphene to warrant fish consumption advisories, levels of toxic chemicals in Lake Superior are low compared to other Great Lakes. Because of the relatively clean waters of the lake and the basin's small industrial base, Lake Superior governments have agreed to set aside the basin as a special demonstration area with a goal of zero discharge and zero emission for nine toxic chemicals. Some of these chemicals have been associated with electric utilities. The governments recognize that electric utilities, industry and residents of the basin will all have a role in the march towards zero. The Lake Superior governments urge the electric utilities to consider 1) the proposed load reduction schedules for Lake Superior, 2) a US inventory of PCB equipment and 3) innovative solutions that bring facilities closer to zero discharge and zero emission.

Key words: binational, electric utilities, Lake Superior, LaMP, mercury, PCBs, special designations, virtual elimination, zero discharge

1. Purpose

The purpose of this paper is to introduce the Binational Program to Restore and Protect the Lake Superior Basin and the role of electric utilities. Utilities are associated with a variety of discharges and emissions, including cooling water and wastewater discharges, stack emissions and solid and hazardous waste management practices which may release toxic chemicals such as PCBs, mercury and dioxin. The Lake Superior governments encourage the electric utilities to consider the load reduction recommendations made by the Lake Superior Binational Forum. These recommendations will be included in a formal public comment period overseen by the United States Environmental Protection Agency (USEPA). The governments especially invite dialog and commitments to voluntary, innovative approaches to reducing toxic chemicals associated with electric utilities.

2. Introduction to the Binational Program

Lake Superior is the cleanest and largest of the Great Lakes. Its watershed is small for its size and has a relatively small population and industrial base (Superior Work Group, 1995a). The watershed is mostly forested and much of the land is in public ownership. Because of the small population and high water quality, the International Joint Commission (IJC) and the Lake Superior governments consider the basin to be a good

Water, Air and Soil Pollution **90**: 31–40, 1996.
© 1996 *Kluwer Academic Publishers.*

demonstration area for an aggressive approach to achieve virtual elimination of toxic and persistent chemicals per the Great Lakes Water Quality Agreement (GLWQA).

In September 1991, the Lake Superior governments responded to an IJC challenge to set aside the Lake Superior Basin as a zero discharge demonstration area. The governments, including the governors of Minnesota, Wisconsin and Michigan, agreed to the principles of a Binational Program to Restore and Protect the Lake Superior Basin. The Zero Discharge Demonstration Project is unique to this Binational Program. While each of the Great Lakes will have a Lakewide Management Plan (LaMP) per the GLWQA, the Lake Superior LaMP will be the mechanism for delivering the Zero Discharge Demonstration Project. The LaMP will also deliver the ecosystem management approach agreed to in the GLWQA, including protection from non-chemical stressors.

The LaMP for each of the Great Lakes will go through four stages as laid out in the GLWQA. Stage 1 is Problem Identification. For Lake Superior, the governments have compiled water, sediment, fish and wildlife data as well as reports from local areas of concern for comparison to the Smith and Smith (1993) "yardsticks" for designated uses of the waters within the Lake Superior Basin. Other than octachlorostyrene (which has no criteria), the zero discharge chemicals exceed water, sediment or fish "yardsticks" for Lake Superior and are highly ranked in the Great Lakes Water Quality Initiative (GLI) list of bioaccumulative chemicals of concern, the Ontario bans and phase-outs list, and the IJC list of Great Lakes critical chemicals (Superior Work Group, 1995a).

The Binational Program has moved into Stage 2, which is focused on load reduction schedules. The Stage 2 document has been drafted and will be noticed in the US Federal Register for public comments. Stage 2 includes Binational Forum recommendations for reduction schedules aimed at mercury, PCBs, pesticides (chlordane, DDT, dieldrin, hexachlorobenzene, mercury and toxaphene) and dioxin, hexachlorobenzene and octachlorostyrene. Some of those schedules are included in this paper.

Stage 3 is where the "rubber meets the road" as the governments, residents and industries implement the reductions needed to meet the schedule. Specific actions from each source of toxic chemicals are needed for Stage 3. Education, voluntary reduction and pollution prevention are the preferred reduction strategies. Some actions that affect the Basin will occur on a larger scale (e.g., Minnesota's new waste combustor rule will eliminate small incinerators statewide). The governments will begin Stage 3 in 1996.

The LaMP deals with several categories of toxic chemicals, but the first priority is the nine chemicals identified in the Zero Discharge Demonstration Project. Specifically, the governments have agreed to a zero discharge and zero emission goal for chlordane, DDT and metabolites, dieldrin, dioxin, hexachlorobenzene, mercury, octachlorostyrene, PCBs and toxaphene. The goal as stated in the binational agreement is "to achieve zero discharge and zero emission of certain designated persistent, bioaccumulative toxic chemicals which may degrade the ecosystem of the Lake Superior Basin."

This goal statement has several significant phrases, the first being the word "goal." The Binational Program's zero discharge goal does not have the force of law, although laws and rules may eventually be associated with the Program. For example, the binational agreement's action plans include some activities from GLI. It is important that all parties keep in mind that zero discharge and zero emission is a goal rather than a law.

The term "zero" is crucial. "Zero" means zero molecules. The same way that a manufacturer strives to have zero defects in a product or an employer strives for zero lost-time injuries to workers, the program strives for zero molecules. From a practical viewpoint, "zero" may mean that the loads are so low that even the most ardent environmentalists agree there are higher priorities than pursuing the last molecule.

The words "discharge" and "emission" are also important. In regulatory programs, a typical discharge is a pipe transferring wastewater from a facility to a waterbody and an emission is a stack transferring waste gases from a facility to the air. Cross media transfers and nonpoint discharges and emissions should not be underestimated, however. For example, the Minnesota Pollution Control Agency estimates that half of the mercury emitted in Minnesota comes from use and disposal of products (MPCA Mercury Task Force, 1994). Since landfills offgas the mercury from disposed products and other toxic chemicals may move off-site, even a landfill can be a source of emissions.

One implication of the choice of Lake Superior for the Zero Discharge Demonstration is the selection of measures of progress. The largest load of the nine chemicals slated for zero discharge and zero emission in the Lake Superior Basin is from atmospheric deposition. A large portion of that atmospheric deposition is from outside the Lake Superior Basin. There are two important considerations regarding atmospheric deposition. First, in order to significantly reduce loads to Lake Superior, the governments must export successful reduction activities from the Binational Program. Second, we can not use traditional indicators such as contaminant levels in fish, sediment or water to measure the progress of the Zero Discharge Demonstration since reduced loads from facilities in the basin will be offset by atmospheric deposition. Instead, we will have to measure change in the behavior of residents and industry.

Responsibility for developing the Binational Program rests with three groups, the Lake Superior Task Force, the Superior Work Group and the Lake Superior Binational Forum. The Task Force is composed of upper government managers, mostly from water quality programs in both the US and Canada. Their role is to set priorities and provide direction. The Work Group is composed of government technical staff. Their task is to carry out the Binational Program goals using the LaMP process established in the GLWQA. The Binational Forum is composed of stakeholders from across the basin. Forum members have included people from mines, pulp and paper mills and small business as well as Native Americans, health professionals, local government staff, environmentalists and educators. Among other responsibilities, the Forum agreed to recommend a series of load reduction schedules to the Lake Superior governments. While these three groups will develop the program with input from the public, it will ultimately be the responsibility of the residents to implement the program.

3. Sources of Toxic Chemicals and Load Reduction Schedules

The nine zero discharge chemicals have a variety of sources. The Superior Work Group has developed use trees for dioxin, hexachlorobenzene, mercury, octachlorostyrene and PCBs. For example, the mercury use tree has four branches for natural sources of mercury, deliberate use, mercury production and storage and mercury as a by-product. These large branches split up into smaller branches, ending with specific sources. Not all of the sources in the use trees are found in the Lake Superior Basin, but the trees are the starting point for estimating the load of chemicals from sources within the Basin.

Beginning with mercury, the Superior Work Group also estimated the loads of the critical chemicals from US and Canadian sources within the Lake Superior Basin. The Lake Superior governments estimate that roughly 2,700 kg/yr of mercury was used, generated or released from Lake Superior Basin sources in 1990 (Superior Work Group, 1995b). Mercury in products accounts for 1330 kg/yr; 940 kg/yr in batteries alone (Superior Work Group, 1995b). Industry accounts for about 800 kg/year, with a single copper smelter accounting for most of that load (Superior Work Group, 1995b). Fuel combustion is third with 400 kg/yr, followed by municipal sources and incineration at about 100 kg/yr each (Superior Work Group, 1995b). These 1990 load estimates are based mostly on Minnesota estimates (MPCA Mercury Task Force, 1994) normalized to the Lake Superior Basin using the 1990 population.

Reductions in mercury used in batteries and the copper smelter shutting down in 1995 have already resulted in at least a fifty percent reduction. Draft load estimates were available to the Binational Forum during its deliberation on mercury load reductions. The Forum recommendations were transmitted to the governments in October 1994 (Figure 1). The Forum has recommended the year 1990 as a base line followed by sixty percent load reduction by the year 2000, eighty percent reduction by 2010 and virtual elimination by 2020. Also, the Forum recommendations apply to atmospheric deposition from outside the basin as well as releases from within the basin.

The mercury recommendations were followed by Forum workshops held on PCBs and pesticides. The Forum transmitted recommendations for load reductions of these two chemical groups to the governments in April 1995. The Forum PCB recommendations are included in Figure 2. The Forum emphasized destruction of PCBs that are in use, storage or in highly contaminated soils or sediments. Their preferred strategy is to remove and destroy PCBs before they are accidentally released to the environment. The Forum also made some recommendations for PCBs that are more widely dispersed in the environment.

The base line year for the accessible PCBs is 1995, with a thirty-three percent reduction recommended by the year 2000, followed by a sixty-five percent reduction in 2010, a ninety percent reduction in 2015 and virtual elimination in 2020. The Forum also transmitted recommendations on pesticides.

Goal: Virtual elimination of mercury within the Lake Superior Basin from all sources within the basin and from all sources outside the basin that are transported into the basin.

Guiding Principles:
- Stress pollution prevention and voluntary means as methods of achieving zero discharge.
- Be aggressive, creative and reward positive behavior.
- Use a full array of mercury reduction methods, including regulations.
- Develop the array of mercury reduction methods in greater detail and report to the Forum and the public for comments before implementation.
- No new uses of products containing mercury should be allowed in the basin.
- Governments should provide incentives to get mercury out of fuel emissions.
- Non-essential uses of mercury should be eliminated.
- Ensure that air deposition is included in the strategy.
- Track the cost to implement reductions. This will provide background information for other projects to virtually eliminate toxic substances.
- Report progress on a quarterly basis, especially on mercury reduction regulations, to the Forum.

Recommendations: Timeline to Eliminate Mercury
- baseline (1990) - base year for loadings of anthropogenic mercury to water, air and other sources
- 5 years (by the year 2000) - 60% reduction
- 15 years (by the year 2010) - 80% reduction
- 25 years (by the year 2020) - Virtual Elimination
  (Note: each of the loading reductions applies to all anthropogenic sources within the basin and from all anthropogenic sources outside the basin that are transported into the basin.)

Fig. 1. Lake Superior Binational Forum Mercury Recommendations October 24, 1994

Goal: The destruction of accessible/in-control PCBs with a 5-10 year window to develop cost effective solutions.

Guiding Principles:
- Encourage development of innovative technology and demonstration projects.
- Emphasize destruction rather than storage (one-step rather than two-step process for destruction).
- In-basin destruction is preferred where appropriate and practical.
- As out-of-control sources become accessible/in-control, management will follow the existing timeline for destruction.
- Promote partnerships between government and regulated industries and the affected communities to accomplish these goals.
- Complete an inventory of both accessible/in-control and out-of-control sources (1995 baseline).
- Remove PCBs at the end of the useful lifespan of equipment.
- Work to remove barriers and obstacles (e.g., bureaucratic and regulatory impediments and jurisdictional differences).
- Use education opportunities to reduce the NIMBY (Not In My Backyard) attitude.

Recommendations:
Timeline (1995 baseline) to Destroy "Accessible/In-Control" PCBs
- 5 years (by the year 2000) - 33%
- 10 years (by the year 2005) - 60%
- 15 years (by the year 2010) - 95%
- 25 years (by the year 2020) - 100% or Virtual Elimination
  Out-of-Control Sources
The Forum recommends the following course of action:
- gather information
- assess available technology
- complete an inventory
- set priorities and timelines for destruction based on risk, cost and technical feasibility prior to 1998

Fig. 2. Lake Superior Binational Forum PCB Recommendations, April 1, 1995

Additional workshops were held on dioxin, hexachlorobenzene and octachlorostyrene. The Forum recently submitted these recommendations to the governments. The Forum was not able to reach consensus on all the dioxin recommendations. The original load reduction recommendation was for virtual elimination within the Basin by 2015 and virtual elimination from sources outside the Basin by 2020. Dissenting members of the Forum recommend an eighty percent reduction of sources from the Basin by 2005, ninety percent by 2015 and a ninety percent reduction of airborne loads from the US and Canada by 2020. In both recommendations, the base line year is 1990. The Lake Superior governments estimate fuel combustion contributes roughly 0.3 g/yr (8%) of the hexachlorobenzene and 2.4 g/yr (0.5%) of the dioxin released from sources in the Lake Superior Basin (Superior Work Group, 1995b).

4. Impact on Electric Utilities

The Lake Superior governments and the Binational Forum emphasize voluntary reductions but recognize the need for incentives to change from existing practices. According to the binational agreement approved by the governors "The prevention of pollution is the preferred approach to environmental protection." The Binational Program also recognized two other strategies for achieving zero discharge, specifically special designations and controls and regulations. Also, the Binational Program is bracketed by other programs that are enforcement oriented such as the GLI.

Each of the strategies identified in the Binational Program has implications for the electric utilities. Fourteen public electric utilities that are located in the Basin. Half of these electric utilities are small municipal facilities, mostly in Minnesota.

4.1. POLLUTION PREVENTION

Pollution prevention is the preferred strategy for achieving zero discharge and zero emission. The Wisconsin Department of Natural Resources is convening a group of Wisconsin, Minnesota and Michigan electric utility representatives, government staff and environmentalists to identify the best pollution prevention opportunities in the basin. An example of a pollution prevention activity already accomplished in the Lake Superior Basin took place at the National Steel Pellet Company in Keewatin, Minnesota. The company, Minnesota Power and mining consultants carried out an energy audit of the facility. Changes in the production line made the process more efficient, lowering the amount of energy needed per ton of pellets.

Chemical feedstocks are another potential pollution prevention target. Potlatch Corporation, a pulp and paper manufacturing facility in Cloquet, Minnesota detected mercury in their effluent after switching to a new bleaching process in March 1994. The mill discharges to Western Lake Superior Sanitary District (WLSSD), which imposed a local limit of 0.3 µg/L mercury in 1995. The District has an effluent limit of 0.024 µg/L mercury based on the Minnesota water quality standard of 0.0069 µg/L.

With the assistance of WLSSD, Potlatch discovered the mercury came from a new source of sulfuric acid needed for the ECF bleaching. The company informed its suppliers that they would need proof of low mercury content in future shipments of chemical feedstocks. Without tracking down and eliminating the source of the mercury, both Potlatch and the District faced the possibility of mercury limit exceedences. Other facilities have switched from mercury cell caustic soda to membrane cell caustic soda. These feedstocks may be "low-hanging fruit" to reduce mercury loading in discharges from a variety of facilities, including electric utilities, mines and pulp and paper mills.

Multimedia inspections are another pollution prevention activity in the Lake Superior Basin. As part of the Lake Superior Partnership, the Minnesota Pollution Control Agency (MPCA), WLSSD and facilities that discharged into WLSSD carried out a series of multimedia inspections. Facility representatives met with the inspection team prior to inspection. The day of the inspection, the entire facility was covered by the whole team, with inspectors later tracking their respective permits in greater detail. Minnesota Power accompanied some of the inspections to look for energy efficiency opportunities. The positive results of the inspections were "one-stop shopping," which was appreciated by the companies and the MPCA found some sites that needed to be brought into the permitting process.

4.2. SPECIAL DESIGNATIONS

The Binational Program identifies two types of special designations with regulatory implications for the US portion of the Lake Superior Basin. Canadian equivalents do not currently exist, although Ontario and Canada are examining possible functional equivalents. The governors of Michigan, Minnesota and Wisconsin committed to "initiate appropriate state procedures" to adopt Lake Superior Outstanding International Resource Water (LSOIRW) and Lake Superior Outstanding National Resource Water (LSONRW).

Under LSOIRW, applications to discharge increased loads of any of the nine chemicals of concern would have to include an antidegradation demonstration, including installation of the best technology in process and treatment. Under LSONRW, any new or expanded discharges of the nine chemicals of concern would be prohibited. The Lake Superior special designations may be adopted as part of the Great Lakes Water Quality Initiative. In addition, the National Wildlife Federation (NWF) and others have petitioned the states and USEPA to adopt a Outstanding National Resource Water designation for the entire basin. Table I describes the special designations.

Currently, Minnesota is considering LSOIRW for the Minnesota portion of the Basin as part of the triennial standards review process. At the time this paper was prepared, two Wisconsin counties and two Native American bands have approved special designation resolutions supporting the NWF petition. Michigan has held public meetings on the LSOIRW designation. However, the Michigan Department of Natural Resources does not have enough agreement between interested parties to submit rule language to the

Michigan legislature at this time. In addition, a suit filed against USEPA by NWF
regarding USEPA's review of Michigan water quality rules will affect Michigan's future
actions regarding special designations.

TABLE I

Lake Superior special designations

Designation	Chemicals	Area	Sources
LSOIRW	nine zero discharge chemicals	US portion of the Lake Superior basin	point source discharges
LSONRW	nine zero discharge chemicals	special areas within the US portion of the Lake Superior basin	point sources discharges
NWF ONRW	critical chemicals in the Lake Superior LaMP, excluding Al, Cd, Cr, Fe, Mn, Ni, Pb and Zn	US portion of the Lake Superior basin	point and nonpoint sources

4.3 CONTROLS AND REGULATIONS

On the US side, the Binational Program incorporates a variety of ongoing regulatory
activities, including the GLI, the pulp and paper cluster regulation, sediment criteria and
remediation and Maximum Achievable Control Technology (MACT) standards. The
Program also requires stormwater analysis in cities over 5,000 people and incorporation
of Toxics Reduction Plans in National Pollutant Discharge Elimination System (NPDES)
permits. Electric utilities are likely to be affected most by nationwide programs that
involve emission standards, but some Binational Program activities may have an impact.
For example, power plants' NPDES permits may include a requirement for Toxics
Reduction Plans (TRPs) under the Binational Program.

Although controls and regulations that require treatment of point source discharges and
emissions are an important tool for achieving the zero discharge and zero emission goal,
that role is limited. Treatment to reduce the release of persistent and bioaccumulative
toxic chemicals is difficult and often results in transferring the chemicals from one
medium to another. For example, WLSSD discovered high levels of mercury in effluent
were due to scrubber water from treating emissions from their sludge and refuse derived
fuel incinerator. The scrubber water was rerouted. Although that loop is now mostly
closed, some of the mercury ends up in ash that is deposited in a WLSSD landfill (Stepun
and Tuominen, 1995). WLSSD is currently working to reduce mercury releases. The
incinerator emissions have dropped from 94 pounds in 1992 to 48 pounds in 1993 due to
improvements in scrubber treatment and source reduction (Coyer, 1995).

One special area of concern for electric utilities is reporting decommissioned PCB filled
or contaminated equipment. In the US, the Toxic Substances Control Act (TSCA)
requires that facilities that use PCB equipment report when that equipment is
decommissioned and the PCBs destroyed. The utilities in the Great Lakes have done a

good job reporting to the USEPA on the status of PCB equipment and have achieved significant reductions (Lingle and Wilson, 1993).

A voluntary inventory of PCBs in use and in storage provides a more accurate picture of the quantities of PCBs remaining on the Canadian side of the Basin (Superior Work Group, 1995b). Table II summarizes the Canadian PCB inventory. Note that Canadian PCBs were reported as liters and have been converted to kilograms assuming a density of 1.15 kg/L. Also, the Canadian inventory includes the total volume of PCB contaminated materials/fluids. (US estimates are reported as "pure PCBs.") Therefore, Canadian and US PCB estimates are not comparable.

The US governments are unable to estimate the quantity of PCBs still in use in the Lake Superior Basin with a high degree of accuracy. TSCA is good for tracking the quantity of PCBs decommissioned, but is not useful to determine the quantity of PCBs remaining. Some Lake Superior utilities have provided detailed inventories of remaining PCB equipment in the Lake Superior Basin (Minnesota Power and Light, 1994). Others, especially the small municipal utilities that characterize the scattered communities of the Basin are unquantified.

TABLE II.

PCB Inventory in the Ontario Portion of the Lake Superior Basin (kg).

Source	Storage	Use
Industrial	134,200	194,830
Utilities	57,880	11,300
Municipal	4,360	14,815
Commercial	22,140	32,700
TOTAL	218,580	253,645

Table III shows the estimated PCB use for utilities and industries on the US side of the Lake Superior Basin (Superior Work Group, 1995b). Information about the PCB inventory is essential for establishing the baseline for the Forum reduction schedule. It is also necessary to know who owns PCBs and in what quantities in order to design incentives to decommission PCB equipment. The state governments surrounding Lake Superior would appreciate comments from utilities on the concept of a PCB inventory and decommissioning incentives.

TABLE III.

Estimates of PCBs in Use in the US Portion of the Lake Superior Basin (kg).

Source	Low End Estimate	High End Estimate
Large utilities: capacitors	10,359	12,161
Large utilities: transformers	50	58
Small utilities	15,737	21,472
Industries	523,885	524,333
TOTAL	550,000	558,000

5. Summary

The Binational Program to Restore and Protect the Lake Superior Basin establishes an ambitious goal of zero discharge and zero emission of nine toxic chemicals, including some that are associated with electric utilities. The best strategy for achieving this goal is to change raw materials or processes to avoid the generation and release of these chemicals. Although additional action is needed to achieve the zero discharge goal, there has already been significant progress by the electric utilities in reducing some of the chemicals. Striving for the zero discharge goal will require effort from the electric utilities as well as their clients and governments. Rather than asking stakeholders to react to a plan drawn up without their input, the Lake Superior governments want involvement from all stakeholders from the beginning. Please examine the Forum recommendations and consider both the practical concerns and the innovative solutions that might influence the reduction schedules. Remember, this is a demonstration project. We are looking for innovative approaches, new partnerships, voluntary actions, suggestions for government policies, and industry expertise to contribute to future rulemaking efforts.

References

Coyer, G: 1995, *Zero Discharge Pilot Project Work Plan*. Western Lake Superior Sanitary District (WLSSD), 63 pp.

Lingle, J.W. and Wilson, D.S.: 1993, *Utility Response to EPA Region 5 PCB Phasedown Program*. Wisconsin Electric Power Company and Decision Focus, Inc., 36 pp.

Minnesota Power and Light: 1994, *Minnesota Power's Response to EPA's Request for "Voluntary PCB Phasedown Program,"* Minnesota Power and Light, 28 pp.

MPCA Mercury Task Force: 1994, *Strategies for Reducing Mercury in Minnesota*, Minnesota Pollution Control Agency (MPCA), 58 pp.

Smith, J. and Smith, I.R.: 1993. *Yardsticks for Assessing the Water Quality of Lake Superior*, Environment Canada Great Lakes Office and Ontario Ministry of the Environment and Energy, 109 pp.

Stepun, J. and Tuominen, T.: July 27, 1995, *Presentation to the Mercury Collaboration Committee*, Western Lake Superior Sanitary District (WLSSD).

Superior Work Group: 1995a, *Current Status of Critical Pollutants: LaMP Volume II- Critical Pollutants, Stage 1: Problem Identification*, 150 pp.

Superior Work Group: 1995b, *DRAFT Loading Reduction Targets for Critical Pollutants: Volume II-Critical Pollutants, Stage 2 Load Reductions/Management Options Program*, 231 pp.

CONSEQUENCES OF PROPOSED CHANGES TO CLEAN WATER ACT THERMAL DISCHARGE REQUIREMENTS

J. A. VEIL[1] and D.O. MOSES[2]

[1] *Argonne National Laboratory, 955 L'Enfant Plaza, SW, Suite 6000,Washington, DC 20024,* [2] *U.S. Department of Energy, PO-62, Washington, DC 20585*

Abstract. This paper summarizes three studies that examined the economic and environmental impact on the power industry of: a) limiting thermal mixing zones to 1,000 feet (~ 305 m), and b) eliminating the Clean Water Act (CWA) §316(a) variance. Both of these proposed changes were included in S. 1081, a 1991 Senate bill to reauthorize the CWA. The bill would not have provided for grandfathering plants already using the variance or mixing zones larger than 305 m. Each of the two changes to the existing thermal discharge requirements were independently evaluated. Power companies were asked what they would do if these two changes were imposed. Most plants affected by the proposed changes would retrofit cooling towers and some would retrofit diffusers. Assuming that all affected plants would proportionally follow the same options as the surveyed plants, the estimated capital cost of retrofitting cooling towers or diffusers at all affected plants ranges from $21.4 to 24.4 billion. Both cooling towers and diffusers exert a 1%-5.8% energy penalty on a plant's output. Consequently, the power companies must generate additional power if they install those technologies. The estimated cost of the additional power ranges from $10 to 18.4 billion over 20 years. Generation of the extra power would emit over 7.3 billion kg per year of additional carbon dioxide. Operation of the new cooling towers would cause more than 94.5 m^3 per second of additional evaporation. Neither the restricted mixing zone size nor the elimination of the §316(a) variance was adopted into law. More recent proposed changes to the Clean Water Act have not included either of these provisions, but in the future, other Congresses might attempt to reintroduce these types of changes.

Keywords. Thermal, mixing zone, §316(a) variance, cooling tower, diffuser.

1. Introduction

1.1 BACKGROUND

In its first session, the 104th Congress has considered reauthorizing or amending many of the major environmental statutes. One example is H.R. 961, a Clean Water Act (CWA) reauthorization bill passed by the House in May 1995 that proposes many changes to water pollution control requirements. Several years earlier (102nd Congress 1991-1992), the Senate considered and widely debated S. 1081, a CWA reauthorization bill that would have significantly tightened water pollution controls. Of particular interest to the electric power industry were two proposed changes to the existing thermal discharge requirements. The bill proposed: a) limiting mixing zones, including thermal mixing zones, to 305 m from the point of discharge, and b) deleting §316(a) from the CWA, which would take away the valuable §316(a) thermal variance provision. No grandfathering provisions were included for facilities currently using a §316(a) variance or a mixing zone larger than 305 m. Both of these changes were thought to have large economic consequences for the electric power industry.

Although S. 1081 was not passed by the Senate, it is uncertain what changes, if any, Congress will make regarding thermal discharges. While the 104th Congress may not

Water, Air and Soil Pollution **90**: 41-52, 1996.
© 1996 *Kluwer Academic Publishers.*

propose new stricter thermal discharge requirements, future Congresses, with different political makeups and interests, might revisit stricter thermal discharge requirements, such as those proposed in S. 1081. This paper summarizes three reports by Argonne National Laboratory that assess the economic and environmental impacts that such legislative changes could have on the power industry, if such changes were enacted. Two of the reports estimate the impacts of deleting the §316(a) variance from the CWA (Veil, 1993a; Veil et al., 1993) and the third report estimates the impacts of limiting thermal mixing zones to 305 m (Veil, 1994).

1.2. USE OF COOLING WATER IN POWER GENERATION

Water is used in many industrial applications for cooling machinery or for condensing steam. The largest industrial user of cooling water is the steam electric power industry. Most power plants use either once-through cooling or closed-cycle cooling. Once-through cooling systems withdraw large volumes of water from a river, lake, estuary, or ocean, pump the water through the condenser, and return it to the same or a nearby body of water. Closed-cycle cooling systems[a] rely on a cooling tower and basin, cooling pond, or cooling lake. Water is withdrawn from the cooling tower basin, pond, or lake; pumped to the condenser; and then returned to the basin, pond, or lake.

The majority of the generating units that would be affected by the proposed changes to the thermal discharge requirements presently use once-through cooling systems. In 1991, the steam electric generating capacity in the United States totaled 568,871 megawatts (MW). Generating units with 250,466 MW (44%) of this capacity used once-through cooling (Edison Electric Institute, 1993).

1.3. EXISTING THERMAL DISCHARGE REQUIREMENTS

The temperature of the discharged cooling water is typically limited by state thermal water quality standards. Some states have established site-specific water quality standards for temperature or have otherwise established water quality standards that take into account existing power plant discharges. There are two mechanisms that can allow adjustments to the thermal limits. Either or both of these mechanisms can be used at a power plant.

First, thermal mixing zones may be allowed, which provide an opportunity for dilution and in-stream cooling of heated discharges prior to measuring compliance with the thermal standards. Each state has different mixing zone size and shape criteria. Most once-through cooling discharges rely on thermal mixing zones. Second, §316(a) of the CWA allows the states or EPA to establish alternative thermal limits if the discharger can demonstrate that the otherwise applicable thermal effluent limits are more stringent than necessary to protect the organisms in and on the receiving water body, and that other, less stringent effluent limitations would protect those organisms.

---

[a] In this report, the term "closed-cycle cooling" means "cooling tower" unless some other type of closed-cycle cooling system is specifically mentioned.

About 75% of the domestic generating capacity using once-through cooling systems operate under §316(a) variances (Edison Electric Institute, 1993).

2. Methodology

Information was collected from a sample of power plants in different parts of the country that use once-through cooling. In separate surveys, selected power companies were asked what each plant would do: a) if it had to meet thermal limits within a 305-m mixing zone, or b) if the §316(a) variance were no longer available. The power companies also were asked to provide cost estimates, when available, for constructing new facilities and equipment to meet the changed requirements. While geographical diversity was sought and achieved, this was not a statistically random sample of the steam electric power industry. Therefore, the results of this study may not be representative of the entire industry. On the other hand, there is no reason to believe that the cost estimates developed in this manner are not a reasonably good gauge of the actual costs to the industry. Responses were received from 13 companies representing 79 plants for the mixing zone study and from 14 companies representing 38 plants for the §316(a) variance study.

The data from the power companies were used to develop capital cost rates in terms of dollars per kilowatt ($/kW)(Veil, 1993a; Veil, 1994). To estimate national capital costs, these cost rates were multiplied by the national affected capacity in megawatts (MW). The affected capacity was assumed to consist of those generating units that were currently using a §316(a) variance and those generating units that would be unable to meet limits based on a 305-m mixing zone. This methodology assumes that the limited sample of plants providing data is proportionately representative of the nationwide power industry.

3. Results

3.1. §316(A) VARIANCE STUDY - CAPITAL COSTS

Approximately 680 units would be affected if the §316(a) variance were lost. These units have a combined generating capacity of roughly 189,000 MW, which represents 33% of the total steam electric generating capacity in the United States. Of those 189,000 MW, approximately 43,000 MW are attributable to nuclear plants and approximately 146,000 MW are attributable to fossil-fuel plants (Edison Electric Institute, 1992). There are a variety of potential alternatives for each affected plant, including:

- seeking relaxed thermal discharge requirements from the state regulatory agency;
- retiring a plant;
- operating only seasonally or at a lower output;
- moving the discharge structure to deeper water;

- adding a diffuser; and
- retrofitting[b] once-through cooling systems with cooling lakes or ponds or cooling towers.

The 14 power companies that provided information reported that they would retrofit cooling towers at nearly all of their 38 plants now operating under §316(a) variances. Figures 1 and 2 show the capital cost estimates provided by the power companies to retrofit cooling towers as a function of power production capacity. Because costs for construction at a nuclear plant are nearly always higher than those for construction at a fossil-fuel plant, data are presented separately for the two fuel types. The reported cost rates ($/kW scaled to 1992 dollars) for fossil-fuel plants range from $32/kW to $346/kW, with an average of $108/kW for 31 plants. The cost rates for nuclear plants range from $102/kW to $234/kW, with an average of $171/kW for 7 plants (Veil, 1993a). Linear regression analysis was performed on the data. The resulting regression equations and correlation coefficients (r) are shown below.

fossil-fuel plants	$y = 0.105x + 2.2$	$r = 0.77$	(1)
nuclear plants	$y = 0.151x + 31.4$	$r = 0.53$	(2)

where y = millions of 1992 dollars and x = MW

Lines representing these two equations are plotted on Figures 1 and 2. The regression line for fossil-fuel plants fits nicely and is statistically very significant (probability < 0.01). On the other hand, the regression line for nuclear plants does not represent the data set precisely. It is not statistically significant (probability = 0.125).[c]

National cooling tower retrofit costs were estimated by multiplying the appropriate cost rates by the affected capacity (146,000 MW for fossil-fuel plants and 43,000 MW for nuclear plants). For fossil-fuel plants, both the average fossil-fuel cost rate ($108/kW) and the slope of the fossil-fuel regression line ($105/kW) were used. Since the slope of the regression line for nuclear plants is not a reliable indicator of the data set, two other approaches -- the average nuclear cost rate ($171/kW) and the median nuclear cost rate ($201/kW) -- were used to develop the national retrofit cost for nuclear plants.

The results of this analysis show that if §316(a) of the CWA were eliminated and all plants currently operating under §316(a) variances were retrofitted with cooling towers, the estimated national capital cost would range from $15.3 billion to $15.8 billion for fossil-fuel plants and from $7.4 billion to $8.6 billion for nuclear plants. The combined

---

[b] Retrofitting means modifying the existing cooling system and installing additional equipment. This may involve major construction and rerouting of piping.

[c] These probabilities represent the probability that the hypothesis (in this case, that the regression equation accurately expresses the distribution of data points) is incorrect. Generally, a probability ≤ 0.05 is considered significant (the hypothesis is accepted), and a probability < 0.01 is very significant.

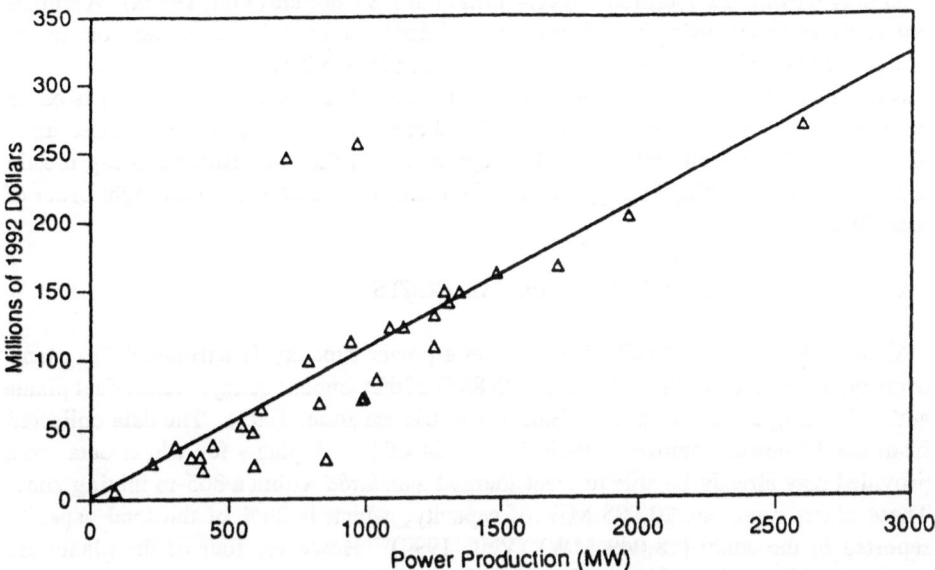

FIGURE 1 Capital Cost to Retrofit Fossil-Fuel Units with Cooling Towers

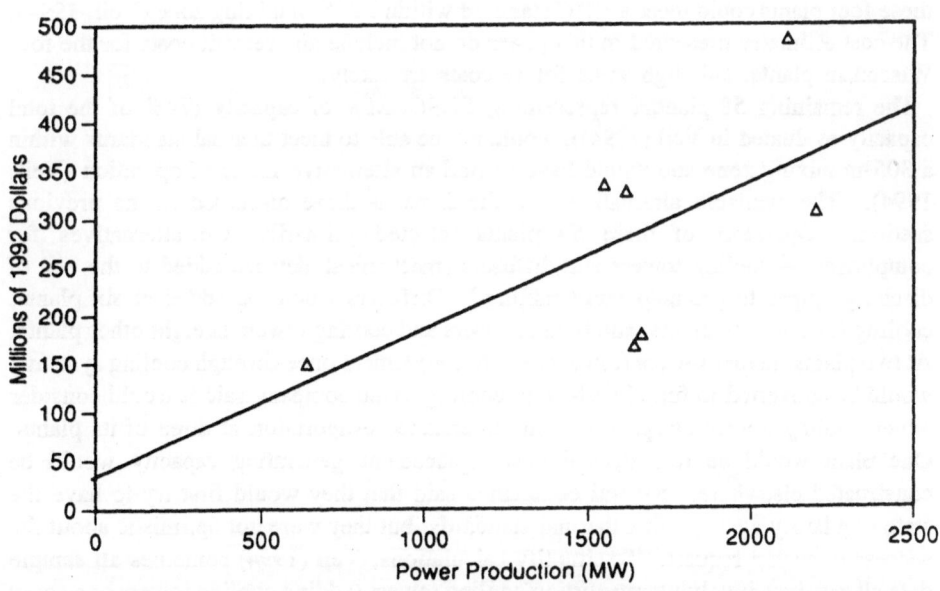

FIGURE 2 Capital Cost to Retrofit Nuclear Units with Cooling Towers

total ranges from $22.7 billion to $24.4 billion in 1992 dollars (Veil, 1993a). A similar but separate study, using a different methodology, estimated the capital cost to the power industry of losing the §316(a) variance at $28.9 billion in 1992 dollars (Edison Electric Institute, 1992). The estimate from Edison Electric Institute (1992) is based on two hypothetical plants, one fossil-fuel and one nuclear, with the costs scaled up to all affected plants. The relatively close agreement of the two estimates using mostly independent methodologies suggests that the estimates are at least in the right order of magnitude.

3.2. MIXING ZONE STUDY - CAPITAL COSTS

About 44% (250,466 MW) of the nation's power capacity is attributable to plants using once-through cooling systems, with 85% of that amount being at fossil-fuel plants and 15% being at nuclear plants (Edison Electric Institute, 1993). The data collected from the 13 power companies indicate that 24 of the 79 plants for which data were provided may already be able to meet thermal standards within a 305-m mixing zone. These plants represent 20,085 MW of capacity, which is 26% of the total capacity reported in the study (78,049 MW) (Veil, 1994). However, four of the plants are located in Wisconsin, which at the time Veil (1994) was published did not have any thermal water quality standards. Therefore, these four plants are reported as being able to meet thermal standards within a 305-m mixing zone. Wisconsin was considering establishing a thermal standard, with 89°F (32°C) being the likely standard. None of these four plants could meet a 32°C standard within a 305-m mixing zone (Veil, 1994). The cost estimates presented in this paper do not include any retrofit costs for the four Wisconsin plants, although some future costs are likely.

The remaining 58 plants[d], representing 57,964 MW of capacity (74% of the total capacity evaluated in Veil (1994)), would not be able to meet thermal standards within a 305-m mixing zone and would have to find an alternative mode of operation (Veil, 1994). The available alternatives are the same as those discussed in the previous section. Operators of these 58 plants selected primarily two alternatives for compliance — cooling towers and diffusers (mechanical devices added to the end of discharge pipes to promote rapid mixing.) Diffusers would be added at six plants, cooling towers at 39 plants, and both diffusers and cooling towers at eight other plants. At two plants, helper towers (towers used to supplement once-through cooling systems) would be converted to full closed-cycle cooling. One company said it would consider either cooling towers or spray systems to enhance evaporation at three of its plants. One plant would be retired, and new replacement generating capacity would be constructed elsewhere. Several companies said that they would first try to have the state regulatory agency relax thermal standards, but they were not optimistic about the success of such a request. To simplify calculations, Veil (1994) combines all sample data alternatives involving retrofitting cooling towers (adding cooling towers or a spray

---

[d] The total of 24 plants meeting the mixing zone and 58 plants not meeting the mixing zone adds to 82 rather than 79 plants. Three of the plants have one set of units that can meet the mixing zone and a different set of units that cannot meet the mixing zone. Those plants have been counted in both categories.

system, converting helper towers to full-time closed-cycle cooling, or adding cooling towers and diffusers) with the one plant that would be retired° into a single category (52 plants). The six plants using just diffusers constitute a second category.

The cooling tower cost rates used to calculate a national cost estimate are the higher end of the ranges from Veil (1994) -- $108/kW for fossil-fuel plants and $201/kW for nuclear plants. For diffusers, the capital cost rates from Utility Water Act Group (1978) were modified to equal $43/kW for fossil fuel plants and $59/kW for nuclear plants (Veil, 1994).

Table I shows the selected alternatives, the percentage of capacity (based on the sample data for this study) anticipated to use each alternative, the estimated capacity (in MW) nationwide that would use each alternative, the cost rates, and the total capital costs. The estimated national capital cost for retrofitting plants that cannot meet thermal standards within a 305-m mixing zone is up to $21.4 billion.

Table I

Calculation of National Cost Estimate (from Veil (1994))

Selected Alternative	Percentage of Capacity Using Alternative*	Affected Capacity Using Alternative (MW)	Cost Rate ($/kW)	Total Cost (million $)
Diffuser				
Fossil-fuel plants	5	12,523	43	538
Nuclear plants	1	2,504	59	148
Cooling towers				
Fossil-fuel plants (mechanical-draft towers)	58	145,270	108	15,690
Nuclear plants (natural-draft towers)	10	25,047	201	5,034
No changes needed	26	65,121	N/A	0
Total	100	250,466	N/A	21,410

* Based on data supplied by power companies for 79 plants.

The cost of retiring a plant and building new generating capacity greatly exceeds the cost of adding a cooling tower. ɪ this case, the cost of installing a cooling tower is used rather than the full cost of building new generating capacity to eep the plant in the data set, and to make the final cost estimate more conservative.

3.3. COSTS ASSOCIATED WITH THE ENERGY PENALTY

Retrofitting cooling towers or diffusers to existing power plants results in a reduction of plant output known as the energy penalty. The energy penalty is caused by increases in turbine back pressure that result in less efficient power generation, additional power requirements for pumping recycled water to the top of a natural-draft cooling tower or operating the fans at a mechanical draft cooling tower, and increased pump head requirements due to the restricted flow through a diffuser. Power companies have several options for dealing with the energy penalty. They can operate the plant at lower net power output, or in some cases, they can run it more frequently or at a higher temperature. The latter option requires that additional fuel be burned to maintain output. In either case, there is an energy cost associated with the retrofitting of cooling towers.

Cooling towers result in an energy penalty for fossil-fuel plants ranging from 1.1 to 4.6%, with most of the data falling between 1.5 and 2.5%. The cooling tower energy penalty for nuclear plants ranges from 1.0 and 5.8%, with the selected data falling between 2 to 3% (Veil et al., 1993). The energy penalty from diffusers is 0.02% for fossil-fuel plants and 0.028% for nuclear plants (Utility Water Act Group, 1978).

The cost of compensating for the energy penalty has two components — the cost of generating replacement energy and the capital cost of building new generating capacity[f]. The replacement energy cost is a function of the cost per kilowatt-hour, historical capacity factors, and the percent energy penalty. Table II summarizes the cost for replacement energy from Veil et al. (1993) and Veil (1994). Annual energy costs range from $370 million to 670 million. The levelized, 20-year costs assume zero real escalation in fuel and variable operating and maintenance costs, a discount rate of 10.5%, and an annual inflation rate of 4%, resulting in a 20-year energy cost ranging from $10 billion to 18.4 billion.

Table II

Nationwide Generation Capacity to Be Replaced Due to Energy Penalty
and Associated Energy and Capital Costs (based on Veil et al. (1993) and Veil (1994))

Category	Replacement Capacity Needed (MW)	Annual Energy Cost ($ million)	Levelized Energy Cost for 20 Years ($ billion)	Capital Cost for Replacement Capacity ($ billion)
Loss of §316(a) Variance	3050 - 4940	420 - 670	11.4 - 18.4	1.4 - 5.3
1000' Mixing Zone	2700 - 4400	370 - 590	10 - 16.2	1.2 - 4.8

---

[f] The percentage of power required to compensate for the energy penalty is relatively small. Some utilities would utilize reserves in their own system or purchase power from other systems. However, if the proposed changes were applied nationwide, the overall energy demand to compensate for the energy penalty would necessitate construction of new generating capacity.

The capital cost is determined on a utility-specific basis considering existing reserve margins, construction schedules for planned facilities, fuel prices, load projections, and the availability and cost of power purchases from other systems. To account for these capital cost issues, Veil et al. (1993) used a range of replacement capacity costs -- $450/kW for a 50-MW combustion turbine unit to $1080/kW for upgrading a 500-MW coal unit to 600 MW (EPRI, 1988; EPRI, 1989). Table II summarizes the energy penalty capital costs estimated in Veil et al. (1993) and Veil (1994). The capital costs for replacement capacity range from $1.2 billion to 5.3 billion.

3.4. ENVIRONMENTAL IMPACTS

Retrofitting cooling towers and diffusers at existing plants using once-through cooling would create secondary environmental impacts. For example, generating additional power to meet the energy penalty would increase carbon dioxide emissions by an estimated 7.3 billion kg per year for plants restricted to a 305-m mixing zone and 8.2 billion kg per year for plants losing the §316(a) variance (Veil et al., 1993; Veil, 1994). Although estimates for other airborne pollutants were not generated, substantial quantities of SO_x and NO_x would also be released. Construction of new generating units to meet the additional capacity required by the energy penalty would cause changes in land use, runoff characteristics, and wildlife habitat, among other direct impacts. Conversion from once-through cooling systems to cooling towers would result in increased evaporation of about 164 m^3 per second for plants not meeting the mixing zone limits and 94.5-176 m^3 per second for plants losing the variance (Veil et al., 1993; Veil 1994). Other potential impacts from cooling towers include cooling tower drift (fog plumes and downwind deposition of water droplets high in total dissolved contaminants), noise, aesthetics, additional discharge of biocides in cooling tower blowdown, and additional solid waste generated as cooling tower basin sludge.

Many of the environmental impacts identified above are not as significant at plants where cooling towers have been designed and built as part of the original installation. This conclusion might not hold true if a large number of new cooling towers were installed as retrofits. Many of the plants that currently operate under §316(a) variances are older plants located in or near urban or suburban areas. Environmental impacts like drift or noise, which are mitigated by the large buffer zones around plants in rural locations, could present serious problems for urban locations. Freezing or fogging from cooling tower plumes could present a safety hazard.

The water consumption issue may be the most critical concern for a retrofitted cooling tower. If a plant is designed to consume a certain volume of water through evaporation, then that volume is factored in from the time a plant is built. However, if a cooling tower is added later, adequate water resources may not be available to accommodate the increased fresh water demand.

4. Discussion

4.1. REGULATORY AGENCY MUST BE SATISFIED THAT 316(A) VARIANCE IS PROTECTIVE

Although potential environmental impacts can be associated with discharges permitted under §316(a) variances, regulatory agencies have the responsibility to ensure that the impacts are minimal or nonexistent. A §316(a) variance is not trivially granted. To receive the variance, the discharger must demonstrate to the satisfaction of the regulatory agency that a discharge that exceeds the otherwise applicable thermal requirements will still protect a balanced, indigenous population in and on the receiving water. The effort required to make this case varies greatly, depending on state requirements and the site-specific potential for impacts. Typically the demonstration involves extensive evaluation of potential impacts and characterization of local aquatic populations. A regulatory agency can reject a demonstration or ask the discharger to study certain issues in more detail.

In 1992, the U.S. Environmental Protection Agency (EPA) evaluated the effectiveness of the §316(a) variance program. EPA found that for the majority of facilities, impacts from thermal effluent have not been found to be large or permanent. Several cases in which severe problems were found may have been the result of inadequate permit limits, rather than facility noncompliance with permit limitations (EPA, 1992).

Considering the oversight authority provided to regulatory agencies, the §316(a) variance program represents an "environmentally safe" form of economic incentive for thermal dischargers. The use of §316(a) and other types of CWA variances can greatly reduce the cost to dischargers (and concomitantly, ratepayers) without increasing the risk to the environment beyond acceptable levels (Veil, 1993b).

4.2. OTHER AFFECTED INDUSTRIES

The steam electric power industry is the primary industrial sector that would be affected by a 305-m mixing zone limit, but other industries that use large volumes of water for cooling (e.g., steel, aluminum, paper, and cement manufacturing industries and waste-to-energy facilities) might also be affected. Impacts of the proposed mixing zone restrictions on these other industries were not estimated. The §316(a) variance is used almost exclusively by the power industry. Consequently, loss of the variance would have relatively little impact on other industrial sectors.

5. Conclusions

• A 1991 Senate bill, S. 1081, which would have prohibited §316(a) variances and restricted thermal mixing zones to 305 m, was not passed by the 102nd Congress. Although CWA reauthorization legislation introduced subsequently does not propose prohibiting §316(a) variances or restricting thermal mixing zones, it remains unclear how Congress will ultimately deal with thermal discharge issues. Until Congress has formally reauthorized the CWA, the potential for changes to the thermal requirements remains.

• Based on data collected from a large sample of power companies representing different geographic regions, most plants currently operating under a §316(a) variance could not meet thermal standards without the variance and most plants using once-through cooling systems could not meet thermal standards within a 305-m mixing zone. The estimated cost to retrofit cooling towers and diffusers at existing power plants would range from $21.4 billion to 24.4 billion. Owners of the retrofitted plants would need to spend an estimated $10 billion to 18.4 billion in additional fuel costs and from $1.2 to 5.3 billion to construct additional generating capacity to overcome the energy penalty.

• The §316(a) variance program has not caused significant environmental degradation and has resulted in considerable cost savings to ratepayers, yet deletion of the variance would have a large negative economic impact on the power industry. The magnitude of the estimated costs of losing the §316(a) variance is similar to the 1993 net electric operating income for major U.S. investor-owned electric utilities ($30.2 billion) (DOE, 1995). There appears to be little justification for deleting the §316(a) variance.

• Any attempt to place statutory restrictions on thermal mixing zones would result in a very large cost to the power industry and perhaps to other industries as well. Policymakers should give careful consideration to these costs, in addition to the increased air emissions, solid wastes, and water evaporation attributable to cooling towers and diffusers, before adopting any national thermal mixing zone restrictions. The potential benefits of a 305-m or smaller mixing zone have not been widely discussed. Unless the potential benefits are believed to be commensurate with the large cost, little justification exists for limiting thermal mixing zones to 305 m from the point of discharge.

Acknowledgements

This work was sponsored by the U.S. Department of Energy, Office of Policy under Contract W-31-109-Eng-38. C. Richard Bozek of Edison Electric Institute and Kristy A.N. Bulleit and Steven J. Koorse of Hunton & Williams (representing the Utility Water Act Group) were instrumental in coordinating data submissions and industry review of the draft reports.

References

Edison Electric Institute: 1992, *Evaluation of the Potential Costs and Environmental Impacts of Retrofitting Cooling Towers on Existing Steam Electric Power Plants That Have Obtained Variances under Section 316(a) of the Clean Water Act*, prepared by Stone and Webster.

Edison Electric Institute, 1993: *Power Statistics Database*, Utility Data Institute.

Electric Power Research Institute (EPRI): 1988, *An Evaluation of Integrated-Gasification-Combined-Cycle and Pulverized-Coal-Fired Steam Plants*, EPRI AP-5950.

Electric Power Research Institute (EPRI): 1989, *TAG Technical Assessment Guide, Electrical Supply — 1989*, EPRI P-6587-L, Vol. 1, Rev. 6.

U.S. Department of Energy, Energy Information Administration (DOE): 1995, *Financial Statistics of Major U.S. Investor-Owned Electric Utilities*, DOE/EIA-0437(93)/1.

U.S. Environmental Protection Agency (EPA): 1992; *Review of Water Quality Standards, Permit Limitations, and Variances for Thermal Discharges at Power Plants*, EPA 831-R92001.

Utility Water Act Group: 1978, *Thermal Control Cost Factors*, prepared by National Economic Research Associates and Stone and Webster Engineering Corp.

Veil, J.A.: 1993a, *Impact on the Steam Electric Power Industry of Deleting Section 316(a) of the Clean Water Act: Capital Costs*. Argonne National Laboratory, ANL/EAIS-4.

Veil, J.A.: 1993b, *Water Environ. & Tech.* **5**(8), 25-26.

Veil, J.A., VanKuiken, J.C., Folga, S., and Gillette, J.L.: 1993, *Impact on the Steam Electric Power Industry of Deleting Section 316(a) of the Clean Water Act: Energy and Environmental Impacts*. Argonne National Laboratory, ANL/EAIS-5.

Veil, J.A.: 1994, *Impact of a 1,000-Foot Thermal Mixing Zone on the Steam Electric Power Industry*. Argonne National Laboratory, ANL/EAD/TM-15.

PART III

ECOLOGICAL / HEALTH RISKS

MULTIPATHWAY HEALTH RISK ASSESSMENT OF POWER PLANT WATER DISCHARGES

C. SEIGNEUR[1], E. CONSTANTINOU[1] AND L. LEVIN[2]

[1]ENSR Consulting and Engineering, 1420 Harbor Bay Parkway, Suite 160, Alameda, California 94502,
[2]Electric Power Research Institute, 3412 Hillview Avenue, Palo Alto, California 94303

Abstract. The chemicals released with water discharges from a fossil-fueled power plant may present health risks through a variety of exposure pathways including ingestion of drinking water, fish consumption and dermal absorption while swimming. The Total Risk of Utility Emissions (TRUE) model provides a framework that allows one to assess the multipathway health risks associated with water discharges from power plants. The TRUE model formulation is presented with particular emphasis on the treatment of water discharges. The application of the TRUE model to a coal-fired power plant is described and the potential health risks by chemical and exposure pathway are discussed.

Key Words: Health risk, power plant, water discharges, drinking water, fish consumption, dermal absorption, ingestion

1. Introduction

The discharge of wastewater from industrial facilities is currently regulated by means of National Pollutant Discharge Elimination System (NPDES) permits for point source discharges and stormwater type permits for non-point source discharges. Concerns over potential health and ecological impacts of some chemicals present in wastewater discharges have prompted the consideration of the analysis of such impacts. To address the potential health effects of water discharges from power plants, it is necessary to quantitatively evaluate the exposure of the human population to the discharged chemicals through a simulation of their fate and transport in surface water, uptake by the human population and associated health effects. To that end, the Electric Power Research Institute (EPRI) sponsored the development of a multimedia health risk assessment model that allows the evaluation of the health risks associated with the chemical releases from power plants into the environment. This model named, "Total Risk of Utility Emissions," (TRUE) is described, with particular emphasis given to the formulation of the surface water transport and fate module and water exposure pathways. A case study that demonstrates the application of the TRUE model to a coal-fired power plant is also presented.

Water, Air and Soil Pollution **90**: 55-64, 1996.
© 1996 *Kluwer Academic Publishers.*

2. Description of the TRUE Model

2.1 OVERVIEW

The TRUE model was designed to address the potential public health impacts of multimedia chemical releases from fossil-fueled power plants. Emissions from the power plant stacks, wastewater discharges and disposal of solid waste are input to the model along with environmental and physiographic characteristics of the study area. Using this information, the model then predicts estimates of the potential carcinogenic and chronic non-carcinogenic health risks to exposed individuals as well as to the local population as a whole.

The TRUE model is particularly useful in assessing the overall effectiveness of installing control equipment that will reduce the release of chemicals in one environmental medium but may, simultaneously, increase the amount of chemicals released into another medium. An example of such a control equipment would be a flue gas scrubber which will reduce stack emissions of sulfur dioxide, soluble gases (e.g., HCl and oxidized Hg), and some particulate-bound chemicals, but may lead to additional water discharges and/or solid waste disposal. The overall risk reduction benefits of a flue gas scrubber can, therefore, be assessed by using the TRUE model to estimate the potential health risks of the multimedia chemical releases from the power plant, before and after installation of the scrubber.

We present in this section an overview of the formulation of the TRUE model. A more detailed description of the model is provided by Constantinou and Seigneur (1993). Particular emphasis is given to the description of the surface water component of TRUE.

2.2 OVERALL MODEL FRAMEWORK

TRUE is composed of several individual modules that handle the fate and transport of chemicals in the environment and the foodchain and predict the levels of exposure and potential health risks to human individuals and populations. The environmental media considered are air, soil, surface water and groundwater. Chemical contamination of air, soil and surface water may result either directly through chemical release into each specific medium (e.g., stack emissions to the atmosphere, water discharges into a river) or indirectly through intermedia transport processes that carry chemicals from one medium to another (e.g., deposition that carries chemicals from the atmosphere to soil or surface water surfaces). Contamination of groundwater may occur only indirectly, through leaching from the overlying soil column.

Transport of chemicals in the foodchain is simulated for a total of five components (i.e., produce, fish, beef, dairy milk and mother's milk). The concentrations in each of these foodchain components are calculated by means of empirically derived, chemical-specific bioconcentration/bioaccumulation factors (Bonfiglio et al., 1994). An exception is made in the aquatic treatment of mercury which is based on a state-of-the-art treatment of the reactions of mercury in water and its uptake by various trophic levels that include phytoplankton, zooplankton, small fish and piscivorous fish (Hudson et al., 1994; Constantinou and Seigneur, 1994).

Human exposure to chemicals present in air, soil, water, and food can be assessed through the characterization of human activities. The location, time, and duration of exposure of a population or an individual determine their exposure. The assumptions used in an individual's exposure calculations, as well as typical values for the parameters involved in these calculations, are presented by Liu (1994) and are as follows: exposure duration = 70 yrs; body weight = 70 kg; inhalation rate = 20 m^3/d; ingestion rate for drinking water = 2 l/d; ingestion rate for soil = 100 mg/d; ingestion rate for fish = 0.037 kg/d, beef = 0.075 kg/d, dairy products = 0.30 kg/d, vegetables = 0.05 kg/d and fruit = 0.028 kg/d.

The amount of a chemical that enters the human body through inhalation, ingestion, and dermal absorption represents the dose of that chemical received by an individual. Three major pathways are considered: (1) inhalation (gases and particles), (2) ingestion (food, drinking water, swimming water, and soil), and (3) dermal absorption (water and soil).

The effects of a chemical on human health are estimated through the use of dose-response models for carcinogenic effects, chronic non-carcinogenic effects, and acute non-carcinogenic effects. These dose-response models are based on toxicological animal data, epidemiological human data and, in limited cases, physiological and pharmacokinetic data.

Twenty-one chemicals (or groups of chemicals) are presently included in TRUE. These include sixteen inorganic chemicals (HCl, HF, NH_3 and HCN, and the following 12 metals: As, Be, Cd, Cr, Cu, Pb, Mn, Hg, Mo, Ni, Se, and V); three organic species (benzene, formaldehyde, and toluene), and two organic groups: polycyclic aromatic hydrocarbons (PAHs) and dioxins/furans. Toxicology profiles for the above-listed chemicals were developed by Von Burg et al. (EPRI, 1994).

The dose-response relationships include reference doses for non-carcinogenic chronic effects and cancer potency slope factors for carcinogenic effects. The basis for selecting these dose-response relationships is discussed by Brown and Schwab (1994). Acute effects were not considered to be an issue for routine emissions from a power plant. A reference dose is defined as the dose, with an appropriate margin of safety, below which no non-carcinogenic health effects are anticipated. If the ratio of the calculated dose divided by the reference dose (hazard index) is less than 1, no chronic health effects are anticipated. For carcinogenic health effects, the lifetime individual risk of developing cancer is calculated by multiplying the dose by a cancer potency slope factor. The cancer potency slope factor corresponds to the estimated risk of developing cancer per unit dose of chemical received. It is, therefore, expressed in inverse units of dose.

The individual modules of TRUE are connected through interfaces that transfer information from one module to another (e.g., chemical loads in overland runoff are transferred to the surface water module to be used in the prediction of chemical concentrations in surface water).

The model study area in TRUE is defined by a radius of up to 100 km around the power plant. The study area is divided into population subregions defined by angular sectors and radial divisions. Environmental calculations, exposure-dose, and health effect calculations for the exposed individuals are performed within each subregion.

2.3 SURFACE WATER FATE AND TRANSPORT MODEL

Liquid-borne chemical wastes resulting from the operation of power plants may be directly released into nearby surface water bodies. Additional pollutant loads to surface water bodies may result from (1) transport of a fraction of the chemicals present on site (e.g., coal pile, solid waste disposal) by overland stormwater runoff and (2) transport of a fraction of chemical mass deposited from the atmosphere by overland runoff.

Surface water models, that simulate the transport and fate of chemicals in surface water bodies (i.e., rivers, lakes, reservoirs, estuaries, and embayments), were reviewed by Villars *et al.* (1993). The major transport and fate mechanisms accounted for in surface water models include advection, dispersion and molecular diffusion, sedimentation and resuspension, adsorption-desorption, chemical decay, and biological degradation. The surface water model selected by Villars *et al.* (1993) to simulate these processes in TRUE is the surface water component of the Water Emissions Risk Assessment Model (WTRISK) (Bolton, 1989).

The surface water transport model of WTRISK is based on the following assumptions:

- Steady-state conditions
- Plug flow with uniform vertical and lateral dispersion for rivers and estuaries
- Well-mixed volume with turnover rate for lakes, reservoirs and embayments
- Water discharges occurring from point sources, runoff, and distributed sources with or without additional flow contributions
- First-order decay or removal (e.g., sedimentation) of chemicals

In WTRISK, the fate and transport of chemicals in a river system is simulated by dividing the river into segments, so-called river reaches, along which the geometric and flow characteristics are assumed to be fairly uniform. The chemical concentrations are calculated at several points along a river reach and a mass balance is performed at the junction between two continuous river reaches to calculate the resulting concentration at the beginning of the downstream reach. The basic equations governing the fate and transport of a chemical along a river reach in the case of a point source with distributed load are summarized below:

$$C(x) = \frac{1}{K}\left(\frac{W_p}{A}\right) + \left[C_0 - \frac{1}{K}\left(\frac{W_p}{A}\right)\right] \cdot \exp\left[-K \cdot \frac{x}{u}\right] \qquad (1)$$

where: x = Downstream distance from the beginning of the reach (L); C(x) = Chemical concentration at point x (ML^{-3}); C_0 = Chemical concentration at x=0 (ML^{-3}) (see Equation (2)); W_p = Distributed chemical load along the reach $(MT^{-1}L^{-1})$; K =

First-order chemical removal constant (T^{-1}); A = River cross-sectional area; and u = Average reach flow velocity (LT^{-1}).

The chemical concentration at the beginning of the reach, C_0, is calculated from the following equation:

$$C_0 = \frac{Q_u C_u + Q_p C_p}{Q_u + Q_p} \tag{2}$$

where: Q_u = Upstream water discharge ($L^3 T^{-1}$); C_u = Upstream chemical concentration (ML^{-3}); Q_p = Point source water discharge ($L^3 T^{-1}$); and C_p = Point source chemical concentration (ML^{-3}).

Equation (1) is slightly modified in the case where the distributed chemical load is accompanied by distributed water discharge.

For the estimation of chemical concentrations in a lake, WTRISK assumes complete mixing and takes into account the processes of decay and sedimentation, modeled as first-order reactions. The equation used for the calculation of the uniform chemical concentration in a lake is as follows:

$$C = \frac{W_{pL}}{Q_L + V_L (K_1 + K_2)} \tag{3}$$

where: C = Uniform chemical concentration in lake (ML^{-3}); W_{pL} = Chemical loading rate into the lake (MT^{-1}); Q_L = Lake exchange flow rate ($L^3 T^{-1}$); V_L = Lake volume (L^3); K_1 = First-order decay constant of chemical in the lake (T^{-1}); and K_2 = First-order sedimentation constant of chemical in the lake (T^{-1}).

3. Case Study: Water Discharges from a Coal-Fired Power Plant

The TRUE model was applied to estimate the potential public health impacts of water discharges from a coal-fired power plant into a river. The boiler of the facility is a 565 MW unit which burns bituminous coal of eastern origin. Low- and medium-sulfur coals are blended together to provide a fuel with approximately 1.5% sulfur content by weight. An electrostatic precipitator (ESP) is used for particulate control.

Bottom ash is removed from the boiler ash hoppers by an ash sluicing system. After a period of dewatering, the bottom ash is loaded on trucks and disposed of in a safe manner off-site. The associated sluice water is collected in a retention pond and the clarified effluent is discharged into a nearby creek which drains into a major river (NPDES outfall x). Fly ash is removed from the ESP hopper by an ash sluicing system as well. The sluiced fly ash is transported via pipeline to an abandoned mine five miles away. Water accumulated on the mine floor is pumped out and sent to a water treatment facility. After clarification, the water is discharged into a nearby creek which drains into the same river (NPDES outfall y).

The concentrations of 14 chemicals were sampled and analyzed in the two liquid discharge streams. These concentrations are presented in Table I.

TABLE I
Liquid Discharges[1]

Chemical	Retention Pond (NPDES x) Discharge Rate (mg/s)	Water Treatment Facility (NPDES y) Discharge Rate (mg/s)	Combined NPDES x and y Discharge Rate (mg/s)
Arsenic	<0.37	<2.70	<3.07
Beryllium	0.17	<1.08	0.17; < 1.25
Cadmium	<0.07	<0.54	<0.61
Chromium (total)[2]	<0.74	<5.40	<6.14
Copper	<1.48	<10.80	<12.28
Lead	0.44	<1.62	0.44; <2.06
Manganese	17.76	70.20	87.96
Mercury	<0.01	<0.11	<0.12
Molybdenum	<3.70	<27.00	<30.70
Nickel	<1.48	<10.80	<12.28
Selenium	<0.37	<2.70	<3.07
Vanadium	<1.48	<10.80	<12.28
Chloride	1110.00	51840.00	52950.00
Fluoride	11.84	135.00	146.84

(1) < : Not detected. The reported value for nondetected chemicals corresponds to the detection limit.
(2) Total chromium was assumed to consist of 5% Cr(VI) and 95% Cr (III).

Only five chemicals were detected: Be, Pb, Mn, chloride and fluoride. Both Be and Pb are listed as carcinogens by the U.S. EPA; however, EPA does not provide any quantitative dose-response relationship for lead. Two individual scenarios were evaluated in the risk assessment: (1) a primary scenario involving detected chemicals only; and (2) a secondary scenario accounting for both detected and non-detected chemicals at 50% of their respective detection limits.

The study area examined in the present application was defined to be the area within a 50 km radius of the power plant. The population density in the area is high. The 1990 U.S. Census database shows a total population of approximately 2.3 million people living within 50 km from the facility. The model study area was divided into 40 subregions that are defined by 8 angular sectors of 45° each, and 5 radial divisions at distances of 10, 20, 30, 40, and 50 km from the facility. A schematic of the study area with its subregion divisions and main hydrologic features is provided in Figure 1.

An average annual total of 104 cm of precipitation is reported to reach the ground. From that, it was estimated that 51 cm return to the atmosphere through evapotranspiration, 34 cm recharge the area's groundwater system through infiltration, and 19 cm recharge the area's surface water system through overland runoff. Where there are surface water-groundwater interconnections, the groundwater may be locally recharging the surface water and vice-versa. No general statement can be made for this mechanism, however, since it depends on the local relationship between surface water and groundwater, and may vary significantly from point to point.

The major surface water bodies in the area include four interconnected rivers (Figure 1). Smaller creeks were not included in the surface water analysis, since they eventually drain into one of the major water bodies mentioned above, and do not significantly contribute to the area's water supply or recreational resources. Water discharges from the plant occur in a creek that drains into river A.

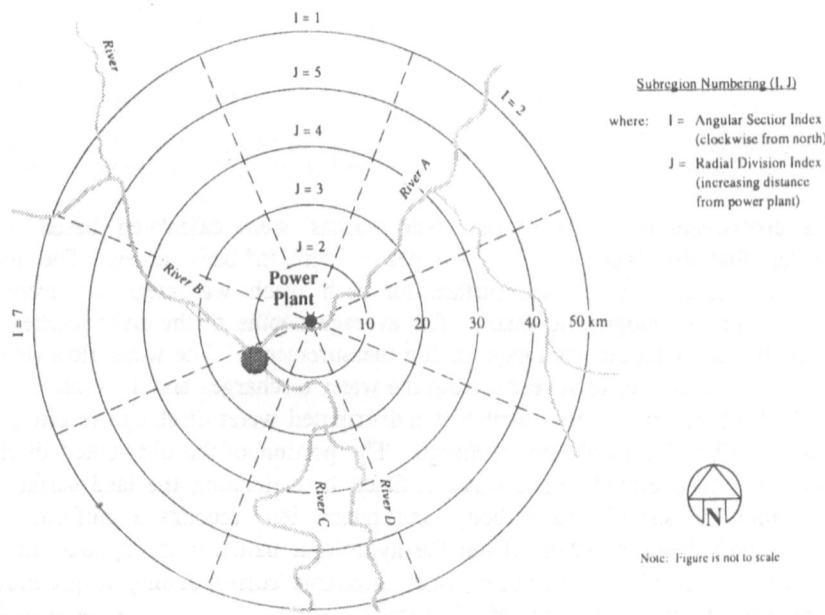

Fig. 1 Model Study Area

Rivers A and B were considered together in a continuous river system composed of nine distinct river reaches (i.e., portions of the river with uniform flow and geometric characteristics). Rivers C and D were divided into five and two reaches, respectively. It should be pointed out that even though the entire water system is analyzed from a hydrologic point of view, only the portion of the A-B river system downstream of the power plant is affected by the liquid discharges evaluated in this case study. The geometric and flow characteristics of the A-B river system are summarized in Table II.

TABLE II
Characteristics of the A - B River System

Reach No./River	Length (km)	Flow Rate (m³/s)	Width (m)	Depth (m)	Cross-Sectional Area (m²)	Velocity (m/s)
1A	18.0	495.6	310	4.7	1435	0.34
2A	14.0	512.2	340	4.7	1576	0.33
3A	6.0	626.8	340	5.0	1675	0.37
4A	16.0	647.4	340	5.0	1675	0.39
5A	5.0	664.9	340	5.0	1675	0.40
6A	22.0	669.8	310	5.0	1525	0.44
7B	8.0	1124.5	360	5.6	1985	0.57
8B	15.0	1154.5	450	5.6	2489	0.46
9B	22.5	1157.3	410	5.6	2265	0.51

The cross-sectional areas of the river reaches were calculated based on the assumption that the river channel is a trapezoid with 45° bank slopes. The average width of the river at the water surface for each reach was measured from U.S. Geological Survey topographic maps. The average depths of the river reaches were estimated by interpolation from gage station measurements. The water flow velocities were then calculated, for each reach, from the water discharges and the cross-sectional areas. Each of the reaches was attributed a distributed water discharge resulting from overland runoff and groundwater recharge. The portion of the distributed discharge corresponding to overland runoff was calculated by estimating the land surface area draining into the specific water body, and taking into account a uniform annual overland runoff depth of 19 cm (from the hydrologic balance) throughout the study area. The portion of the distributed water discharge corresponding to groundwater recharge was calculated using an annual average groundwater recharge depth of 84 cm (i.e., by assuming that the total amount of water infiltrating through the ground eventually returns to the surface by recharging the area's surface water bodies), and considering the same drainage areas used in the overland calculations. At reaches being recharged by areas extending beyond the study area, additional distributed discharges were added as necessary for the total flow at water station locations to match the measured annual average water discharges. In addition to the distributed water discharges, point discharges also had to be specified at the beginning of each one of the considered river systems as well as at two additional points along river A, four additional points along river B, and two additional points along river C. These point water discharges were calculated based on annual average measurements at an upstream station and an estimation of the additional amount of water draining into the river from the surrounding areas located between that water station and the particular points.

4. Results of Case Study

The maximum risks in the study area resulting from the power plant's liquid emissions were calculated to occur in subregion (7,1) (i.e., northwest sector, 0-10 km), the population of which was assumed to consume water and fish from reach 5 of the A-B river system (i.e., the reach immediately downstream of the point of discharge). Two scenarios were considered. In the first scenario, only chemicals detected during the sampling program were included. In the second scenario, all chemicals were included, with half the detection limit (DL) being used for chemicals that were not detected.

4.1 PRIMARY SCENARIO (DETECTED CHEMICALS ONLY)

Carcinogenic Risk.
- The maximum total carcinogenic risk was calculated to be 2.3×10^{-8} (i.e., 0.023 in a million).
- This risk was primarily due to beryllium ingestion from drinking water (71.2%). Beryllium ingestion from fish consumption followed with a contribution of 28.4%. Dermal absorption due to contact with water had an insignificant contribution of 0.4%.

Non-Carcinogenic Risk:
- The maximum total non-cancer hazard index (HI) was calculated to be 0.027.
- This HI was primarily due to ingestion of chlorine compounds from drinking water (44.3%) and fish consumption (43.7%). Exposure to chlorine compounds due to dermal contact with water had a smaller contribution of 12%.
- The cumulative HI from the other four detected chemicals (i.e., beryllium, lead, manganese, and fluorine) was insignificant compared to that of chlorine compounds (i.e., 6×10^{-4} and was primarily due to ingestion).

4.2 SECONDARY SCENARIO (INCLUDING NON-DETECTED CHEMICALS AT 1/2DL)

Carcinogenic Risk:
- The total carcinogenic risk in this secondary scenario increased to 0.3 in a million (i.e., an increase by a factor of 13 from the primary scenario).

● The major contributor to this risk was ingestion of arsenic due to drinking water (35%) and fish consumption (32%). Exposure to arsenic and beryllium due to dermal contact with water had an insignificant contribution of 1%.

Non-Carcinogenic Risk:
● The total maximum HI in this secondary scenario was calculated to increase by an insignificant amount from the primary scenario (i.e., 0.028 versus 0.027), since the primary contributing chemical to non-cancer risk remained chlorine compounds, which were detected in the liquid discharges.

● The cumulative HI from all other chemicals was calculated to be 1.4×10^{-3} and was primarily due to ingestion (84%). The contribution of the dermal absorption pathway due to contact with water was 16%.

5. Conclusion

We have described the formulation of a multimedia health risk assessment model, the TRUE model, with particularly emphasis on the treatment of water discharges. The TRUE model is particularly useful in assessing the effect of transferring some chemical discharges from one medium to other media through the installation of emission control equipment (e.g., a scrubber that will reduce air emissions but may simultaneously increase water discharges or solid waste disposal).

The TRUE model was applied to the simulation of the potential health risks associated with water discharges from a coal-fired power plant. No significant health risks were estimated, even when non-detected chemicals were included in the analysis.

Acknowledgements

This work was funded under project WO-3081-01 from the Electric Power Research Institute, Palo Alto, California.

References

Bolton, J.G.: 1989, *User's Guide to the Water Emissions Risk Assessment Model (WTRISK)*, Working Draft. WD-3725-3-EPRI, Rand Corporation, Santa Monica, California, U.S.A.
Bonfiglio, M., Zale R., Ruffle B., Hawkins E. and Anderson P.: 1994, *Environ. Software*, **9**, 115-131.
Brown, S.L. and Schwab B.: 1994, *Environ. Software*, **9**, 161-179.
Constantinou, E. and Seigneur C.: 1993, *Environ. Software*, **8**, 231-246.
Constantinou, E. and Seigneur C.: 1994, *Computer Techniques in Environmental Studies V*, **1**, 255-265, P. Zannetti, ed., Computational Mechanics Publications, United Kingdom.
Electric Power Research Institute (EPRI): 1994, *Electric Utility Trace Substances Synthesis Report*, **4**, Appendix P, Toxicology Profiles, EPRI TR-104614-V4, Palo Alto, California, U.S.A.
Hudson, R.J.M., Gherini S.A., Watson C.J. and Porcella B.P.: 1994, *Mercury Pollution: Integration and Synthesis*, 473-523, Lewis Publishers.
Liu, D.: 1994, *Environ. Software*, **9**, 153-160.
Villars, M., Gerath M. and Galya D.: 1993, *Environ. Software*, **8**, 135-152.

WATERSHED RISK ANALYSIS MODEL
FOR TVA's HOLSTON RIVER BASIN

C.W. CHEN[1], J. HERR[1], R.A. GOLDSTEIN[2], F.J. SAGONA[3], K.E. RYLANT[4], and
G.E. HAUSER[5]

[1] Systech Engineering, 3180 Crow Canyon Place #260, San Ramon, CA 94583
[2] Electric Power Research Institute, P.O. Box 10412, Palo Alto, CA 94303 [3] TVA, Chattanooga, TN 37402
[4] TVA, Muscle Shoals, AL 35660 [5] TVA Engineering Laboratory, Norris, TN 37918

Abstract. The Electric Power Research Institute has launched a research project to develop a conceptual risk analysis framework for watershed management of point and nonpoint source pollution. The research leads to the design of an engineering model to 1) process and translate water quality data (coliform, BOD, DO, suspended solids, temperature, sediment, etc.) into decision variables (suitability for water contact sports and swimming, fish spawning, fish survival, human consumption of fish, and freedom from algal nuisance, etc.) and 2) predict water quality improvements from proposed management alternatives. Actual development of the model is being carried out with the Tennessee Valley Authority (TVA) for the Holston River watershed. The effort includes model construction by importation of GIS map files, stringing together existing watershed and reservoir models, calibration of the model, and selection of decision variables and water quality check points. The model calculates hydrology, waste load, water quality and suitability of fish habitats at headwaters. The base case results and improvements after best management alternatives will be compared to the data observed by TVA's River Action Team. The final product will be a user friendly tool that stakeholders can use to find a cost effective method of improving water quality, including market-based pollution trading.

Keywords: watershed, water quality, point source vs nonpoint source pollution, decision variables

1. Introduction

Decision making processes for water quality management are changing from the command and control approach of the federal government to consensus building at the local level. Instead of rigid technology based treatment requirements for individual discharges, stakeholders at the local level are involved in setting goals and strategies to address all point and nonpoint discharges on a watershed scale. They need to identify where the water quality problems occur and how one can best find a cost effective solution, including through pollution trading. The emphasis in judging water quality is also shifting away from using concentrations of specific chemical constituents and toward determining whether general goals are achieved, such as suitability for human use and aquatic life.

The model discussed in this paper has been designed to accurately simulate the physics of water quality in a watershed and satisfy the needs of stakeholders and decision makers.

2. TVA System

The TVA system includes tributary watersheds, rivers, and reservoirs. The system is managed from many perspectives including power generation, flood control, recreation,

Water, Air and Soil Pollution **90**: 65-70, 1996.
© 1996 *Kluwer Academic Publishers.*

navigation, and water quality.   Ecosystem health requires good water quality for spawning habitat at headwaters, high D.O. and low suspended sediment in rivers, low nutrient and sediment inflows to the reservoirs, etc.   Water quality issues thus vary in space.

This paper will deal with the Holston River watershed.   Modeling starts from a tributary watershed, moving gradually downstream to rivers and reservoirs. An investigation is being made in this watershed to determine how farming practices can be improved to reduce sediment and nutrient loads.   These types of nonpoint source pollution impair aquatic habitat in headwater streams, lower D.O. in rivers, and contribute to siltation and eutrophication in the TVA reservoirs.

3. Holston River Watershed

The Holston River's headwaters are in the Appalachian Mountains of Virginia, Tennessee, and North Carolina.   The river flows about 240 km southwest and meets the Tennessee River near Knoxville, TN.   Most of the watershed is used for agriculture.

The model represents the watershed as catchment, river, and reservoir objects linked together.   The spatial data for the model is imported from Digital Elevation Maps (DEMs) which are available from the USGS over the Internet.   The DEM data is processed to determine the extents and physical properties of rivers, catchments, and reservoirs.   Each object is linked to the object downstream.   The physical model is displayed as a map in the Windows environment on a PC.   The Holston watershed is divided into 71 catchments, with 56 river segments and 5 reservoirs as shown in Figure 1.

The model calibration began at one of the uppermost sections of the Holston watershed, as shown highlighted in Figure 1.   The area encompasses the Middle and South Forks of the Holston River from their headwaters to their confluence.   The watershed of the Middle Fork has an area of 606 km^2.   Elevations within the watershed vary from 520 to 1,220 meters above sea level and the average slope is 11%.   About 40% of the land in the Middle Fork watershed is used for pasture and crops, less than 10% is urbanized, and the remainder is forest.   The 826 km^2 South Fork watershed is more rugged, with elevations from 520 to 1,750 meters above sea level and an average slope of 16%. About 20% of the South Fork watershed is used for agriculture and less than 1% is urbanized.

The physical representation of the South and Middle Fork watersheds was generated from a DEM in the same manner as the full Holston River watershed.   The watershed is divided into 48 catchments, each with its own river segment, as shown in Figure 2.

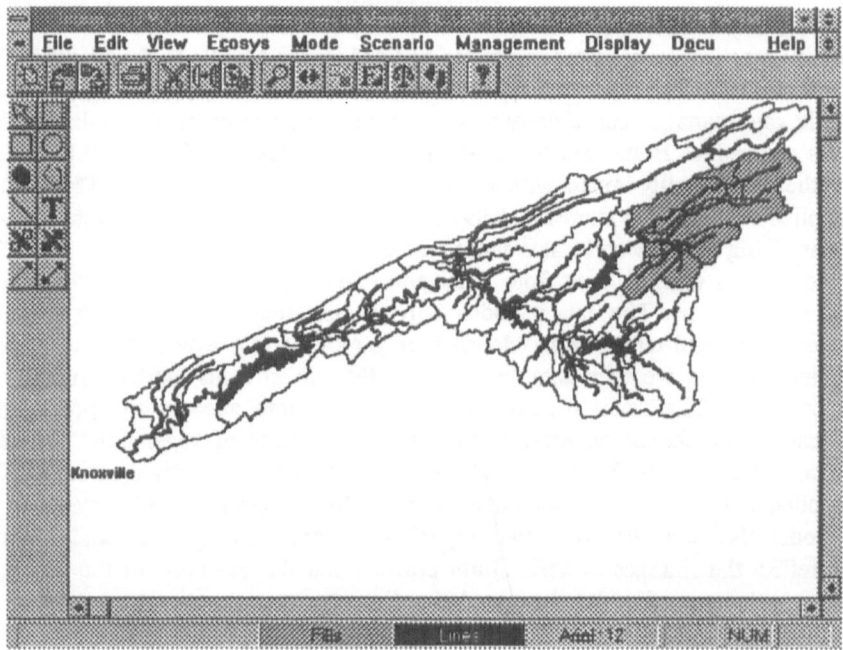

Fig. 1. Model representation of the Holston River watershed with the South and Middle Forks highlighted

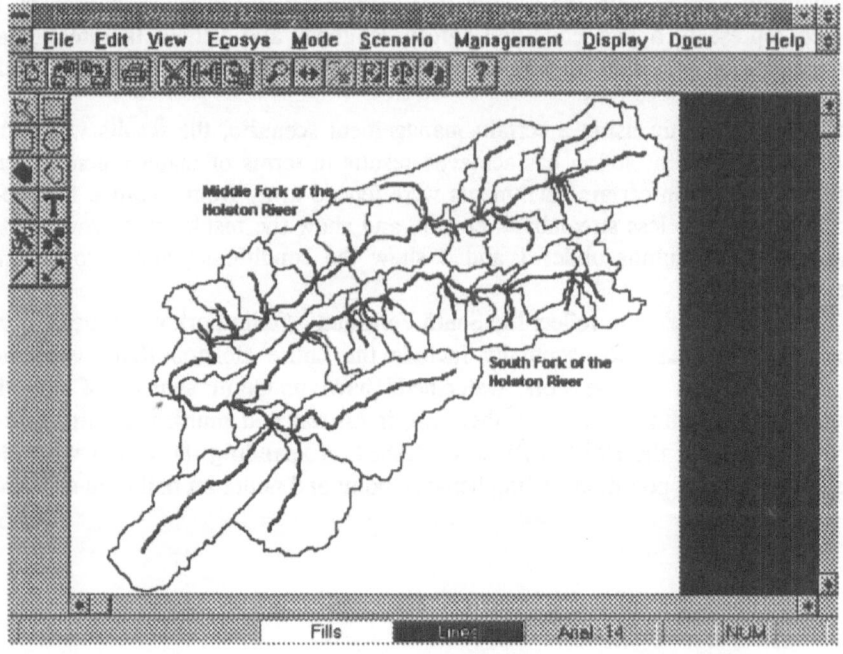

Fig. 2. Model representation of the South and Middle Fork Holston River watersheds

4. Model Simulations

Agricultural practice studies were conducted by TVA in the Middle Fork watershed from 1986 through 1993. The practices included reduction of overgrazing, fencing off streams from cattle, and management of dairy waste. These prevented erosion of sediment from the banks and reduce BOD load to the streams. The watershed model will be used to evaluate the water quality impact with and without the improved practices. The model is being applied to the period from 11/1/1991 to 10/31/1993, when additional studies were made quantifying the flow and water quality in the South and Middle Forks.

The model uses weather data from Kingsport, Tennessee compiled by the National Climatic Data Center. The data includes daily precipitation, maximum and minimum temperature, dewpoint temperature, cloud cover, atmospheric pressure, and wind speed. The model uses the weather data to calculate flow in streams, surface runoff, and subsurface flow. The water quality simulation takes into account the precipitation quality, catchment chemistry, erosion potential, and effects of vegetation. Figure 3 shows a comparison of model results and observed field data. The observed stream flow was compiled by USGS and the water quality data is from a special TVA study.

The model also simulates time varying BOD, nitrogen, phosphorus, and sediment loads to reflect the changes in agricultural practice and the condition of the watershed without such changes. Simulated suspended sediment, ammonia, dissolved oxygen, and temperature will be translated into an estimated Index of Biotic Integrity (IBI), which is a parameter used by TVA to evaluate ecosystem health. Figure 4 shows how results can be displayed in terms of decision variables.

In that example, the characteristics of the local fish population are displayed in a red/yellow/green format, with the criteria that correspond to the colors chosen by the user. Red represents a bad condition, green is good, and yellow is intermediate. Displaying decision variables in this format helps stakeholders identify water quality issues.

After the model is run using a certain management scenario, the results will reflect whether the management action has achieved results in terms of management criteria. For example, if cows in certain catchments were fenced away from streams, the model simulation would have less streambank erosion and show the resulting improvement in fish habitat. The columns under 1 and 2 show the conditions under two different management scenarios.

After the model is set up to reflect the South Fork and Middle Fork watersheds, it will be expanded downstream to eventually include the entire Holston River watershed. While the South and Middle Fork watersheds have nonpoint sources of pollution associated with agricultural land use, there are industrial and municipal point sources farther downstream in the Holston River watershed. Expanding the model will allow analysis of the effect of pollution trading between point and nonpoint pollution sources.

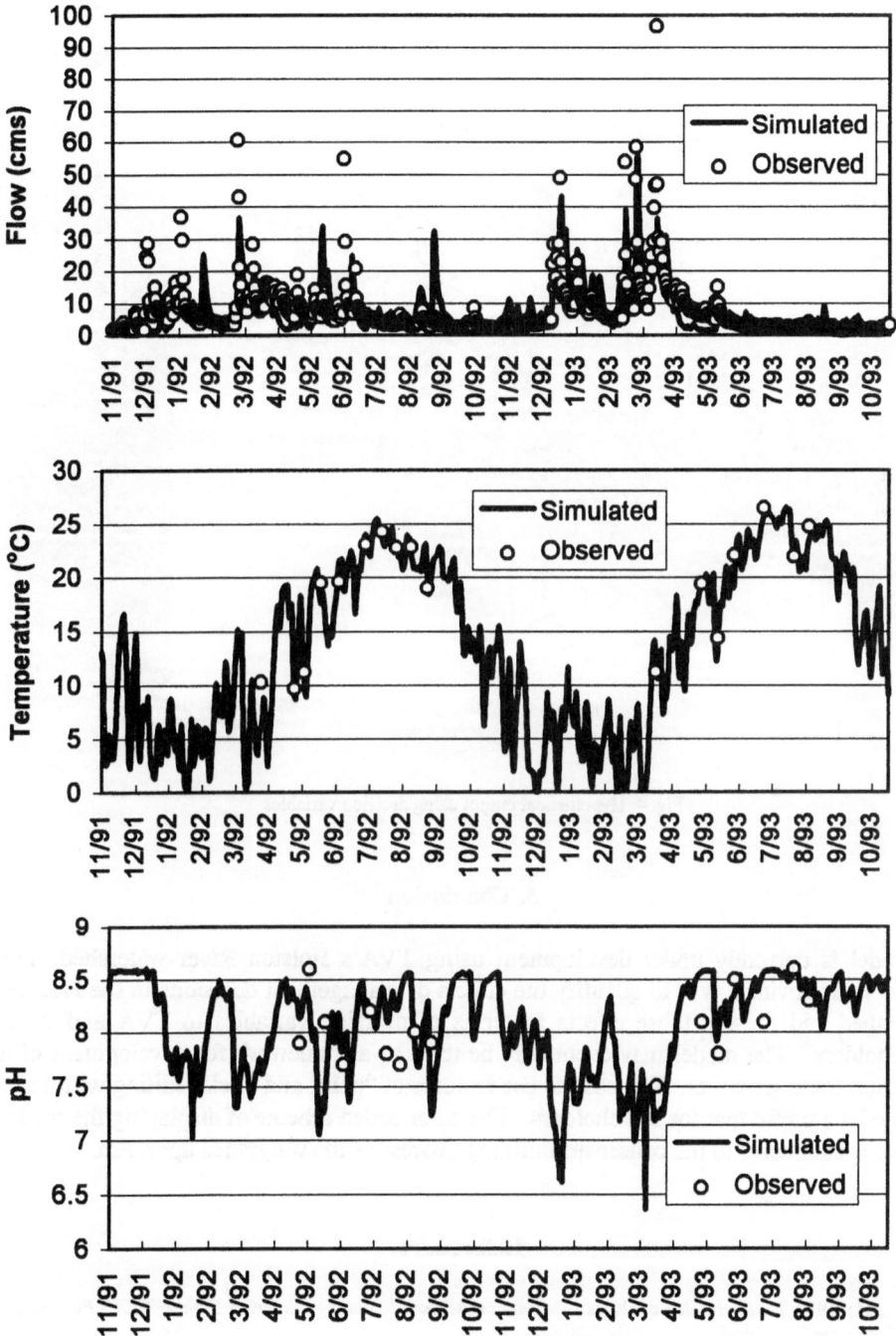

Fig. 3: Flow, temperature, and pH in the Middle Fork of the Holston River

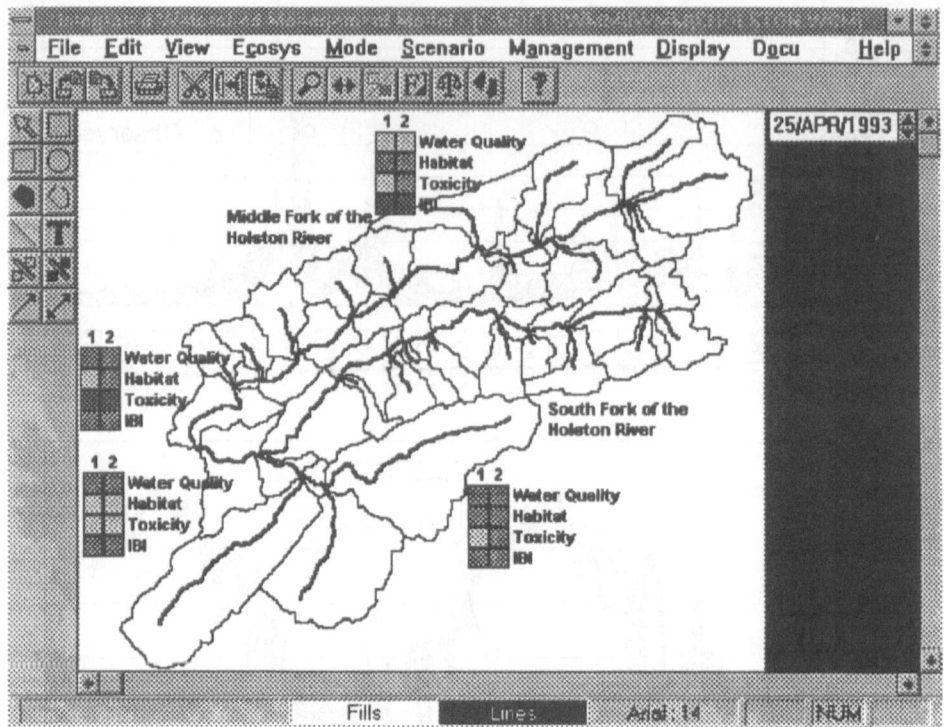

Fig. 4: Hypothetical output using decision variables

5. Conclusion

A model is currently under development using TVA's Holston River watershed. The model will provide a way to quantify the effects of management decisions in the Holston watershed and present those effects in terms of decision variables to TVA and other stakeholders. The model may eventually be used as a framework for development of a pollution trading market. The easy to use features of "point and click" editing will make the model a useful tool for stakeholders. The color coded scheme of displaying the model results is conducive to the consensus building process of the watershed approach.

References

Avera, J.B.: 1994, *Determination of Oxygen Demanding Material Loadings Entering Eight of TVA's Reservoirs*, Tennessee Technological University, 236 pages.

Chen, C.W., Herr, J., Gomez, L.G., and Golstein, R.A.: 1995, *Watershed Risk Analysis Tool*, presented at the First International Conference Water Resources Engineering Div./ASCE, et. al., San Antonio, TX (August 1995).

Tennessee Valley Authority (TVA): 1994, *Holston Watershed 1994 Stream Assessment*, 338 pages.

Tennessee Valley Authority (TVA): 1995, *TVA Integrated Watershed Approach Holston River Basin*, 48 pages.

INFERRING ECOLOGICAL RISK FROM TOXICITY BIOASSAYS

S. FERSON[1], L.R. GINZBURG[2] and R.A. GOLDSTEIN[3]

[1]*Applied Biomathematics, 100 North Country Road, Setauket, NY 11733 USA.* [2]*Ecology and Evolution, SUNY at Stony Brook, Stony Brook, NY 11794 USA.* [3]*Electric Power Research Institute, Palo Alto, CA 94303 USA*

Abstract. Results from toxicological bioassays can express the likely impact of environmental contamination on biochemical function, histopathology, development, reproduction and survivorship. However, justifying environmental regulatory decisions and management plans requires predictions of the consequent effects on ecological populations and communities. Although extrapolating the results of toxicity bioassays to potential effects on the ecosystem may be beyond the current scientific capacity of ecology, it is possible to make detailed forecasts at the level of a population. We give examples in which toxicological impacts are either magnified or diminished by population-dynamic phenomena and argue that ecological risk assessments should be conducted at a level no lower than the population. Although methods recently proposed by EPA acknowledge that ecological risk evaluations should reflect population-level effects, they adopt approaches from human health risk analysis that focus on individuals.

Keywords: ecological risk, ecotoxicology, populations, ecosystems, bioassays, population-level endpoints

1. Introduction

Bartell (1993) has suggested that human health risk assessments are a subset of ecological risk assessments in the obvious sense that humans are one of the species in the natural world. Guidance promulgated by EPA this year likewise prescribes methods for ecological risk evaluation that are slightly modified from human health risk analysis. The analogy between human health and ecological risk is misleading, however, in at least one important respect. In human health risk assessments, concern is focused on individuals. A single death, or even a single instance of disease, is a cause for some alarm. In ecological risk assessments for which the focal species is not humans, individual-level impacts do not generally elicit the same concern unless the species is endangered or otherwise specially protected. As is acknowledged in EPA guidance, it is population-level effects that are of interest for planning mitigation and management strategies.

For instance, a study showing a 10% decrease in sperm motility in a species of fish at a contaminated site may, by itself, be a questionable justification for expensive remediation. Only by translating this result to the population level do we understand what the consequence of the impact really is. Depending on the life history of the particular species at the particular site, such an effect might result in a grave consequence such as the local extinction of the species, or it might correspond to no population-level effect whatever. When it results in some intermediate impact on the population, making the translation allows us to see just what the expense of remediation buys in terms that are comprehensible to ecologists and lay people alike.

Water, Air and Soil Pollution **90**: 71-82, 1996.
© 1996 *Kluwer Academic Publishers.*

In the early days of environmental protection, regulators considered an impact ecologically significant if it could be *detected* above statistical noise. This was a conscious decision to be prudent in the face of our generally poor understanding of how natural ecological processes work. One of the results of this approach was the development of hazard quotients which compare the likely exposures of an environmental contaminant to levels that would be expected to yield detectable impacts. The many problems with such quotients have been widely discussed (ORNL, 1986; Bartell *et al.*, 1992; Suter *et al.*, 1993; Landis and Yu, 1995) and we will not reiterate them here. We suggest that the current state of ecology as a science can support a considerably more refined approach. In this paper, we argue that it is relatively simple to extend the existing methodology to the population-level using population dynamic models and risk analysis tools. This approach provides an important increase in the ecological relevance of an assessment with only a modest redirection in the empirical effort it requires.

2. Are ecosystem-level assessments impossible?

The difficulty of extrapolating from laboratory toxicity bioassays of some environmental contaminant to its eventual ecological consequences has been the focus of much discussion recently (ORNL, 1986; Cairns, 1992; Bartell *et al.*, 1992; Suter *et al.*, 1993; Landis and Yu, 1995; Hughes *et al.*, 1995). There is general agreement on the need to extend toxicological assessments to higher levels of biological organization (populations, communities and ecosystems), but no agreement on how to accomplish this task. The most significant limitation impeding the development of test protocols at higher levels is the state of ecological theory. In order to make assessments at the community or ecosystem levels, ecologists need to understand the interactions among species and trophic levels. Despite on-going research in the theory of trophic interactions, there are no generally accepted models that can be used as basis of ecosystem-level impact assessment.

The problem is not that there is no theory; the problem is that there are too many theories. Several ways have been suggested to model the relationship between a prey population and the predator population it supports. Most ecological simulations of food chains or the dynamics of trophic interactions are based on the Lotka-Volterra predation model developed in the 1920's (Lotka, 1925; Volterra, 1926) or a straightforward generalization (Rosenzweig and MacArthur, 1963) of it. Recently, a number of studies have challenged the basic assumptions of the traditional approaches (Arditi and Ginzburg, 1989; Matson and Berryman, 1992; Ginzburg and Akçakaya, 1992; Hanski, 1991). These studies have generated wide scientific discussion (Abrams, 1994; Sarnell, 1994; Diehl *et al.*, 1994; Akçakaya *et al.*, 1995; McCarthy *et al.*, 1995). Some theoreticians have combined the competing models of trophic interaction in equations of general form (DeAngelis *et al.*, 1975; Hassel and Varley, 1969) although ecologists do not even agree on which generalization to use.

The inadequacy of the current state of ecological theory for community-level and ecosystem-level risk assessments is evidenced by the recent controversy in the ecological literature about the correct form for modeling trophic interactions. So long as ecologists cannot even agree on the very basic assumptions of food web and food chain models, any risk assessment protocol developed at the community-level would be on shaky ground. As much as we would like to be able to evaluate ecological risks at the ecosystem level, we may be practically constrained to the population level.

3. Are population-level assessments necessary?

This section describes several examples that illustrate the potential complexity of population dynamic responses to toxicants. This complexity arises from the demographic structure of natural populations, including density dependence and age or developmental stage distribution. Ignoring this complexity can lead to both over-regulation of relatively innocuous substances and under-regulation of comparatively injurious contaminants.

3.1 CASCADING AND CONFLICTING EFFECTS

In the past, toxicity assessments were usually focused on acute mortality effects, and regulatory decisions were based solely on such results. Given that a field population experiences fluctuations in its natural mortality rate as a result of environmental and demographic stochasticity, how should one interpret the result of a bioassay? What does a certain increase in mortality due to exposure to a toxicant really mean for the population as a whole? How small must a change in mortality be to be ecologically unimportant? How should a change in the *variance* of mortality be interpreted?

With the recent use of chronic and 'life table response' (Caswell, 1989) experimental protocols, as well as comparative field studies, effects on survivorship, growth and reproduction are now monitored. In fact, there are many variables that could be selected for measurement, including growth rate or body size of individuals, survival rates, scheduling of mortality, frequency or scheduling of spawning, number, number or viability of eggs per brood, number of broods, maturation time, susceptibility to disease, and any number of other possibly significant variables. Which variables are the most important? When different variables yield results of different magnitudes, how should they be integrated? What if different variables yield contradictory results? For instance, how should one interpret an increase in mortality accompanied by an increase in fecundity?

Assessing effects at the level of the population solves both kinds of problems. Focusing on the population gives us a way to interpret the many measured variables and any potentially disparate findings. The *meaning* of a small increase in mortality rate due to a contaminant can be determined by projecting its consequence in terms of the total population's future abundance and vitality. When there are conflicting effects such as decreasing survivorship but increasing fecundity, they can be integrated in a straightforward way in a population-level assessment that takes both birth rate and death rate into account to predict population size. Even seemingly extraneous toxicant effects such as a change in sex ratio can be incorporated into the assessment.

It is important to recognize the necessity of performing a population projection to make this determination. It can be difficult to predict population-level consequences of impacts expressed at the organismal level. For instance, it is well known in population biology that inconspicuous and apparently minor impacts on individuals can sometimes cascade through population dynamics into significant effects at the level of the population. Conversely, it is also possible that seemingly major impacts on individuals translate into negligible population-level consequences once the normal population feedbacks have been taken into account. Furthermore, when a toxicant has conflicting effects such as decreasing survivorship but increasing fecundity, it is impossible to say whether the change in fecundity balances the change in survivorship without a specific population-level analysis.

3.2 TEMPORALLY LAGGED EFFECTS

The consequences of environmental stress on a population may be delayed in time, sometimes by many years, as a result of the age structure of a population. For instance, if a chemical contaminant disrupts development of larval fish so that they can never reproduce, the population consequences of a contamination event may occur long after the contaminant itself has disappeared. There may also be more complicated kinds of delayed effects. For instance, the recovery of the bluegill sunfish *Lepomis macrochirus* population at Hyco Reservoir after contamination by selenium was predicted (EPRI, 1991) to be cyclic, at first overshooting then undershooting its carrying capacity (Figure 1). Without a population-level assessment, one would not have predicted the population crash in

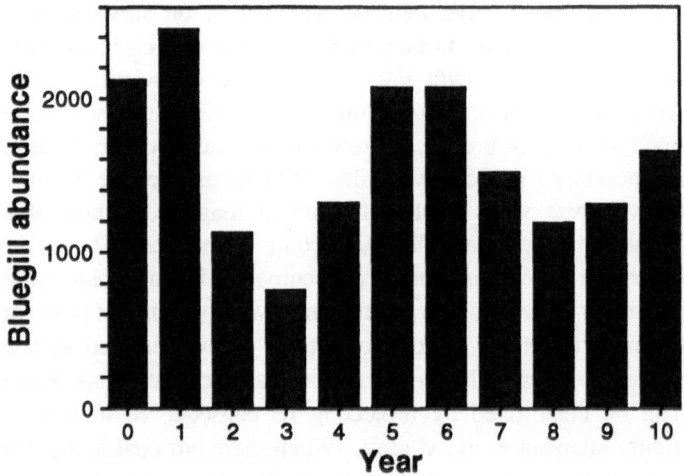

Fig. 1. Predicted cyclic recovery of a bluegill population after selenium poisoning.

years 2 and 3, and would likely have been unprepared for any public outcry it might precipitate. A population-level assessment shows that the decline in those years is a normal part of the population dynamics of this species.

3.3 COMPENSATORY PHENOMENA

In natural populations, overcrowding induces changes in vital rates such as increased mortality or decreased reproduction that keep the population near or below a size called the carrying capacity. There is much debate in ecology about how often real populations actually experience such overcrowding effects, and generally about how often population dynamics is governed by density dependence, of which overcrowding is an example. Whenever real populations in the natural world are regulated by such phenomena, they will have steady state population sizes. The existence of such a steady state implies that some impacts of toxicants and other stresses may be absorbed and produce little effect at the level of the population. For instance, if a toxicant is found to kill 10% of juveniles exposed to it in a toxicity bioassay, but in the natural population 90% of juveniles will die anyway as a normal result of the density dependence that governs the population, the toxicant may have a negligible effect at the population level. Such phenomena are known as compensation in the regulatory literature.

We describe one, perhaps extreme, example of compensation. Based on 11 years of observation, McFadden *et al.* (1967) described the demography of brook trout *Salvelinus fontinalis* at a site in Michigan which we summarize in Table I.

TABLE I

Demography of brook trout

Age-group	Abundance (individuals)	Fecundity (individuals/female)	Survivorship (annual fraction)
0	4471	0	0.420
I	2036	39	0.170
II	287	129	0.0860
III	14	301	0.0230
IV	0	455	0.0
Coefficients of variation			
Survivorships of age-group 0	0.17		
Survivorships of older groups	0.26		
Fecundities	0.25		
Correlations		Fecundities	Survivorships of older groups
Survivorships of age-group 0		−0.776	−0.476
Survivorships of older groups		0.549	
Density dependence follows the Ricker function (α=0.3 and β=0.000015)			

We used the abundances in the table as the initial distribution for a 25-year simulation of the effects of demographic (Akçakaya, 1991) and environmental stochasticity. Fecundities and survivorships were modeled as lognormal variables having the indicated means, interannual coefficients of variation, and correlation structure. We used the Ricker function (Ricker, 1975) to model brook trout density dependence and fitted the Ricker parameters α and β to the McFadden data. We computed the population size as the sum of individuals in age groups I, II, III and IV (but not group 0) and estimated the risk that the population will be lower than some threshold size at the end of 25 years using RAMAS/age (Ferson and Akçakaya, 1990). The result is displayed as the dotted line in Figure 2.

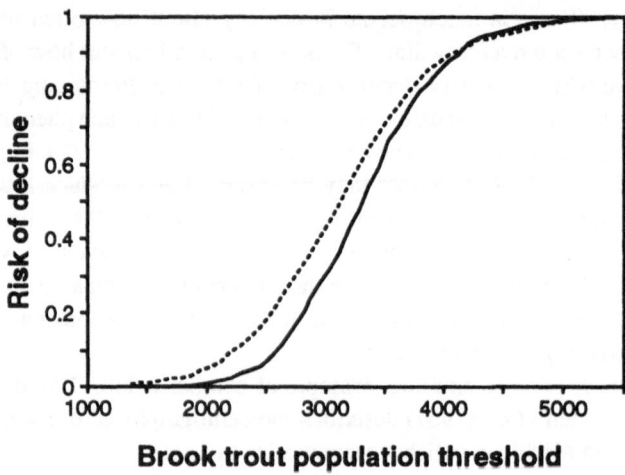

Brook trout population threshold

Fig. 2. Risks of brook trout population decline for normal (dotted) and stressed (solid) conditions.

Surprisingly, when we simulated stressed conditions by decreasing fecundities by 20%, we observed a general *lowering* of the risks of population decline (shown as a solid line in the figure). Although counterintuitive, this reaction to stress is not considered unusual for salmonids (Michael Power, pers. comm.) and seems to be a normal feature of their life history. Only after the stress has increased past some breaking point is the population negatively affected. In this case, even reducing mean fecundities by 50% will not have much of an effect on the risk of decline. But when the reduction is 75%, the population crashes and the risks of decline increase to unity. This result would never be predicted without a population-level assessment. It illustrates that the life history and ecology of a species can strongly influence the effect of a toxicant.

We do not expect that such counterintuitive results will be common. No doubt, small impacts will usually have fairly small effects, and large impacts will typically have commensurately large effects. But compensation will tend to reduce the magnitude of effects. In fact, as we have argued elsewhere (Ginzburg *et al.*, 1990), one can make a conservative estimate of the effect by ignoring compensatory phenomena when there is a poor empirical understanding of it. On the other hand, when there are reasonably good data about compensation, making use of them in a population-level assessment may be able to make up for poor information about toxicity. Because of how density dependence works, we may actually be more certain about the population-level effects than we are about the effects on the vital rates used as inputs.

3.4 DEPENSATORY PHENOMENA

In life history studies, density dependence is commonly thought of as the phenomenon that produces overcrowding effects. But another form of density dependence produces effects that have been called *under*crowding (Allee, 1938), in which the vitality of a population is lessened at smaller densities. Undercrowding can result from any of a number of natural mechanisms collectively known as Allee effects. In the regulatory literature, this phenomenon is known as depensation. Population-dynamic depensation (Allee effects on vital rates), while not ubiquitous in nature, does appear in a few percent of species studied, and is believed to be responsible for the collapse of at least one commercial fishery (Myers, et al., 1995). Depensation may also occur in toxicity responses directly. Indeed, Allee's original experiments (Allee and Bowen, 1932) with goldfish suggest that organisms at high density can withstand the toxic effects of a poison (colloidal silver) better and longer than similar organisms at low population density. In other words, the toxicity depends not only on the concentration of the toxicant, but on the population size as well. This suggests that the effect of even a constant level of contamination will vary in time as the population fluctuates under changing environmental conditions. Consequently, forecasting the eventual effect of a toxicant requires one to conduct a population-level assessment, without which experimental toxicity data are quite ambiguous.

4. What endpoints can be used at the population level?

Having decided to focus on population-level effects, one has a choice of several kinds of summaries to consider. One measure is the rate of population growth or decline. In the case of stochastic population dynamics, this classical measure is of limited usefulness, since it does not represent the inherent and inescapable variability that affects all natural biological systems. Other possible measures of a population's dynamics include the risk of population decline to some level over a given time horizon. Because this is a probabilistic notion, such a measure encapsulates the natural variability of vital processes and can express the consequences of impacts due to contaminants against the scale of background variability. In this section, we discuss three possible summaries for assessment purposes and consider the relative merits of each.

4.1 ASYMPTOTIC GROWTH RATE

A traditional measure of population growth or decline is the asymptotic rate of population growth, classically given the symbol λ. It is the ratio between the population sizes at successive time steps after all transient effects have died away. For discretely structured populations, it can be computed as the dominant eigenvalue of the transition matrix. If λ

is greater than one, the population is growing; if it is less than one, the population is declining. If we assumed the vital rates were constant over time, the magnitude of λ would be a measure of the growth of the population and we could use the equation

$$N_{t+\tau} = \lambda^\tau N_t$$

to predict the future size of the population at time $t+\tau$ from its size N_t known at any time t (so long as t is large enough that the population has equilibrated). Caswell (1989) has argued that λ is a reasonable synopsis of the current state of the population. The argument is that λ measures the robustness of a population, summarizing its capacity for future growth in a single number. There are simple formulas to estimate λ's sensitivities (Caswell, 1978) and elasticities (de Kroon *et al.*, 1986) to small changes in the vital rates. Using these formulas, an analyst can improve data collection strategies for better estimating λ and design efficient management or mitigation actions to increase the population's robustness. In ecotoxicology, several researchers have used λ for summarizing a population's dynamics, and have interpreted changes in λ when designing and assessing the success of management strategies (Pesch *et al.*, 1987; Pesch *et al.*, 1991; Caswell, 1995).

Although λ has been widely used by biologists and managers to assess impacts of various kinds on populations, it can actually be a poor measure of a population's health. Even if λ is very large, the population can still decline whenever the distribution in the various age or stage classes is far from stable structure. For instance, if all individuals present are post-reproductive, the population will be extinct within a generation, no matter how large λ is. In fact, after one time step, the population can be either much smaller or much larger than would be predicted by λ. Because the initial distribution of abundance plays such an important role, λ by itself is not a good predictor of the near-term population dynamics. Likewise, λ is not a good measure of the long-term dynamics either. Its implicit assumption that the vital rates are constant over time is simply not a tenable one, and vital rates often vary dramatically from season to season.

The central problem with λ is that it cannot express the stochastic variation of biological populations growing under fluctuating environmental conditions. In natural settings, such variation can be as large as or even larger than the population trends themselves. We have argued that, in the case of stochastic population dynamics, λ is a measure of limited usefulness, since it does not represent the natural variability that affects all natural biological populations (Burgman *et al.*, 1993). Surely, any toxicity assessment that pretends ecological relevance must recognize and express natural variability.

4.2 AVERAGE ABUNDANCE

A possible alternative to the asymptotic growth rate is simply the population abundance after some time period. Abundance is trivially simple to compute, requiring only repeated multiplication of the abundance distribution by transition matrices. At the end of a specified time period, the population size is referred to as the terminal abundance. In general, a stochastic model of population dynamics will yield a distribution for this terminal abundance. Such a distribution gives a complete answer to the question of how large (or small) the population will be after a specified period of exposure or impact. The average of the distribution, or any measure of its central tendency, gives a single scalar number that characterizes the response of the population to the treatment.

The abundance will depend, of course, on the length of the time horizon and on the starting population size and its initial distribution in the various age classes or developmental stages. So long as the time period is fixed—and long enough for impacts of concern to become evident—and the starting abundance is the same for the normal and stressed populations, the terminal average abundance is a summary of population-level effects that is at least as reasonable as, and perhaps better than, the asymptotic growth rate. Since it is also easier both to compute and to explain to lay people than λ, it suggests itself as a convenient summary of population-level response to a toxicant.

4.3 RISK OF POPULATION DECLINE

The risk of population decline measures the probability that the population will fall below a given threshold. In many cases, there is no particular single threshold that is obviously best to use in this calculation, so the risk is often presented as a function for a range of thresholds. This function is sometimes called the quasi-extinction risk (Ginzburg et al., 1982). Bridges et al. (1996) illustrates the use of quasi-extinction risk in toxicological assays.

The risk of population decline can be calculated either for falling below the threshold at least once during the simulation or for ending up below the threshold at the final time step of the simulation. In both cases, risk can be expressed as a curve (such as the two in Figure 2) which estimates the exceedance probabilities for population size. These risk curves summarize the dynamics of the population in a fully comprehensive way.

Although there are no simple formulas for sensitivities and elasticities of the risk of population decline like those for λ, perturbation analysis (Uryasev, 1994) permits the calculation of comparable quantities for the risk summary directly in the Monte Carlo simulation without additional computational cost.

5. Importance of the time horizon

Computing the average abundance or the risk of decline requires the specification of a time horizon. We feel that the decision about the time horizon should be explicitly stated by those making the assessment, rather than being buried as a hidden assumption in the analysis as it sometimes is. These summaries force us to state the time scale over which we make a forecast and prevent us from pretending that we can predict a population's infinite-time behavior, which is neither practically achievable nor even desirable in principle. The primary problem with such asymptotic predictions is that they are often irrelevant. If a population declines to extinction within the next decade, it simply doesn't matter what the asymptotic-time behavior of the population would have been.

Elementary reliability considerations also suggest that short-term time horizons should be favored over long-term or asymptotic time frames. Short-term predictions are generally more reliable than long-term predictions which must extrapolate far beyond the domain over which observations have been made. All this is not to say that we should disavow long-term perspectives. It is surely prudent to worry about long-term impacts, especially if the toxicant accumulates or has delayed effects. We merely emphasize that one cannot expect to draw trustworthy conclusions about consequences dozens or hundreds of years into the future based on data that span five or fewer years.

In some cases, the generation time of the focal species will suggest an appropriate time horizon over which an assessment should be made. In other cases, the time horizon will be determined by extra-scientific factors having to do with legal requirements or political circumstances. For example, the licensing period for a facility may be a set number of years irrespective of the generation times of the species that may be affected. When the choice for the time horizon is not obvious, it is always possible to make computations for several different time horizons and present the results as functions of time. When this is done, the explicitness of statements about time can be properly preserved.

6. Conclusions

We make the following five conclusions:

(1) Methods developed for human health risk assessments which focus on individuals are generally inappropriate for ecological risk assessments when the focus is shifted to entire populations in ecological communities.

(2) Although it would be desirable to extrapolate the results of toxicity bioassays to predict their consequent effects on entire ecosystems, doing so would generally require scientific understanding far beyond current ecology. Because single species population dynamics is much better understood than interspecific interactions, assessments may be practically limited to the population level. Making more comprehensive assessments will require further methodological development in ecology.

(3) The consequences of some putative environmental impact can only be interpreted in the context of population dynamics which can magnify or attenuate effects on individuals. Attention to population-level effects can distinguish impacts that are severe from those that are only transitory or turn out to be minor in magnitude.

(4) A risk-analytic summary such as the risks of population decline, rather than any simple hazard quotient or scalar estimate of population growth, is the appropriate endpoint for ecological risk assessment. Such risks should be compared against background risks arising from normal environmental fluctuations to determine the portion of risk attributable to the impact of the stressor.

(5) The time horizon over which the assessment is made should be finite and explicit in the analysis. Asymptotic or infinite-time predictions are unreliable because of ever changing environmental and political conditions.

Acknowledgements

We thank Tom Dillon and Todd Bridges of the Waterways Experiment Station of the U.S. Army Corps of Engineers for motivating this discourse. We also thank Michael Power of the University of Manitoba who gave advice on salmonid life history. This work was supported by the U.S. Army Corps of Engineers and the Electric Power Research Institute.

References

Abrams, P.A.: 1994, "The fallacies of 'ratio-dependent' predation." *Ecol.*, **75**, 1842-1850.

Akçakaya, H.R.: 1991, "A method for simulating demographic stochasticity." *Ecol. Modelling*, **54**, 133-136.

Akçakaya, H.R., Arditi, R., Ginzburg, L.R.: 1995, "Ratio-dependent predation: an abstraction that works." *Ecol.*, **76**, 995-1004.

Allee, W.C.: 1938, *Cooperation Among Animals with Human Implications*. NY: Henry Schuman.

Allee, W.C., Bowen, E.: 1932, "Studies in animal aggregations: mass protection against colloidal silver among goldfishes." *J. of Experimental Zoology*. **61**, 185-207.

Arditi, R., Ginzburg, L.R.: 1989, "Coupling in predator-prey dynamics: ratio-dependence." *J. of Theor. Biol.*, **139**, 311-326.

Arditi, R., Saïah, H.: 1992, "Emprical evidence and the theory of ratio-dependent consumption." *Ecol.* **73**, 1544-1551.

Bartell, S.M.: 1993, "Ecological risk assessment." Keynote Address, presented at the ASTM Third Symposium on Environmental Toxicology and Risk Assessment: Aquatic, Plant, and Terrestrial, Atlanta, GA.

Bartell, S.M., Gardner R.H., O'Neill, R.V.: 1992, *Ecological Risk Estimation*. Boca Raton, FL: Lewis.

Bridges, T.S., Wright, R.B, Gray, B.R., Gibson, A.B., Dillon, T.M.: 1996, "Chronic toxicity of Great Lakes sediment to *Daphnia magna*: elutriate effects on survival, reproduction and population growth." *Ecotoxicology* [in press].

Burgman, M.A., Ferson, S., Akçakaya, H.R.: 1993, *Risk Assessment in Conservation Biology*. London: Chapman & Hall.

Cairns, J.: 1992, "Paradigms flossed: the coming of age of environmental toxicology." *Environ. Toxic. and Chem.* **11**, 285-287.

Caswell, H.: 1978, "A general formula for the sensitivity of population growth rate to changes in life history parameters." *Theor. Popul. Biol.* **14**, 215-230.

Caswell, H.: 1989, "The analysis of life table response experiments. I. Decomposition of treatment effects on population growth rate." *Ecol. Modelling*, **46**, 221-237.

Caswell, H.: 1995, "Demography meets ecotoxicology: untangling the population level effects of toxic substances," *Quantitative Ecotoxicology: A Hierarchical Approach*, M. Newman (ed.), Chelsea, MI: Lewis.

Caswell, H.: 1989, *Matrix Population Models: Construction, Analysis and Interpretation*. Sunderland, MA: Sinauer.

de Kroon, H., Plaiser, A., van Groenendael, J., Caswell, H.: 1986, "Elasticity: the relative contribution of demographic parameters to population growth rate." *Ecol.* **67**, 1427-1431.

DeAngelis, D.L., Goldstein, R.A., O'Neill, R.V.: 1975, "A model for trophic interaction." *Ecol.*, **56**, 881-892.

Diehl, S., Lundberg, P.A., Gardfjell, H., Oksanen, L., Persson, L.: 1994, "*Daphnia*-phytoplankton interactions in lakes: is there a need for pragmatic consumer-resource models?" *Amer. Natural.*, **142**, 1052-1061.

EPRI: 1991, *Use of RAMAS to Estimate Ecological Risk: Two Fish Species Case Studies*. Palo Alto, CA: Electric Power Research Institute. EPRI EN-7176.

Ferson, S., Akçakaya, H.R.: 1990, *RAMAS/age User Manual: Modeling Fluctuations in Age-structured Populations*. Setauket, NY: Applied Biomathematics.

Ginzburg, L.R., Akçakaya, H.R.: 1992, "Consequences of ratio-dependent predation for steady state properties of ecosystems." *Ecol.*, **73**, 1536-1543.

Ginzburg, L.R., Ferson. S., Akçakaya, H.R.: 1990, Reconstructibility of density dependence and the conservative assessment of extinction risks.
Conservation Biol. **4**, 63-70.

Ginzburg, L.R., Slobodkin, L.B., Johnson, K., Bindman, A.G.: 1982, "Quasiextinction probabilities as a measure of impact on population growth." *Risk Analy.* **21**, 171-181.

Hanski, I.: 1991, "The functional response of predators: worries about scale." *Trends in Ecol. and Evol.*, **6**, 141-142.

Hassel, M.P., Varley, G.C.: 1969, "New inductive population model for insect parasites and its bearing on biological control." *Nature*, **223**, 1133-1136.

Hughes, J.S., Biddinger, G.R., Mones, E., (eds.): 1995, *Envir. Toxic. and Risk Assessment - Third Volume, ASTM STP 1218*, Philadelphia: American Society for Testing and Materials.

Landis, W.G., Yu, M.-H.: 1995, *Intro. to Envir. Toxic.* Boca Raton, FL: Lewis.

Lotka, A.J.: 1925, *Elements of Physical Biology*, Baltimore: Williams and Wilkins. Republished as *Elements of Mathematical Biology*, New York: Dover, 1956.

Matson, P., Berryman, A.: 1992, "Ratio-dependent predator-prey theory." *Ecol.* **73**, 1529.

McCarthy, M.A., Ginzburg, L.R., Akçakaya, H.R.: 1995, "Predator interference across trophic chains." *Ecol.*, **76(4)**, 1310-1319.

McFadden, J.T., Alexander, G.R., Shetter, D.S.: 1967, "Numerical changes and population regulation in brook trout Salvelinus fontinalis." *J. of the Fisheries Research Board of Can.*, **24**, 1425-1459.

Myers, R.A., Barrowman, N.J., Hutchings, J.A., Rosenberg, A.A.: 1995, "Population dynamics of exploited fish stocks at low population levels." *Science.* **269**, 1106-1108.

ORNL: 1986, *User's Manual for Ecological Risk Assessment.* Oak Ridge, TN: Oak Ridge National Laboratory, March 1986. ORNL-6251, ESD Pub. No. 2679.

Pesch, C.E., Munns, W.R., Gutjahr-Gobell, R.: 1991, "Effects of a contaminated sediment on life history traits and population growth rate of *Neanthes arenaceodentata* (Polychaeta: Nereidae) in the laboratory." *Environ. Toxicol.* **10**, 805-815.

Pesch, C.E., Zajac, R.N., Whitlach, R.B., Balboni, M.A.: 1987, "Effect of intraspecific density on life history traits and population growth rate of *Neanthes arenaceodentata* (Polychaeta: Nereidae) in the laboratory." *Marine Biology.* **96**, 545-554.

Ricker, W.E.: 1975, *Computation and Interpretation of Biological Statistics of Fish Populations.* Bulletin 191, Ottawa: Dept. of Fisheries and Oceans.

Rosenweig, M.L., MacArthur, R.H.: 1963, Graphical representation and stability conditions of predator-prey interactions. *American Naturalist.* **97**, 209-223.

Sarnelle, O.: 1994, "Inferring process from pattern." *Ecol.*, **75**, 1835-1841.

Suter, G.W.II, Barnthous, L.W., Bartell, S.M., Mill, T., Mackay, D., Paterson, S.: 1993, *Ecological Risk Assessment.* Boca Raton, FL: Lewis.

Uryasev, S.: 1994, "Analytic perturbation analysis of discrete event dynamic systems." Proceedings of the Fourth International Conference on Computer Integrated Manufacturing and Automation Technology, Troy, NY, Los Alamitas, CA: IEEE Computer Society Press.

Volterra, V.: 1926, "Fluctuations in the abundance of a species considered mathematically." *Nature.* **118**, 558-560.

EVALUATION OF CONTAMINATED GROUNDWATER CLEANUP OBJECTIVES

CANDY ARQUIETT[1], MARK GERKE[2], IRENE DATSKOU[1]

[1]Center for Risk Management, Oak Ridge, TN 37830*, [2]University of Tennessee, Knoxville, TN 37996

*Oak Ridge National Laboratory, managed by Lockheed Martin Energy Research Corp. for the U.S. Department of Energy under contract number DE-AC05-96OR22464.

Abstract. The U.S. Department of Energy's (DOE's) Environmental Restoration Program will be responsible for remediating the approximately 230 contaminated groundwater sites across the DOE Complex. A major concern for remediation is choosing the appropriate cleanup objective. The the cleanup objective chosen will influence the risk to the nearby public during and after remediation; risk to remedial and non-involved workers during remediation; and the cost of remediation. This paper will discuss the trends shown in analyses currently being performed at Oak Ridge National Laboratories' (ORNL's) Center for Risk Management (CRM). To evaluate these trends, CRM is developing a database of contaminated sites. This paper examines several contaminated groundwater sites selected for assessment from CRM's database. The sites in this sample represent potential types of contaminated groundwater sites commonly found at an installation within DOE. The baseline risk from these sites to various receptors will be presented. Residual risk and risk during remediation will be reported for different cleanup objectives. The cost associated with remediating to each of these objectives will also be estimated for each of the representative sites. Finally, the general trends of impacts as a function of cleanup objective will be summarized.

Keywords: cleanup objectives, groundwater, public risk, worker risk, cost, hypothetical homesteader risk

1. Introduction

Over the past several years, policy makers have increasingly pointed out the need for an integrated global approach to environmental management that allows funding to be allocated in ways that are both environmentally responsible and economically viable. Recent debates in Congress over environmental regulatory reform have also underscored the need for policies that carefully scrutinize both the costs and benefits associated with various cleanup objectives. On a practical level, agencies like the U.S. Department of Energy (DOE), which are tasked with environmental cleanup, have had to balance increased demands for environmental restoration with decreased budgets—a conflict that has further heightened the necessity for integrated global planning to ensure that limited resources are allocated to achieve the greatest risk reduction possible to humans and the environment.

In order for the DOE to prioritize effectively, it needs to gain an overview of the costs and benefits associated with various cleanup objectives at an installation or across a group of installations. A multi-attribute model incorporating cost and risk used at Oak Ridge National Laboratory's (ORNL's) Center for Risk Management offers a good means of effectively prioritizing environmental projects. Using groundwater plumes at the Savannah River Site (SRS) as an example, this paper describes risk/cost analysis and how it can be used. For a more detailed description of the methods used in this analysis, the reader should refer to *Environmental Restoration Programmatic Environmental Impact Statement Methodology for Estimating Human Health Risks* (ORNL, 1995a).

Water, Air and Soil Pollution **90**: 83–92, 1996.
© 1996 *Kluwer Academic Publishers.*

2. Purpose

The purpose of this paper is to demonstrate the benefits of performing risk/cost assessments of various cleanup objectives. These assessments provide decision makers with a more comprehensive overview of potential remediation projects while 1.) helping to ensure that limited resources are allocated in the most efficient manner possible, 2.) pinpointing the remediation technologies driving the potential risk and cost associated with different cleanup objectives and thereby indicating where alternative technologies are needed, and 3.) offering a defensible tool to use for renegotiating a risk-based compliance agreement with the regulatory agencies.

3. Scope

Ideally, prioritization of environmental restoration projects should be based on a site-wide analysis that examines a number of variables such as public risk, worker risk, cost, technology feasibility, and stakeholder involvement; however, an installation-wide analysis of the SRS is beyond the scope of this paper, which is intended as an example. Instead, for the purposes of illustration, this paper focuses on the A- and M-, TNX, and D- Areas of localized groundwater contamination at the SRS. The analysis provided in this document is solely intended to demonstrate the usefulness of evaluating different cleanup objectives. This report is not intended to provide actual remediation guidance for these groundwater plumes which are currently being analyzed for remediation by the site. Any attempt to remediate the groundwater without first evaluating its relationship to the rest of the facility and the installation as a whole would result in unnecessary expenditures (e.g., if contamination from other release sources continued to migrate to the groundwater). This study looks at how different cleanup objectives can result in varying risks and costs. It does not look at how the groundwater sites relate to other sites within the installation.

4. Cleanup Objectives

This paper examines two representative cleanup objectives: 1.) containment and 2.) treatment; and evaluates some of the technologies typically associated with these objectives. The cleanup objectives discussed in this report and their associated technologies are described as follows:

- **Containment.** For the containment cleanup objective, the groundwater is contained by the use of slurry walls. At the slurry walls, the contaminated water is collected with extraction wells and treated by ex-situ air strippers. Groundwater use is restricted for this cleanup objective.

- **Treatment.** The treatment cleanup objective entails removal of only the contaminants from the area through treatment without the removal of the contaminated media. The groundwater is treated with vapor extractors without being removed from the aquifer.

The containment and treatment objectives are compared to each other and to a no action scenario under which no attempt is made to mitigate or prevent offsite releases of contaminants.

Information from the *Methodology for Selecting Technologies and Estimating Waste Volumes, Costs, and Labor Requirements for Environmental Restoration* is used to evaluate these objectives.

5. Site Description

Groundwater at SRS has been contaminated by the disposal of soluble solids and liquids at waste sites. Sites evaluated in this study include the A- and M-Area, D-Area, and TNX Area (Figure 1). Some liquids were placed in seepage basins that allowed them to seep into the ground. Some solid materials and liquids were buried in landfills where rainwater seeped through the waste and leach materials down to the groundwater. Thus the groundwater is contaminated with radionuclides, metals, and organic chemicals. The descriptions given below are from the *Savannah River Site 1993 Annual Environmental Report*.

5.1. A- AND M-AREAS

The administration and manufacturing areas, A and M, are located in the northwest portion of SRS. A-Area houses administrative and research facilities. The latter includes the Savannah River Technology Center (SRTC, known as the Savannah River Laboratory, SRL, until 1992), which conducts environmental research and management projects. M-Area produces nuclear fuels, targets, and other reactor components.

Surface elevations across these areas range from approximately 350 to 380 feet above mean sea level (MSL). Surface drainage is toward Tims Branch, approximately 5,000 feet to the southeast, and toward valleys to the northwest and southwest that lead to the Savannah River. The nearest SRS boundary is approximately one-half mile to the northwest.

The water table beneath A-Area and M-Area slopes to the southeast, south, and southwest towards Tims Branch and other discharge points. Geohydrology information from these areas indicates that most of the water of the upper saturated zone migrates downward into lower water-bearing zones.

Most of the contamination near A-Area and M-Area is caused by chlorinated volatile organic compounds used in the manufacturing operations at M-Area. These compounds are common contaminants of groundwater throughout the United States. Surface waste sites associated with the A- and M-Area groundwater plume include the: A-Area Burning/Rubble Piles, A-Area Ash Pile, A-Area Coal Pile Runoff Containment Basin, A-Area Metal Burning Pit, M-Area Hazardous Waste Management Facility, Metallurgical Laboratory Seepage Basin, Miscellaneous Chemical Basin, and Motor Shop Oil Basin.

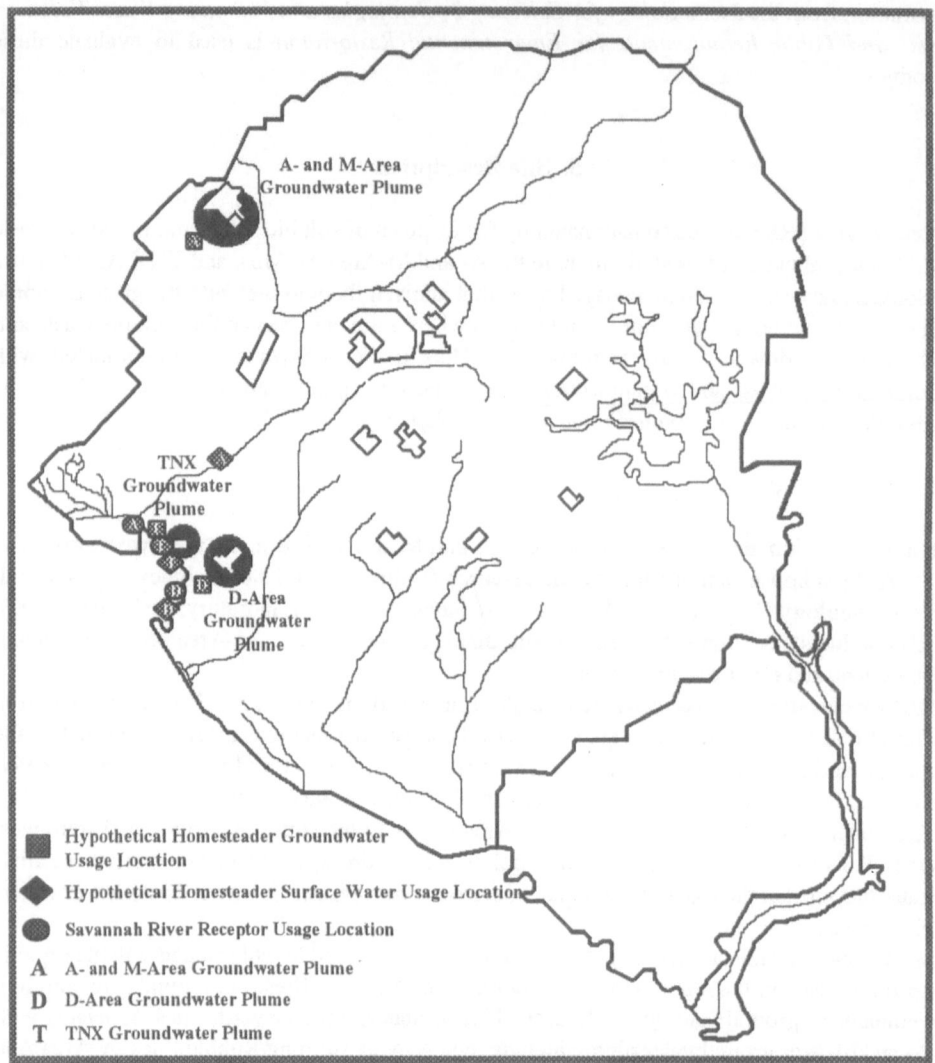

Fig. 1. Locations of Groundwater Plumes and Potential Receptors.

5.2. D-AREA

A heavy water production plant in D-Area began operation in 1953 to produce heavy water to moderate and cool the site's reactors. The facility stopped production in 1981 because of sufficient supply of heavy water. The facilities currently operating in D-Area include a heavy water reprocessing facility, a coal-fired power plant, and a laboratory facility that analyzes process effluent samples. Sources of contamination include: D-Area Burning/Rubble Pit, D-Area Coal Pile Runoff Containment Basin and Ash Basins, and D-Area Oil Disposal Basin.

5.3. TNX AREA

The TNX Area is situated in the southwestern portion of SRS one-quarter mile east of the Savannah River. The area served as a pilot-scale testing and evaluation facility which supported chemical processes for fuel and target manufacturing areas, Separations Area, and most recently the Defense Waste Processing Facility (DWPF). Wastewater generated in support of this initiative was discharged to unlined earthen basins through a network of process sewers. Sources of contamination include the: TNX Burying Ground, Old TNX Seepage Basin, and New TNX Seepage Basin.

5.4. RECEPTORS

The study analyzes three receptor scenarios, which are described briefly as follows:

- **Hypothetical Homesteader.** The Hypothetical Homesteader is evaluated to give an upper bound estimate of onsite risks. Under this most conservative of risk scenarios, DOE does not keep the installation under institutional controls. It is assumed that an agricultural receptor resides on or near each contaminated area at the installation. This hypothetical receptor raises all the agricultural products that he consumes; uses groundwater for all household and agricultural purposes; and fishes, swims, and boats in the surface water body closest to the site.

- **Savannah River Receptor.** The Savannah River receptor scenario is designed to evaluate risks from direct exposure to the Savannah River. This receptor uses the Savannah River for all agricultural, household, and recreational activities. Institutional controls are in place for this receptor.

- **Remediation worker.** Because exposure, injuries, and illnesses are likely to occur during remediation, this scenario is evaluated to provide an estimate of the physical hazards and exposure risk faced by workers. Remediation Workers are assumed to be employees who work at the release source 8 hours per day, 250 days per year until remediation is completed. The remediation worker is limited to a seven year period at the site to maintain a realistic scenario.

Table I summarizes the possible exposure pathways for each of these receptors. See *Environmental Restoration Programmatic Environmental Impact Statement Methodology for Estimating Human Health Risks* for further information on assumptions for these receptors (ORNL, 1995a).

Table I
Matrix of Exposure Routes for each Receptor

Exposure Pathway/ Exposure Route	Receptor Type		
	Hypothetical Homesteader	Savannah River Receptor	Remediation Worker
Groundwater:			
Household Use	●	—	—
Irrigation - crops	●	—	—
Crops - animals	●	—	—
Ingestion - crops	●	—	—
Ingestion - animals	●	—	—
Surface water:			
Household Use	—	●	—
Irrigation - crops	—	●	—
Crops - animals	—	●	—
Ingestion - crops	—	●	—
Ingestion - animals	—	●	—
Fishing/Shell fishing	●	●	—
Recreational	●	●	—
Air:			
Inhalation - vapors	●	—	●
Inhalation - particulates	●	—	●
Direct Exposure			
Soil (incidental ingestion)	●	—	—
Direct radiation	●	—	—

● Exposure via this route is evaluated.
— Exposure via this route is not evaluated.

For this study, risks are reported as the potential for cancer incidence to the maximally exposed individual (MEI) within each receptor scenario. This study also evaluates physical hazard incidents (e.g., heat stress or construction injury) for workers. Physical hazards occur as a direct result of remediation activities.

5.5. CONTAMINANTS OF POTENTIAL CONCERN

Contaminants of potential concern are based on information provided in *U.S. Department of Energy Eastern Area Programs Sitewide Risk Prioritization for Savannah River Site* (ORNL, 1995b). For a complete list of the contaminants evaluated and the concentrations used, please refer to Volume 2 of this report.

6. Results

This section presents the results of the study performed for the representative cleanup objectives for the A- and M-Area, D-Area and TNX Area groundwater plumes. The results assume that surface waste areas from which contamination could potentially migrate have been remediated. In actuality, this may not be true; therefore, the results of this study should not be used to evaluate actual cleanup options for the site.

The results are for the containment and treatment cleanup objectives and the technologies typically associated with these cleanup objectives. These two cleanup objectives are compared to a no action scenario. This section provides an integrated analysis of public risk, worker risk and physical hazard, and remediation cost.

Table II shows maximum post-remedial risks to the varying receptors for each of the cleanup objectives. The Hypothetical Homesteader and Savannah River Receptor risks reported are for the MEI. Worker risks and physical hazards are for the entire worker population. To compare these numbers, one must remember that the Hypothetical Homesteader and Savannah River Receptor values must be multiplied by the population that could potentially be exposed under a given scenario.

Cost values for each of the cleanup objectives are also given in this table. Cost and physical hazard estimates are based on the total person-hours spent at the site and the cost per person-hours and physical hazard per person-hour, respectively, for each activity. The total person-hours are directly related to the size of the site, but not always to the total volume.

Table II
Results

Plume	Cleanup Objective	Hypothetical Homesteader Risk	Savannah River Receptor Risk	Remediation Worker Risk	Remediation Worker Physical Hazard	Cost (Thousands of Dollars)
A- and M-Area	No Action	1	4×10^{-09}	N/A	N/A	N/A
	Containment	2×10^{-03}	4×10^{-13}	1×10^{-08}	190	$319,000
	Treatment	2×10^{-01}	4×10^{-11}	7×10^{-08}	120	$1,200,000
D-Area	No Action	4×10^{-04}	4×10^{-11}	N/A	N/A	N/A
	Containment	2×10^{-08}	4×10^{-15}	2×10^{-13}	35	$40,000
	Treatment	4×10^{-06}	4×10^{-13}	8×10^{-11}	4	$68,000
TNX Area	No Action	1×10^{-03}	2×10^{-10}	N/A	N/A	N/A
	Containment	5×10^{-06}	4×10^{-13}	3×10^{-11}	17	$17,200
	Treatment	5×10^{-04}	2×10^{-10}	3×10^{-11}	0.06	$610

Contaminants of concern (COCs) for the Hypothetical Homesteader at the A- and M- Area Groundwater Plume include, in descending order of contribution to total risk: trichloroethylene, chloroform, tetrachloroethylene, 1,1,2-trichloroethane, uranium-238, 1,1-dichloroethylene, dibromochloromethane, uranium-234, tritium, uranium-235, uranium metal, carbon tetrachloride, and benzene. For this same plume, the Savannah River Receptor COCs were: 1,1-dichloroethylene and 1,2-dichloroethane.

At the D-Area Groundwater Plume, the COCs for the Hypothetical Homesteader, in descending order of contribution include: tritium, trichloroethylene, and tetrachloroethylene. The Savannah River Receptor COCs for this plume were: trichloroethylene and tritium.

COCs for the Hypothetical Homesteader at the TNX Area Groundwater Plume include: carbon tetrachloride, potassium-40, uranium-234, uranium 238, chloroethene, trichloroethylene, tritium, and radium-226. For the Savannah River Receptor, COCs from this plume include: potassium-40, carbon tetrachloride, uranium-234, uranium-238, chloroethene, trichloroethylene, and tritium.

The groundwater exposure pathway drives the risk for the Hypothetical Homesteader. Although the surface water exposure pathway was also evaluated for this receptor, contribution from it to the total risk is insignificant. For the Savannah River Receptor, the surface water exposure pathway was the only pathway evaluated. It should be noted from Table I that the exposure pathways for the Hypothetical Homesteader from surface water are different than those of the Savannah River Receptor.

For the Hypothetical Homesteader by the A- and M-Area Groundwater Plume, many of the COC reach the groundwater well within the first 150 years. Uranium and its isotopes are an exception to this, reaching the well after about 600 years. COCs do not show up in significant levels at Savannah River Receptor for at least 2,000 years.

At the D-Area Groundwater Plume, all of the COCs reach the Hypothetical Homesteader within the first 150 years. Transport time to the Savannah River Receptor for all COCs is less than 300 years.

COCs contributing the most to the Hypothetical Homesteader risk at the TNX Groundwater Plume reach the receptor within the first 50 to 100 years, with the exception of uranium and its isotopes. COC to the Savannah River Receptor take anywhere from 150 years for potassium-40 to 3,000 years for uranium-234, and -238.

7. Discussion

Figure 2 shows the sum of the groundwater plumes' post-remediation risk on a logarithmic scale for the Hypothetical Homesteader and the Savannah River Receptor for each of the cleanup objectives evaluated. This figure shows that the risk reduction of the containment option is greater than that of the treatment option. The main cause of this difference is that the treatment of groundwater is very inefficient in most cases. It is difficult to remove contaminants from soil because these contaminants will adhere to the soil both chemically, through reaction with the organic matter in the soil, and mechanically, in the crevices of the soil. Because the containment option also impedes contaminants from traveling away from the site by the use of a slurry wall and extraction wells, it is much more efficient at reducing risks.

Figure 3 shows the total Remediation Worker risk and physical hazard as well as the remediation cost in thousands of dollars of all the groundwater plumes for each cleanup objective. This figure shows that the total worker risk and the cost associated with the containment option is less than that of the treatment option. The total person-hours required to apply the technologies at these sites is greater for treatment than for containment because of the complexity of the in situ vapor extraction technology. However, the Remediation Worker's physical hazard totals are almost the same for both treatment levels. This is due to the large amount of construction required for containment, mainly caused by the erection of slurry walls to contain the plume.

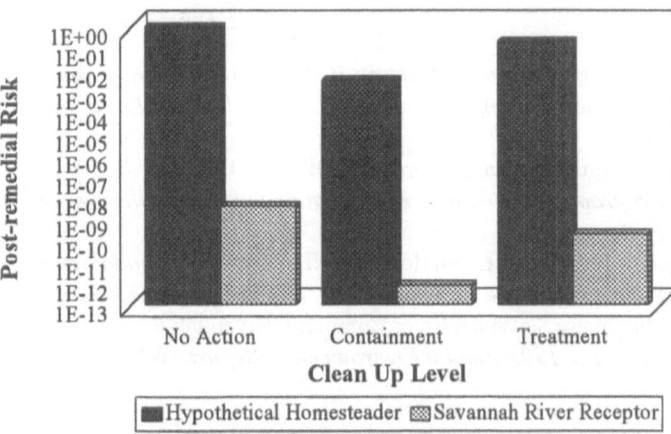

Fig. 2. Post-remedial Risk to the Hypothetical Homesteader and Savannah River Receptor for the Different Cleanup Objectives

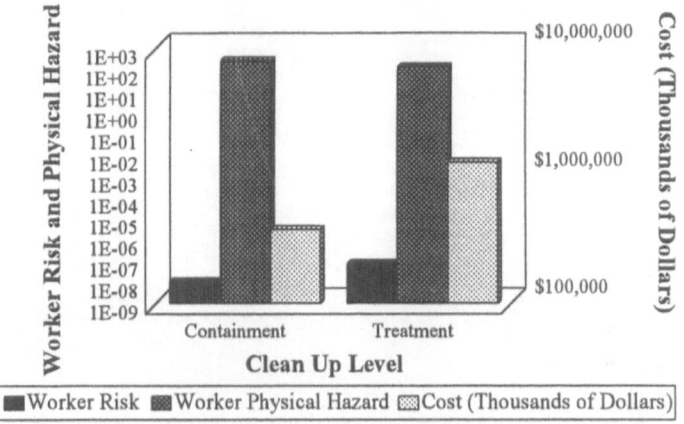

Fig. 3. Cost and Remediation Worker Risk and Physical Hazards for the Different Cleanup Objectives

Another factor affecting the risk apparent from this study is the effect of contaminant type and distance from the receptor on the time to exposure. The A- and M-Area Groundwater plume is much farther from the Savannah River Receptor than is the D-Area or TNX Area Groundwater plumes. Because of this, the contaminants driving the risk to the Savannah River Receptor from the A- and M-Area are exclusively volatiles that travel quickly through the groundwater. The risk from this same source to the Hypothetical Homesteader, who is directly by the release source, is also affected by uranium and its isotopes, a slow moving contaminant. The risk from these isotopes appears much later than those from the volatiles that contribute risk.

References

Oak Ridge National Laboratory (ORNL): 1995a, *Environmental Restoration Programmatic Environmental Impact Statement Methodology for Estimating Human Health Risks*. ORNL-6865.

Oak Ridge National Laboratory (ORNL): 1995b, *U.S. Department of Energy Eastern Area Programs Sitewide Risk Prioritization for Savannah River Site*. (Document Number Forthcoming).

Pacific Northwest Laboratories (PNL): 1994, *Methodology for Selecting Technologies and Estimating Waste Volumes, Costs, and Labor Requirements for Environmental Restoration*.

Westinghouse Savannah River Company, Savannah River Site (WSRC-SRS): 1994, *Savannah River Site 1993 Annual Environmental Report*. WSRC-TR-94-077.

ASSESSING SELENIUM CYCLING AND ACCUMULATION IN AQUATIC ECOSYSTEMS

GEORGE L. BOWIE [1], JAMES G. SANDERS [2], GERHARDT F. RIEDEL [2], CYNTHIA C. GILMOUR [2], DENISE L. BREITBURG [2], GREGORY A. CUTTER [3], and DONALD B. PORCELLA [4]

[1] Tetra Tech, Inc., 3746 Mt. Diablo Blvd., Suite 300, Lafayette, CA 94549 [2] The Academy of Natural Sciences, Benedict Estuarine Research Center, 10545 Mackall Rd., St. Leonard, MD 20685 [3] Old Dominion University, Department of Oceanography, Norfolk, VA 23529 [4] Electric Power Research Institute, P.O. Box 10412, Palo Alto, CA 94304

Abstract. We conducted a joint experimental research and modeling study to develop a methodology for assessing selenium (Se) toxicity in aquatic ecosystems. The first phase of the research focused on Se cycling and accumulation. In the laboratory, we measured the rates and mechanisms of accumulation, transformation, and food web transfer of the various chemical forms of Se that occur in freshwater ecosystems. Analytical developments helped define important Se forms. We investigated lower trophic levels (phytoplankton and bacteria) first before proceeding to experiments for each successive trophic component (invertebrates and fish). The lower trophic levels play critical roles in both the biogeochemical cycling and transfer of Se to upper trophic levels. The experimental research provided the scientific basis and rate parameters for a computer simulation model developed in conjunction with the experiments. The model includes components to predict the biogeochemical cycling of Se in the water column and sediments, as well as the accumulation and transformations that occur as Se moves through the food web. The modeled processes include biological uptake, transformation, excretion, and volatilization; oxidation and reduction reactions; adsorption; detrital cycling and decomposition processes; and various physical transport processes within the water body and between the water column and sediments. When applied to a Se-contaminated system (Hyco Reservoir), the model predicted Se dynamics and speciation consistent with existing measurements, and examined both the long-term fate of Se loadings and the major processes and fluxes driving its biogeochemical cycle.
Key words: selenium, biogeochemical cycling, bioaccumulation, toxicity experiments, simulation model

1. Introduction

Selenium (Se) is an environmental toxicant that has been responsible for significant ecological damage in several aquatic ecosystems. Se discharged in coal fly ash pond effluent has caused fishery declines in several power plant cooling lakes (Cumbie *et al.*, 1978; Lemly, 1985). Assessment of Se toxicity is difficult since exposure occurs simultaneously through multiple oxidation states and chemical forms, each with different chemical, biological, and toxicological properties. In addition, aquatic organisms play a major role in the biogeochemical cycling of Se and are therefore partly responsible for determining chemical speciation and exposure.

We conducted a joint experimental research and modeling study to develop a methodology for assessing Se toxicity in aquatic ecosystems (Porcella *et al.*, 1991; Bowie and Grieb, 1991). The objectives were to determine how the various forms of Se are taken up by aquatic organisms, what roles organisms play in biogeochemical cycling, how Se moves through aquatic food webs, how toxic effects are exerted on aquatic organisms, and how to integrate this information into a predictive modeling framework. The first phase of

Water, Air and Soil Pollution 90: 93-104, 1996.

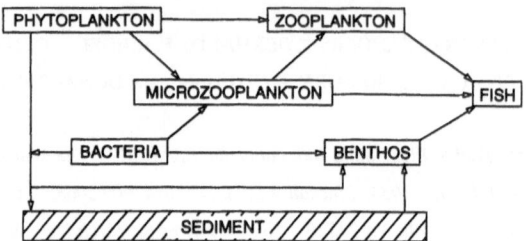

Fig. 1. Potential pathways of carbon and selenium cycling in a model freshwater ecosystem.

the research focused on Se cycling and accumulation (Sanders *et al.*, 1992) and provided the scientific basis and data necessary to parameterize the model. The model provides a conceptual framework that helps focus the experimental research, as well as integrating the experimental results and other data in the literature into a predictive framework that summarizes our current understanding of Se cycling and accumulation in aquatic ecosystems.

2. Experimental Research

Experiments were designed to investigate the biological pathways by which Se is transported in freshwater systems, the transformations that occur via biological uptake and incorporation, and the potential for impact to various trophic levels. The research was based on the simplified food web shown in Figure 1 and investigated the potential toxicity of both dissolved Se species and particulate Se in food. Laboratory studies were designed to detect the toxic effects of Se on the trophic components of a generic lake ecosystem, to determine the rates of transfer and transformation within the food web, and to ascertain the rates and mechanisms of accumulation. The effects of organisms on Se biogeochemical cycling processes were also an important part of the research. All experiments were designed to determine which components and processes must be incorporated into the model framework and to provide estimates of the important rate parameters.

Along with the laboratory studies, additional experiments were performed using natural communities from freshwater ecosystems. These experiments were targeted at lower trophic levels and were designed to compare results of Se uptake, transfer, and transformation from laboratory experiments with similar results from actual ecosystems.

Our laboratory phytoplankton experiments utilized three phytoplankton species, the chlorophyte, *Chlamydomonas reinhardtii*, the cyanophyte, *Anabaena flosaquae*, and the diatom, *Cyclotella meneghiana*, as representative of natural freshwater communities. In general, the three species responded in similar fashion to Se, with minor differences in uptake and transformation observed (Riedel *et al.*, 1991). All three species actively took up inorganic Se compounds, with short-term uptake rates for selenite much more rapid than for selenate. Passive adsorption of selenite to cell walls was an important component of initial uptake. Over the longer term, however, selenite and selenate incorporation were similar, and largely proportional to the concentration of dissolved Se. Uptake of an organo-Se compound was even more rapid, but the reactive compound tested (selenomethionine) probably occurs at low concentrations in natural systems and represents only a fraction of the dissolved organo-Se pool. Toxicity to Se did not occur at environmentally realistic concentrations (Riedel et al., 1991).

Examination of Se interactions with natural phytoplankton communities gave similar results. Toxicity to selenate, however, was observed at much lower concentrations than in laboratory cultures (Riedel et al., unpublished). Transformation of Se species, including volatilization of organo-Se compounds, did occur, but at low levels that suggest little environmental effect (Riedel et al., unpublished).

Bacteria exhibited Se uptake and incorporation patterns similar to those seen for phytoplankton, with short-term uptake rates for selenite much higher than those for selenate. Unlike phytoplankton, over the longer-term, incorporation of selenite continued to be more rapid. Bacteria actively volatilized Se both in culture and in natural waters (Sanders et al., 1992). Bacterioplankton are probably the predominant mediators of Se volatilization in natural communities. Bacteria, together with phytoplankton, are largely responsible for much of the biogeochemical cycling of Se that occurs in natural systems (Riedel et al., unpublished). The effects of microbial communities on redox reactions, detrital Se regeneration, and sediment Se accumulation were measured in several experiments.

In our zooplankton experiments, both protozoans (Sanders and Gilmour, 1994) and cladocerans (Goulden and Riedel, unpublished) actively incorporated Se from food; protozoans from bacteria, and cladocerans from phytoplankton. While some Se is incorporated from the surrounding medium, it appears that foodborne uptake is more important than waterborne uptake. Se assimilation efficiencies are relatively high, averaging approximately 35 percent for protozoans (Sanders and Gilmour, 1994) and 44 to 72 percent for cladocerans (Goulden and Riedel, unpublished); somewhat higher efficiencies have been demonstrated for marine organisms (Reinfelder and Fisher, 1991).

Our fish experiments to date have focused on fathead minnows and measured tissue accumulation from food and water, as well as changes in the Se content of eggs during successive breeding events. Fish incorporate Se primarily from their food; however, some accumulation from water also occurs and the two sources appear to be additive (Breiburg and Riedel, unpublished). Elevated tissue burdens are reflected throughout the organism, including reproductive tissues. Se contents of eggs varies greatly among females and among clutches produced by individuals (Breitburg and Riedel, unpublished).

Overall, biological interactions with Se are dominated by lower trophic levels, primarily phytoplankton and bacterioplankton. Se toxicity to lower trophic levels occurred at very high concentrations, one to two orders of magnitude above those encountered in natural ecosystems. The exception was selenate toxicity to natural phytoplankton, which occurred at levels that have been observed in some contaminated systems (Riedel *et al.*, unpublished). Se availability to higher trophic levels was regulated primarily by the Se content of food items; less Se was taken up directly from the surrounding medium. Toxicity to higher trophic levels, as with the phytoplankton and bacterioplankton, was only seen at relatively high concentrations. The most important biological effect appears to be inhibition and/or failure of reproduction in fish. Because of the high concentrations of Se contained in sediments and the potential for many fish to forage on benthic infauna, future experiments will focus on Se transfer from benthos to fish.

3. Biogeochemical and Food Web Transfer Models

3.1. BIOGEOCHEMICAL CYCLE

The Se cycle is modeled by partitioning all Se in the ecosystem into several compartments representing different oxidation states or chemical forms in water, sediments, and aquatic organisms. The model components are organic selenides [Se(-II)], elemental Se [Se(0)], selenite [Se(IV)], and selenate [Se(VI)]. Most of the oxidation states (organic selenides, selenite, and selenate) can occur in either dissolved forms, particulate forms, or in the cells or tissues of living organisms. Elemental Se can occur in bacterial cells, sediments, and particles suspended in the water column. The particulate forms are diverse and include precipitated elemental Se, organic selenides and other forms incorporated in algal and bacterial cells, selenite and selenate adsorbed to the surfaces of plankton cells, selenite and selenate adsorbed to clay particles or inorganic precipitates such as iron and manganese oxides, and various Se forms incorporated in and adsorbed to detrital organic material. Since the particulate forms vary widely in their fate and chemical behavior, they must be treated differently in the model.

The biogeochemical cycling and food web transfer models are based on a dynamic mass balance approach. A thermodynamic equilibrium approach is not appropriate for modeling Se since the system is never in equilibrium due to the slow kinetics of the oxidation reactions and the continual regeneration of soluble reduced organic forms by biological activity (Cutter, 1991; Cutter, 1992). Mass balance equations are used to quantify the processes that transform Se among different oxidation states and among the water, sediments, and biotic community. Figure 2 summarizes the major compartments and processes modeled. A separate equation is required for each Se compartment and spatial segment in the model, resulting in a large system of simultaneous differential equations coupled by the biogeochemical and toxicokinetic processes linking the Se compartments and the transport processes linking the spatial segments.

Water Column Processes   Sediment Processes

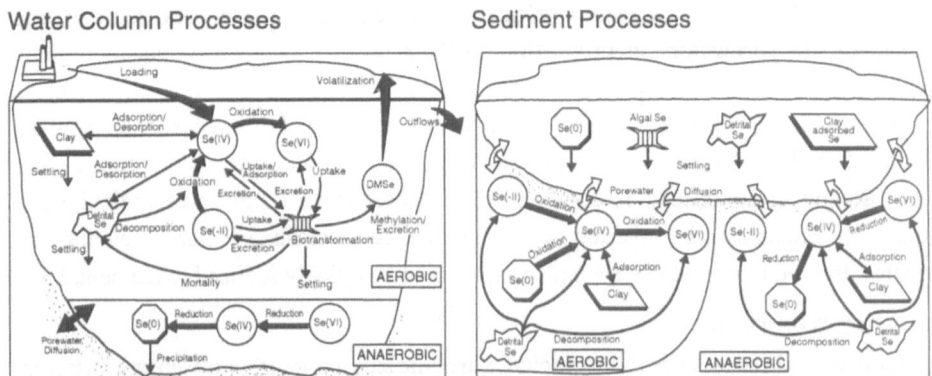

Fig. 2. Selenium biogeochemical cycling processes simulated in the model.

The equations include terms describing the redox reactions among dissolved organic selenides, selenite, selenate, and elemental Se; decomposition of particulate organic Se in the water column and sediments; uptake, biotransformation, and excretion of the three dissolved forms of Se by organisms; production and volatilization of dimethyl selenide by microbial and algal excretions; settling of particulate organic Se, elemental Se, and Se adsorbed to organic particles and clays; diffusive exchange with sediment porewaters; consumption of particles by detrital grazers; and various external loading and transport processes. Uptake, excretion, egestion, and consumption fluxes are calculated in the food web model equations.

3.2. SEDIMENTS

Sediments are modeled as an active layer of constant thickness overlying a deeper inactive layer. The active layer is assumed to be anaerobic, with a thin oxidized microzone at the sediment-water interface. Sediment accumulation is modeled dynamically and depends on deposition rates of clays, detritus, and algae, adjusted for decomposition of sediment organic material and processing by the benthic community. Surface accumulation results in a corresponding deep burial flux that transports Se to the deep inactive layer where it can no longer participate in biogeochemical cycling. The active sediment layer is made up of three physical components—organic particles, inorganic particles, and porewaters—each with different Se speciation.

The inorganic particles include precipitated elemental Se and selenite and selenate adsorbed to settled clays. The organic particles are derived from settling plankton and organic detritus and consist largely of organic selenides, but also include selenite and selenate taken up by aquatic organisms (and incorporated in fecal pellets) and selenite and selenate adsorbed to phytoplankton and microbial cell surfaces. The organic and inorganic particles are modeled separately since only organic components are subject to decomposition processes.

Organic sediment decomposition releases dissolved organic selenides, selenite, and selenate to the porewaters in proportion to their relative concentrations in organic particles. The dissolved porewater forms are then subject to redox reactions and diffusive exchanges with the overlying water column, with minor contributions also coming from the slow burial of dissolved Se in bottom waters by sediment deposition (porewater burial). Reducing conditions are assumed in the porewaters, so dissolved selenite and selenate are reduced and converted to particulate elemental Se. Similarly, selenite and selenate adsorbed to settled clays are also assumed to be quickly reduced to elemental Se.

3.3. FOOD WEB TRANSFER

The food web is user-defined and includes trophic components representing different types of fish, benthos, zooplankton, microzooplankton, algae, macrophytes, and bacteria. The Se in each organism is described by four mass balance equations simulating the dynamics of organic selenides, selenite, selenate, and elemental Se. Each oxidation state is treated separately since uptake kinetics, elimination kinetics, assimilation from food, and toxic effects differ between forms (Besser *et al.*, 1993; Ingersoll *et al.,* 1990; Kleinow and Brooks, 1986). Elemental Se is involved only in bacterial biotransformation reactions, bacterial consumption by microzooplankton, and benthic uptake from ingested sediments. The other three forms—organic selenides, selenite, and selenate—are involved in both waterborne and foodborne uptake by all organisms. The modeled processes include uptake from water and food, biotransformations between forms, elimination processes, and growth dilution.

Uptake from water is modeled as two components: adsorptive uptake to the cell exterior and internal uptake into the organism. Adsorption is of concern primarily for the smallest trophic components like bacteria and algae. Adsorptive uptake is assumed to occur instantaneously according to linear distribution coefficients. Internal uptake can include both passive and active uptake mechanisms, and is modeled as the product of the waterborne concentration, organism biomass, uptake rate coefficient, and temperature correction factor, with either first-order kinetics or Michaelis-Menten saturation kinetics for uptake.

Foodborne uptake is modeled as the product of the consumption rate, consumer biomass, toxicant concentrations in food items, and toxicant assimilation efficiencies during digestion. Food consumption rates are modeled as a saturable function of food supply using a Michaelis-Menten formulation. Consumption is partitioned between different food types based on dietary preferences.

Elimination and biotransformation processes are modeled as a function of the internal toxicant pools in the organisms using first-order kinetics. Elimination is partitioned into nonvolatile soluble excretions, volatile soluble excretions, and fecal material for biogeochemical modeling purposes.

Growth dilution is modeled similar to elimination as a function of the internal toxicant pools using first-order kinetics dependent on net growth rates. Net growth is determined from user-specified population dynamics and calculated fluxes for consumption, predation, nonpredatory mortality, and plankton settling. Almost all processes are temperature adjusted.

3.4. SPATIAL CONSIDERATIONS

Spatial differences in biogeochemical cycling and toxicant exposure are simulated using a three-dimensional modeling approach that divides the water body into a series of horizontal segments and vertical layers, and calculates the movement of materials between spatial segments using a three-dimensional transport equation. The transport processes include advection, dispersion, inflows, outflows, and external loadings. The model is flexible and can be set up using either a one-, two-, or three-dimensional grid, or by representing the whole water body as a mixed system with no spatial resolution. Each spatial segment or layer that contacts the bottom has a sediment layer associated with it. A separate mass balance equation must be solved for each spatial segment, each Se species, and each biogeochemical medium (dissolved, suspended particulates, porewaters, sediment particles, various biotic groups).

For all dissolved, suspended, and planktonic Se forms in the water column, the model integrates the equations for one Se form at a time using an implicit Crank-Nicholson technique that solves the equations for all spatial segments simultaneously. The equations for Se in the sediments, porewaters, benthic organisms, and fish are solved using an explicit numerical scheme since these components are not subject to hydrodynamic transport.

4. Model Application

We applied the model to a contaminated lake with an extensive database to test its ability to predict Se dynamics and speciation, and to analyze the major processes and fluxes that drive the biogeochemical cycle and control the fate of Se loadings. The system was Hyco Reservoir, a power plant cooling reservoir in North Carolina that experienced fishery declines during the mid-1970s as a result of Se contamination from ash pond effluent (CP&L, 1986-93). Implementation of dry fly ash handling in 1990 significantly reduced Se loadings and initiated recovery of the fisheries. We simulated the 8-year period from 1985 to 1992. The first three years coincided with a Se speciation study in the reservoir (Cutter, 1991), and the last three years covered the initial recovery period.

Model rate parameters for biogeochemical processes and biological uptake and excretion were obtained from our experimental research (Sanders et al., 1992) and other data in the literature. Carolina Power & Light Co. monitoring data (CP&L, 1986-93) were used to characterize Se loadings, water quality conditions, aquatic populations, and Se concentrations in biota, water, and sediments. Se speciation measurements in water, sediments, and suspended particles were provided by Cutter (1991). Hydrologic data were obtained from USGS monitoring stations. For purposes of modeling, the Hyco ecosystem

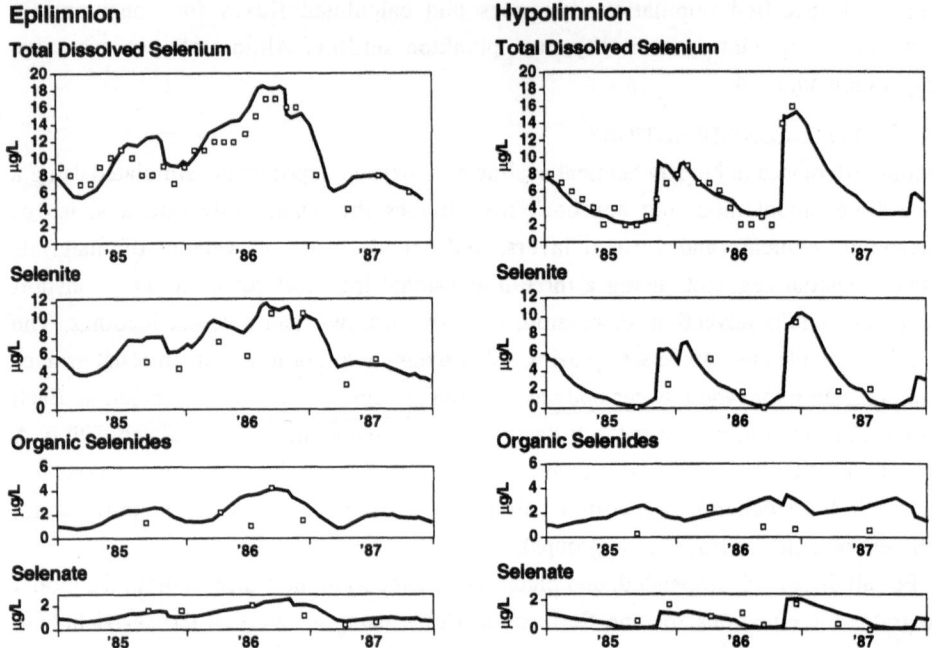

Fig. 3.   Se speciation predictions in Hyco Reservoir from 1985 to 1987. Lines are model predictions, symbols
are data. Speciation data are from Cutter (1991) and total dissolved Se data are from CP&L (1986-88).

was divided into two spatial segments representing the epilimnion and hypolimnion, and
six trophic components representing bacteria, phytoplankton, microzooplankton,
zooplankton, benthic animals, and fish.

4.1. CONCENTRATION PREDICTIONS

Figure 3 compares the model predictions of Se speciation in the water column with
available field data. Seasonal variations in total dissolved Se are driven by variations in
loading rates, reservoir flushing, and the development of anoxic conditions in the
hypolimnion. Selenite, the predominant form in ash pond loading, dominates the
speciation except during periods of hypolimnetic anoxia when fast reduction reactions
convert selenite and selenate to insoluble elemental Se that rapidly settles (Cutter, 1991).
Under aerobic conditions, organic selenides produced by biological cycling processes
such as excretion and decomposition represent about 20 percent of the dissolved Se, and
selenate produced by oxidation of selenite represents about 10 percent.

Figure 4 compares the model predictions of total Se in water and organisms with field
data. Se concentrations in the water column, phytoplankton, and zooplankton respond
rapidly to the reduced loading resulting from dry ash handling initiated in 1990. However,
the benthic community and fish respond slower due to the persistence of Se in sediments
and the importance of the benthic food web in bluegill Se accumulation. Consumer

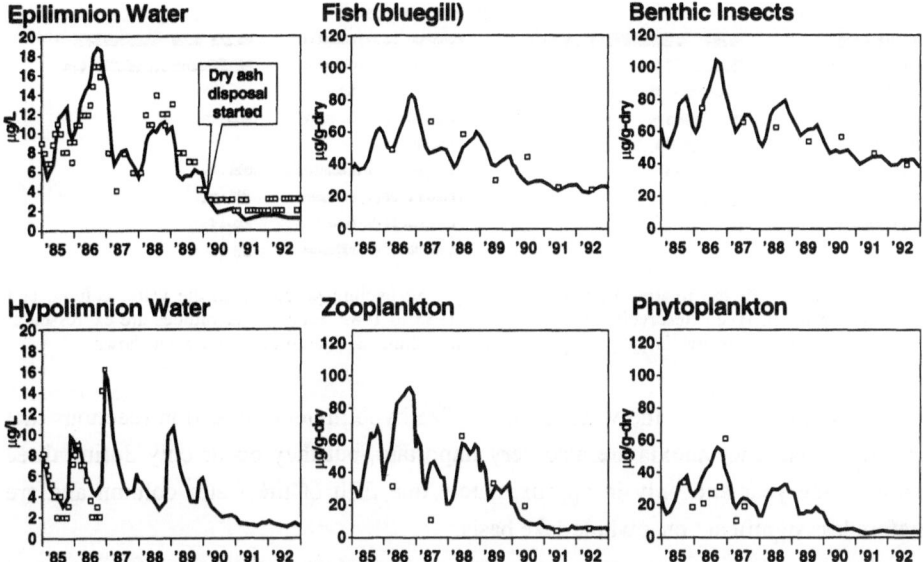

Fig. 4.   Total Se predictions in water column and food web components of Hyco Reservoir from 1985 to 1992. Field data are from CP&L (1986-93) and Cutter (1991). Food web results are for the epilimnion.

organisms obtain most of their Se from food and therefore reflect the Se dynamics of their diet.

4.2. LONG-TERM FATE PREDICTIONS

Analysis of Se pools in the lake shows that approximately 97 percent of the Se resides in the sediments, with only 3 percent in the water column and less than 0.1 percent in the biota. Flux calculations were performed with the model to determine the major processes governing the fate of Se loadings. During 1985 to 1989, over 50 percent of the loadings were deposited in the sediments, about 45 percent was flushed out with outflows, and less than 5 percent volatilized across the lake surface from the dimethyl selenide excretions of algae and bacteria. These estimates are consistent with the field data (Cutter, 1991) and the results of our laboratory experiments (Sanders et al., 1992).

4.3. BIOGEOCHEMICAL CYCLING FLUXES IN WATER COLUMN

The model was also used to analyze the internal Se cycling processes in Hyco during the 1985 to 1989 pre-remediation period. The major process fluxes in the water column and sediments are summarized in Figure 5. The flux magnitudes are expressed as percentages of the maximum flux in each medium to facilitate comparison, and represent whole lake averages.

In the water column (Figure 5a), selenite loading from ash pond effluent is the major flux driving the biogeochemical cycle. The internal cycling is dominated by diffusion of selenite into sediment porewaters, and biological uptake of selenite followed by excretion of organic selenides back to the water column. These fluxes are higher than the oxidation

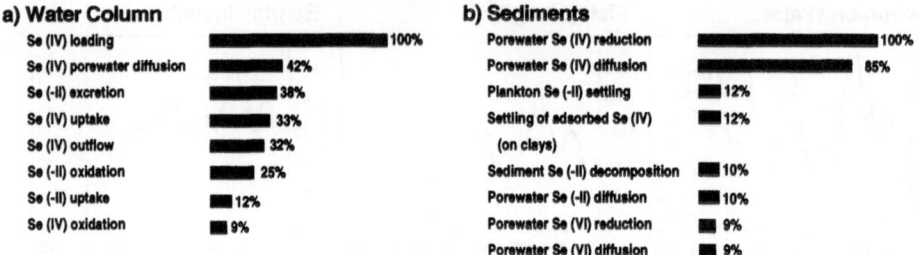

Fig. 5.   Comparison of the relative importance of major Se cycling processes in the water column and
sediments of Hyco Reservoir predicted during 1985 to 1989. Process magnitudes are expressed as
percentages of the maximum process flux in each medium. Smaller processes are not shown.

and reduction reactions in the water column. The hypolimnetic reduction reactions that
occur during summer anoxia are also very important, but they occur only during three
months of the year and only in approximately one fifth of the water column and are
therefore less significant on a whole-lake basis.

Biological uptake and excretion processes are important components of the internal
biogeochemical cycle that convert selenite and selenate to organic selenides. This alters
the Se speciation, but does not contribute significantly to Se removal. Approximately 10
percent of the organic excretions of algae and bacteria were volatile compounds, resulting
in a net volatilization loss of less than 5 percent of the loading.

In contrast, the diffusion of selenite into sediment porewaters is an extremely
important removal process in the water column, as important as outflow fluxes associated
with reservoir flushing. Porewater diffusion of selenite and selenate account for more than
45 percent of the loading, transporting these dissolved forms into the sediments where
they are subsequently deposited. The diffusion and deposition is driven by microbial
organisms in the sediments which, through dissimilatory reduction reactions, strip selenite
and selenate from the porewaters, converting them to insoluble elemental Se (Steinberg
and Oremland, 1990). This creates a large diffusion gradient that maximizes the flux into
the sediments and maintains a continual supply of selenite and selenate for microbial
respiration.

Note that even though the organisms represent an extremely small portion of the total
Se pool (less than 0.1 percent), they are extremely important in driving the
biogeochemical cycle, in controlling speciation in the water and sediments, and in
governing the long-term fate of Se loading to the lake.

4.4. BIOGEOCHEMICAL CYCLING FLUXES IN SEDIMENTS

In the sediments (Figure 5b), diffusion and subsequent microbial reduction of selenite in
porewaters is the most important process driving Se accumulation. The parallel processes
with selenate are approximately 10 percent of the selenite fluxes due to the lower
abundance of selenate in the water column. Settling of algal Se, detrital Se, and selenite
adsorbed to clays are each only 10 to 15 percent of the porewater reduction fluxes.

Settling of elemental Se produced during hypolimnetic anoxia contributes less of the total sediment Se accumulation than the above processes since it occurs only during three months of the year in one fifth of the water column.

Decomposition of organic detrital Se, derived originally from algal and detrital settling, is the most important Se removal process in the sediments. This creates porewater concentrations of dissolved organic selenides that exceed those in the overlying water column, producing an upward diffusion flux that recycles organic Se back to the water. This can be a major source to the water column during lake recovery periods after external loads are eliminated. However, during loading periods (pre-1990), the outward flux of organic selenides is dwarfed by the inward flux of selenite and selenate. As a result, the sediment Se pool accumulates over time, and its speciation is dominated by elemental Se.

5. Conclusions

The joint research/modeling effort has been of considerable value. The experimental research provided the scientific understanding of Se cycling and bioaccumulation processes that allowed us to construct the model. It also gave us the data necessary to parameterize these processes. The model integrated this information into a predictive framework that allowed us to extrapolate from the laboratory to the field. The model results also prompted new questions and helped shape the following experiments, so the research and modeling activities were interactive.

The results presented here for Hyco Reservoir demonstrate that the model predicts Se exposure and food web accumulation, and follows the important biogeochemical processes controlling the fate of Se loading to lakes. Although the results are specific to Hyco, the model is designed to be applicable to other systems. The results concerning the magnitudes and relative importance of the various process fluxes are site-specific, but several important conclusions can be drawn from our results that are applicable to most systems:

1. The sediments are extremely important in both the biogeochemical cycling of Se and perhaps in the accumulation of Se in fish through the benthic food web. The sediments are the major repository of the Se loadings, and a source of exposure to benthic organisms.

2. The organisms play a major role in the Se cycle. Biological processing fluxes are high and are among the major determinants of Se speciation in the water column. Microbial activity in the sediments drives much of the sediment Se accumulation.

3. The lower trophic levels—phytoplankton and bacteria—bioconcentrate Se several orders of magnitude above the concentrations in water. The higher trophic levels—zooplankton, benthos, and fish—obtain most of their Se from food. As a result, planktonic organisms respond rapidly to changes in waterborne Se exposure, while

benthos and fish respond much slower due to continued exposure through the benthic food web. The sediments respond slowly to changes in loadings.

4. Volatilization losses are small in Hyco and probably of minor importance in most lakes. However, they could be important in shallow, highly productive systems.

Acknowledgments

This work was funded by the Electric Power Research Institute under RP2020-11.

References

Besser, J. M., Canfield, T. J., and LaPoint, T. W.: 1993, *Environ. Toxicol. Chem.* **12**, 57-72.
Bowie, G. L. and Grieb, T. M.: 1991, *Water, Air, and Soil Pollut.* **57-58**, 13-22.
Breiburg, D. L. and Riedel, G. F., Unpublished data.
CP&L: 1986-93, *Roxboro Steam Electric Plant: 1985-92 Annual Environmental Monitoring Reports.*
Cumbie, P. M. and Van Horn, S. L.: 1978, Paper No. 32, Proc. Ann. Conf. S.E. Fish and Wild. Agencies, pp. 612-624.
Cutter, G. A.: 1991, *Selenium Biogeochemistry in Reservoirs. Vol. 1: Time Series and Mass Balance Results.* Electric Power Research Institute, Report EPRI EN-7281.
Cutter, G. A.: 1992, *Mar. Chem.* **40**, 65-80.
Goulden, C. E. and Riedel, G. F.: Unpublished data.
Ingersoll, C. G., Dwyer, F. J., and May, T. W.: 1990, *Environ. Toxicol. Chem.* **9**, 1171-1181.
Kleinow, K. M. and Brooks, A. S.: 1986, *Comp. Biochem. Physiol.* **83C**, 61-69.
Lemly, A. D.: 1985, *Reg. Toxicol. Pharmacol.* **5**, 465-486.
Porcella, D. B., Bowie, G. L., Sanders J. G., and Cutter, G. A.: 1991, *Water, Air, and Soil Pollut.* **57-58**, 3-11.
Riedel, G. F., Ferrier, D. P., and Sanders, J. G.: 1991, *Water, Air, and Soil Pollut.* **57-58**, 23-30.
Riedel, G. F., Sanders, J. G., and Gilmour, C. C.: Unpublished manuscript.
Reinfelder, J. R. and Fisher, N. S.: 1991, *Science* **251**, 794-796.
Sanders, J. G., Riedel, G. F., Gilmour, C. C., Osman, R. W., Goulden, C. E., and Sanders, R. W.: 1992, *Selenium Cycling and Impact in Aquatic Systems: Annual Progress Report for 1991.* The Academy of Natural Sciences, Benedict Estuarine Research Laboratory, for Electric Power Research Institute, RP-2020-11.
Sanders, R. W. and Gilmour, C. C.: 1994, *Applied Environ. Microbiol.* **60**, 2677-2683.
Steinberg, N. A. and Oremland, R. S.: 1990, *Applied Environ. Microbiol.* **56**, 3550-3557.

Comparison of *Daphnia magna*, rainbow trout and bacterial-based toxicity tests of Ontario Hydro aquatic effluents

D. W. RODGERS, J. SCHRÖDER[1] AND L. VEREECKEN SHEEHAN

Ontario Hydro Technologies, 800 Kipling Ave, Toronto, ON, CANADA M8Z 5S4; [1]present address - Beak Consultants Limited, 14 Abacus Road, Brampton, ON, CANADA L6T 5B7

Abstract. Over a one year program of intensive monitoring of effluents from Ontario Hydro's nuclear, fossil and hydroelectric generating facilities, the *Daphnia magna* and rainbow trout, *Oncorhynchus mykiss*, acute toxicity tests correlated well, with 61% of the toxic effluents toxic to both species. If the effluent was toxic to only one of the test species it was generally toxic to *D. magna*, with from 23 to 57% of the toxic effluents toxic to *D. magna* only. The greater sensitivity of *D. magna* to boiler blowdown effluent likely resulted from a combination of the low conductivity of boiler blowdown effluent and the smaller size and greater surface to volume ratio of *D. magna* relative to rainbow trout. *D. magna* were also particularly susceptible to oil/water separator samples, with the daphnids frequently observed to be caught at the surface/water interface. These observations suggest that an accumulation of organic material at the air/water interface was responsible for the mortality of *D. magna*. In subsequent tests, we also examined the relationship between the *D. magna* acute toxicity test and a bacterial-based assay (Toxi-Chromotest®) for several toxic effluents from Ontario Hydro stations to determine if bacterial-based tests could provide similar information in less time with smaller sample volumes. The *D. magna* acute toxicity test did not correlate well with the bacterial-based Toxi-chromotest®. In particular, many of the samples which were toxic to *D. magna* were not toxic to the Toxi-chromotest® assay. The poor correlation between the *D. magna* and Toxi-chromotest® likely relates to both the relatively low toxicity of many of the effluent samples, and the fact that in many cases toxicity likely resulted from relatively simple combinations of inorganic toxicants. Accordingly, the Toxi-Chromotest® assay would not seem suitable as a surrogate for the *D. magna* acute toxicity test for our effluents.

Key Words: effluent toxicity; *Daphnia magna*, rainbow trout, Toxi-chromotest®, bacterial-based toxicity test

1. Introduction

Throughout North America and the world, industries such as electrical utilities are increasingly required to include effluent toxicity testing in their operations. Under the Municipal Industrial Strategy for Abatement (MISA) program of the Ontario Ministry of Environment and Energy (OMEE), Ontario Hydro conducted a one year program of intensive monitoring of effluents from its nuclear, fossil and hydroelectric generating facilities (Ontario Hydro, 1992), and is now engaged in a three year program to monitor and control effluents which were toxic in the monitoring program. The present regulations (OMEE, 1995) require monthly or quarterly sampling of specific effluent streams using acute lethality toxicity tests with *Daphnia magna* (Environment Canada 1990a) and rainbow trout, *Oncorhynchus mykiss* (Environment Canada 1990b), followed by chronic toxicity testing using a fathead minnow, *Pimephales promelas*, larval growth

Water, Air and Soil Pollution 90: 105-112, 1996.
© 1996 *Kluwer Academic Publishers.*

stream. Boiler blowdown may be removed either continuously, or intermittently on the basis of feed water parameters such as conductivity.

• Neutralization Sump - Water Treatment Plant (WTP) effluent. This effluent originates from the water purification process which provides demineralized water for the steam/water cycle in both nuclear and fossil stations. It includes waters used for blowdown of clarifiers, backwash of filters, and regeneration of ion exchange resins.

• Oil/Water Separator effluent. This consists of effluent from systems which collect and treat drainage from areas where oil could be spilled or leaked in either nuclear and fossil stations.

• Radioactive Liquid Waste effluent. This effluent is comprised of drainage from a variety of sources within Ontario Hydro's nuclear stations, ranging from floor drains to parts washers, which may contain radioactive materials.

From October through December, 1994, a further series of toxicity tests were conducted on sample streams which were toxic in the initial MISA tests. Again D. magna tests were conducted according to the Environment Canada and OMEE protocols (Environment Canada, 1990b). Because of the large number of samples being processed and their relatively low toxicity (LC50's in initial MISA testing generally > 40% effluent), the D. magna acute toxicity tests were conducted with duplicate samples of 100% effluent. The toxicity of these samples was also evaluated using the bacterial colorimetric assay, Toxi-chromotest® (EPBI- Version 3.0). The test is based on the ability of toxic materials to inhibit de novo synthesis of the inducible enzyme β-galactosidase in a special strain of Escherichia coli, which is highly sensitive to a wide spectrum of toxic materials (EBI 1993). The activity of induced β-galactosidase is determined by hydrolysis of a chromogenic substrate, which was measured photometrically (615 nm) after 90 min incubation. The Toxi-chromotest® tests were conducted with duplicate samples of a serial dilution of 100% effluent (100%, 50%, 25% and 0%). In addition to toxicity tests on the samples as received, toxic samples were treated to reduce toxicity, and additional tests conducted on the treated samples. Treatments for the respective streams were:

• Neutralization Sump - pH of the sample was adjusted to pH 6-7 using 1 N HCl or NaOH as appropriate.

• Oil/Water Separator - sample was passed through an activated charcoal column (Calgon Filtrasorb 400)

• Radioactive Liquid Waste - sample was passed through either an activated charcoal column or a metal chelating resin column (Rohm and Haas IRC-718).

The details of the treatments are given elsewhere (Rodgers et al., 1996).

Throughout the tests, the response of the test organisms to a standard toxicant (2-chlorophenol for D. magna and mercuric chloride for the Toxi-chromotest® test) was consistent. In the single concentration tests, samples were considered toxic to *D. magna* if more than 50% than of the animals in each duplicate test died within 48 h. In the Toxi-chromotest®, samples were considered toxic if the 100% effluent produced a 50% or greater reduction in color of both duplicate samples.

3. Results and Discussion

Throughout the MISA tests, there were marked differences in the toxicity of the particular effluents. For example, greater than 80% of the boiler blowdown samples were toxic to one or both of the test organisms, yet less than 20% of the oil/water separator samples were toxic (Figure 1). Overall, there was reasonably good correspondence between the two tests, as 61% of the toxic samples were toxic to both *D. magna* and rainbow trout. However, the degree of correspondence differed among effluents (Figure 1), with values ranging from 80% (neutralization sump) to 29% (oil/water separator). If the effluent was toxic to only one of the test species, it was generally toxic to *D. magna*, with from 23 to 57% of the toxic effluents toxic only to *D. magna* (Figure 1). In contrast, from 0 to 14% of the toxic effluents were toxic only to rainbow trout.

The correspondence between the *D. magna* acute toxicity test and the bacterial-based Toxi-chromotest® was much weaker than that between the *D. magna* and rainbow trout test (Table 1). In particular, many of the samples which were toxic to *D. magna* were not toxic to the Toxi-chromotest® assay. All toxic samples of radioactive liquid waste and chlorinated sewage treatment effluent were toxic only to *D. magna*. Similarly, of the 4 samples of oil/water separator effluent which were toxic to *D. magna*, only one was also toxic to the Toxi-Chromotest™ test. A reasonable correspondence between the tests was observed only for neutralization sump effluent, where all samples toxic to *D. magna* were also toxic to the bacteria and one sample was toxic to the Toxi-chromotest® but not *D. magna*. However, all of the neutralization sump samples which where toxic to both *D. magna* and the Toxi-chromotest® were highly acidic (pH ≤ 2.5).

The substantial differences apparent in both the toxicity of the different effluent streams and the relative sensitivity of the test organisms among effluent streams likely correspond to variation in both the chemical composition of the specific effluent streams and the biology of the test organisms. Although *D. magna* was generally more sensitive than rainbow trout to many Ontario Hydro effluents, this may not apply to all industries. In Ontario's MISA program, over all industrial sectors, the two species were generally similar in sensitivity, with 24% of samples lethal to *D. magna* and 20% lethal to rainbow trout (Lee *et al.*, 1995). Thus, further examination of the specific characteristics of Ontario Hydro's effluents is required to account for the apparent sensitivity of *D. magna* to these effluents.

The greater sensitivity of *D. magna* to boiler blowdown effluent likely resulted from a combination of the effluents low conductivity (mean less than 10 milliSiemens/m) (Ontario Hydro, 1992), and the smaller size and greater surface to volume ratio of *D. magna* relative to rainbow trout. This interpretation is consistent with the observation that *D. magna* were unable to survive the duration of the acute toxicity test (48 h) in deionized water, while trout were able to survive in deionized water for 96 h (Rodgers, 1994).

and survival test and a *Ceriodaphnia dubia* reproduction and survival test (Environment Canada 1992a, 1992b).

Although debate continues regarding the relative merit of single species tests versus multiple species testing procedures (Cairns 1986), industries such as utilities are continually faced with the problem of relating the various tests and test procedures. Further, these tests require substantial volumes of effluent (ie. determination of a rainbow trout LC50 requires 50 - 100 L of effluent (Environment Canada, 1990b)) and considerable time (from 2 d (48 h) for the *D. magna* acute toxicity test (Environment Canada, 1990a) to 7 d for the chronic tests (Environment Canada 1992a, 1992b)). Consequently, there is considerable demand for tests which provide similar information in less time with smaller sample volumes, and an array of bacterial-based toxicity tests have been developed to meet this demand (Bitton and Dutka 1986). In this paper, we compare the results of *D. magna* and rainbow trout tests during the one year MISA monitoring program, and review our subsequent studies comparing a bacterial-based toxicity test with the *D. magna* acute toxicity test.

2. Materials and Methods

From June 1, 1990 to May 31, 1991, Ontario Hydro conducted a one year program of monitoring of effluents from its nuclear, fossil and hydroelectric generating facilities, as required by the OMEE's MISA program (Ontario Hydro, 1992). Toxicity testing was integral to this program, and over the one year period, more than 700 effluent samples were tested. At each station, effluent from the designated sample streams was collected monthly or quarterly using grab or pump samplers. The toxicity samples were collected in several ≈ 10 L plastic pails lined with food grade polyethylene bags and transported to the test laboratory within 24 h of collection. Acute toxicity tests with both *Daphnia magna* (48 h - 20°C) and rainbow trout (96 h - 15°C) were conducted by serial dilution of the effluent within 5 days of sample collection, following Environment Canada and OMEE protocols (Environment Canada 1990a, 1990b). Total hardness, pH, conductivity and dissolved oxygen concentration of the effluents were analyzed at the beginning and end of the test; in addition, an array of substances were analyzed in each effluent stream as part of the MISA program (Ontario Hydro, 1992). The results of the toxicity tests were summarized and the LC50 calculated by the binomial, probit, or Spearman-Karber methods as appropriate. For this paper, samples were considered toxic if the calculated LC50 was less than or equal to 100% effluent.

In these tests, there were appreciable numbers of toxic samples for the following effluents from nuclear and fossil stations:

• Ash Transport System effluent. This consists of water which may contact either bottom ash only, or both bottom ash and fly ash at fossil stations.

• Boiler Blowdown effluent. This stream is derived from the condensation side of the steam/water cycle that drives the turbine/generator units to generate electricity in both nuclear and fossil stations. Corrosion products and impurities, which tend to concentrate in the boilers, are removed by the boiler blowdown

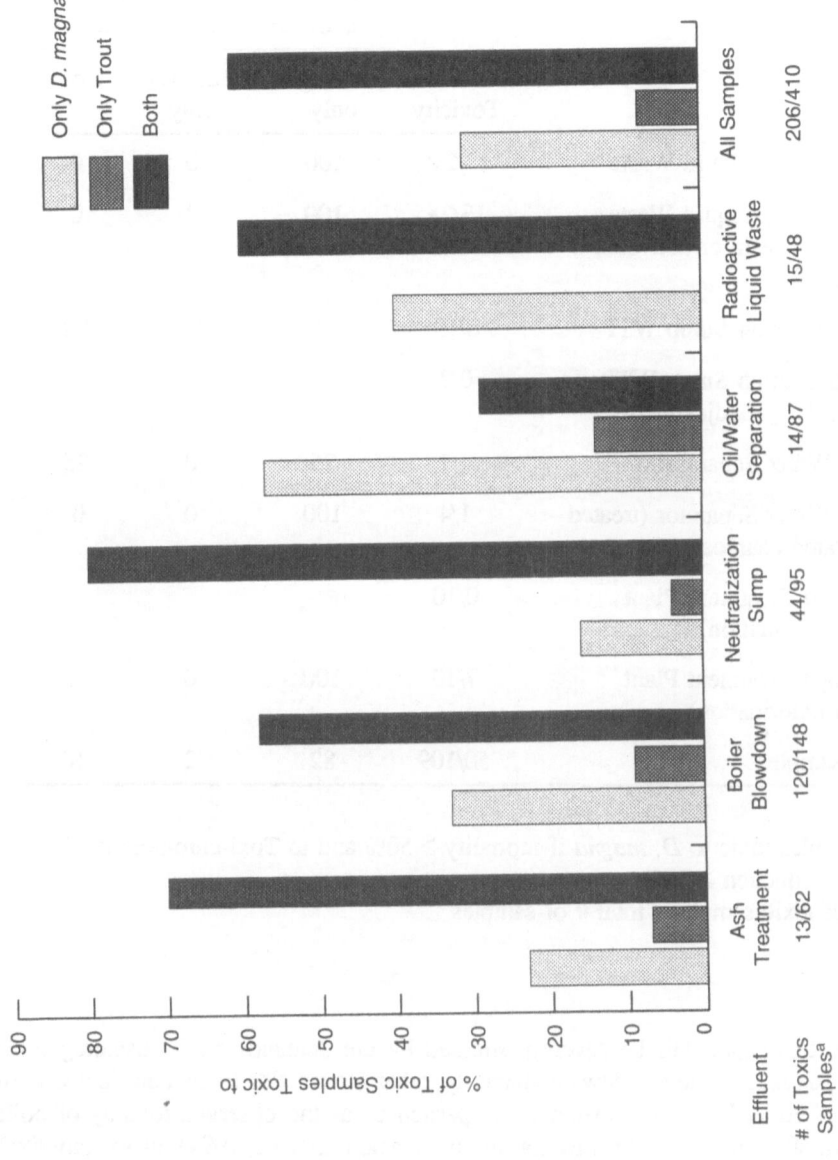

Fig 1: Toxicity of Ontario Hydro MISA Samples to *Daphnia magna* and Rainbow Trout

Table I: Toxicity[a] of Ontario Hydro effluents to *Daphnia magna* and Toxic-chromotest[®]

| | | % of toxic samples toxic to | | |
	Toxicity	*D. magna* only	Bacteria only	Both
Radioactive Liquid Waste	17/24[b]	100	0	0
Radioactive Liquid Waste (treated - resin or activated charcoal)	13/28	100	0	0
Neutralization Sump WTP	8/19	0	12	88
Neutralization Sump WTP (treated - pH adjustment)	0/7			
Oily Water Separator	4/7	75	0	25
Oily Water Separator (treated - activated charcoal)	1/4	100	0	0
Sewage Treatment Plant (pre-chlorination)	0/10			
Sewage Treatment Plant (post-chlorination)	7/10	100	0	0
All samples	50/109	82	2	16

[a] Samples toxic to *D. magna* if mortality $\geq$ 50% and to Toxi-chromotest[®] if color reduction $\geq$ 50%
[b] # of toxic samples / total # of samples

Although *D. magna* will be severely stressed by the demands on its osmoregulatory systems, *per se*, in waters of low conductivity, other factors also likely contributed to the toxicity of boiler blowdown effluent. In particular, as the observed toxicity of boiler blowdown effluent was reduced by Ca^+ or humic acid (Rodgers, 1994), its toxicity likely resulted from an interaction of the low ionic content and slightly elevated concentrations of metals, especially Zn and Cu. This interpretation is consistent with numerous observations and reports that very low concentrations of many metals are toxic to fish and invertebrates in waters of low ionic concentration, and that this toxicity is reduced with addition of Ca^+ or humic acid (Spear, 1981; Meador, 1991; Playle *et al.*, 1991).

D. *magna* were also particularly susceptible to oil/water separator samples, as just under 60% of the toxic samples of this effluent were toxic to *D. magna* only. In these tests, daphnids were frequently observed to be caught at the surface/water interface early during the test and to subsequently expire. These observations suggest that an accumulation of organic material at the air/water interface was responsible for the mortality of *D. magna*, likely through a combination of direct physical trapping, interference with respiration caused an oil coating, and by exposure to locally elevated concentrations of organic chemicals. In the detailed parallel chemical analysis of the MISA effluent streams, however, no specific organic compound or compounds were consistently detected at anything approaching toxic concentrations, and the measured concentrations of oil and grease levels were low (generally 1-6 mg/L) (Ontario Hydro, 1992). Although the toxic compounds remain to be identified, in the subsequent series of tests, passage through an activated charcoal filter removed the toxicity from toxic oil/water separator samples to *D. magna* in 3 of 4 cases.

Although radioactive liquid waste effluent was also more toxic to *D. magna*, we are less certain as to the causes of toxicity to either species. Rather, as this effluent stream collects drainage from a variety of sources within Ontario Hydro's nuclear stations, ranging from floor drains to parts washers, the toxic factors likely varied considerably among stations and with time. Further discussion of our continuing studies to identify and eliminate the toxic components of this stream is contained elsewhere (Rodgers *et al.*, 1996).

Conversely, the response of *D. magna* and rainbow trout to neutralization sump effluent and ash treatment effluent was much more uniform, with 80% and 70%, respectively, of the toxic samples toxic to both species. For both effluents, toxicity was usually related to extremes of pH (pH ≤ 3 or pH ≥ 10) (Ontario Hydro, 1992). As these extreme pH values are outside the zone of tolerance of both species (Dillon *et al.*, 1984), it is likely that mortality was pH-related. This was substantiated in our subsequent tests, where none of the toxic neutralization sump effluent were toxic to *D. magna* after adjustment of pH (Table 1).

Unfortunately, the Toxi-Chromotest™ assay appears much less sensitive than the *D. magna* acute toxicity test for our samples, and would not seem suitable as a surrogate for this test. The general lack of correspondence between the *D. magna* and the Toxi-chromotest® likely relates to both the relatively low toxicity of many of the effluent samples (LC50 ≥ 40% effluent) and the fact that in many cases toxicity likely resulted from relatively simple combinations of toxicants. Previous studies have noted that bacterial-based toxicity tests appear to be less sensitive than the *D. magna* and fish acute toxicity tests for many inorganic toxicants, but may be of similar sensitivity for many organic compounds and complex mixtures (Qureshi *et al.*, 1982; Munkittrick *et al.*, 1991).

Acknowledgments

We thank P. Wiancko, R. Maruska, M. Greenall, J. Reeves (deceased) and J.Sharma for their help and support throughout this study and D. Poirier (OMEE) for his help and constructive comments on various aspects of this work. We also acknowledge the many people from Ontario Hydro Stations who provided samples in a timely and efficient manner.

References

Bitton, G., Dutka, B.J.: 1986, *Toxicity testing using microorganisms Vol 1*, Boca Raton, Florida: CRC Press Inc.

Cairns, J. Jr.: 1986, *Bioscience* 36, 670-672.

Dillon, P.J., Yan, N.D., Harvey, H.H.: 1984, *CRC Critical Reviews in Environmental Control*, 13: 167-194.

Environmental Bio-Detection Inc. (EBI): 1993, *The Toxi-chromotest Version 3.0 - instructions for use*, Brampton. Ont.

Environment Canada: 1990a, *Biological test method: Reference method for determining acute lethality of effluents to Daphnia magna*, Environmental Protection Publications, Conservation and Protection, Reference Method EPS 1/RM/14.

Environment Canada: 1990b, *Biological test method: Reference method for determining acute lethality of effluents to rainbow trout*. Environmental Protection Publications, Conservation and Protection, Reference Method EPS 1/RM/13.

Environment Canada: 1992a, *Biological test method: Test of larval growth and survival using the cladoceran Ceriodaphnia dubia*. Environmental Protection Publications, Conservation and Protection, Reference Method EPS 1/RM/21.

Environment Canada: 1992b, *Biological test method: Test of larval growth and survival using fathead minnows*. Environmental Protection Publications, Conservation and Protection, Reference Method EPS 1/RM/22.

Lee, J.T., Westlake, G.F., Abernathy, S.G., Poirier, D.G., Mueller, M.C.: 1995, *Proceedings of at the 21st Annual Aquatic Toxicity Workshop: October 3-4, 1994, Sarnia Ont.*, Canadian Technical Report of Fisheries and Aquatic Sciences, 2050, p20.

Meador, J.P.: 1991, *Aquatic Toxicol.* 19, 13-32.

Munkittrick, K.E., Power, E.A., Sergy, G.A.: 1991, *Environ. Toxicol. Water Qual.* 6, 35-62.

Ontario Ministry of Environment and Energy (OMEE): 1995, *Effluent Monitoring and Effluent Limits - Electric Power Generation Sector*, Ontario Regulation 215/95 made under the Environmental Protection Act.

Ontario Hydro: 1992, *A review of Ontario Hydro's MISA monitoring results*. Ontario Hydro, NOCS-IR-07291-OUO/TISD-07291-0001.

Playle, R.C., Gensemer, R.W., Dixon, D.G.: 1991, *Environ. Toxicol. Chem.* 11, 381-391.

Qureshi, A.A., Flood, K.W., Thompson, S.R., Janhurst, S.M., Innis, C.S., Rokosh, D.A.: 1982, In: Pearson *et al.* (eds) *Aquatic Toxicology and Hazard Assessment: Fifth Conference*, American Society for Testing and Materials, ASTM STP 766, p179-195.

Rodgers, D.W.: 1994, *Chemical Speciation and Bioavailability* 6, 23-26.

Rodgers, D.W., Vereecken Sheehan, L., Evans, D.W.: 1996, *Water, Air and Soil Pollution*

Spear, P.A.: 1981, *Zinc in the aquatic environment: chemistry, distribution and toxicology*, National Research Council of Canada (NRCC) NRCC No. 1789.

USE OF THE PISCES DATABASE:
POWER PLANT AQUEOUS STREAM COMPOSITIONS

Greg P. Behrens
Doug A. Orr
Robert G. Wetherold
Radian Corporation
8501 N. MoPac Blvd.
Austin, Texas 78759

and

Barbara Toole O'Neil
Electric Power Research Institute
3412 Hillview Avenue
Palo Alto, California 94303

Abstract

The Power Plant Integrated Systems: Chemical Emissions Studies (PISCES) Database sponsored by the Electric Power Research Institute is a powerful tool for evaluating and comparing the level of trace substances in power plant process streams. In this paper, data are presented on the level of several selected trace metals found in a few of the aqueous streams present in power plants. In addition, a brief discussion of other features of the Database is presented.

Keywords: trace elements, power plants, ash pond, discharge, aqueous streams, PISCES Database

1.0 Introduction

The Electric Power Research Institute (EPRI) is conducting a project to assist member utilities with managing and controlling chemical substances in the process streams of power generating systems in light of current regulatory and health effects issues. This project, entitled Power Plant Integrated Systems: Chemical Emissions Studies (PISCES), involves the collection and validation of information on chemical substances in process and discharge streams from power generating systems. One of the results of the EPRI research has been the development of a relational Database to store this information. The Database development represents one element in the overall EPRI effort to 1) define the source, pathways, and concentration of potentially hazardous pollutants; 2) establish the health and environmental risks associated with these chemical substances; 3) determine the performance of available technologies for their measurement and control, if warranted; and, 4) develop a model for estimating their distribution and discharge.

Water, Air and Soil Pollution **90**: 113-122, 1996.
© 1996 *Kluwer Academic Publishers.*

The PISCES Database contains information from public literature sources on the chemical composition of process streams in conventional (steam-electric) fossil-fuel generating systems. In addition to public literature, data from the PISCES Field Chemical Emissions Monitoring (FCEM) Program have been entered into the Database. Examples of literature and field test reports from which data were collected are listed at the end of this paper. Related information, such as plant and control device operating conditions, solid by-product leachate compositions, sample collection and analytical procedures is also compiled in the Database. Currently, nearly 180,000 individual records have been validated, qualified, and entered into the Database.

2.0 Database Description

The PISCES Database is managed on an Oracle™ framework. It consists of multiple tables linked by key attributes. Of the total number of records, approximately 42% are associated with solid streams (such as coal and ash), 31% are liquid streams, and 27% are gas stream data. In reviewing reports, the cited stream names are entered (e.g., blow-down, Run 1) and a standard stream name (e.g., cooling tower blowdown) is assigned. About 200 standard stream names have been used to permit multi-plant comparisons.

The majority of data is for coal-fired power plants. About 5% of the records pertain to oil and gas units. Current data entry efforts are focussed on expanding the existing liquid stream information. Approximately 50,000 new liquid stream records have been identified by various literature searches for entry this coming year.

One of the uses of the Database is to extract information for all plants meeting certain search criteria and assume that the resulting data distribution is representative of the total population. With over 2,400 individual steam-electric systems present in the US, the issue of representativeness must always be considered when data are used to forecast conditions at sites which have not been tested. Summary statistics are also subject to bias. Table I presents summary information from the Database showing the number of plants that have liquid stream information for five substances [arsenic (As), boron (B), chromium (Cr), mercury (Hg), and selenium (Se)] found at coal-fired units. For the substances selected in this paper, data are available from about 2 to 3 percent of the total number of units. In addition, most plants have multiple measurements of a substance in several streams. Although the Database has a relatively large amount of liquid phase information, when separated into individual streams, the number of records is relatively small. Figure 1 presents a breakdown, by stream types, for the five selected substances. Some of the streams have limited information. Some of the data gaps are obvious selections for targeted information gathering.

Table I

Aqueous Phase Data Records for Selected Trace Substances

Analyte	Total Records	Number of Plants	Average Records per Plant	Maximum Records per Plant
Arsenic	879	125	7	44
Boron	322	79	5	21
Chromium	944	133	7	68
Mercury	695	109	6	45
Selenium	799	116	7	44

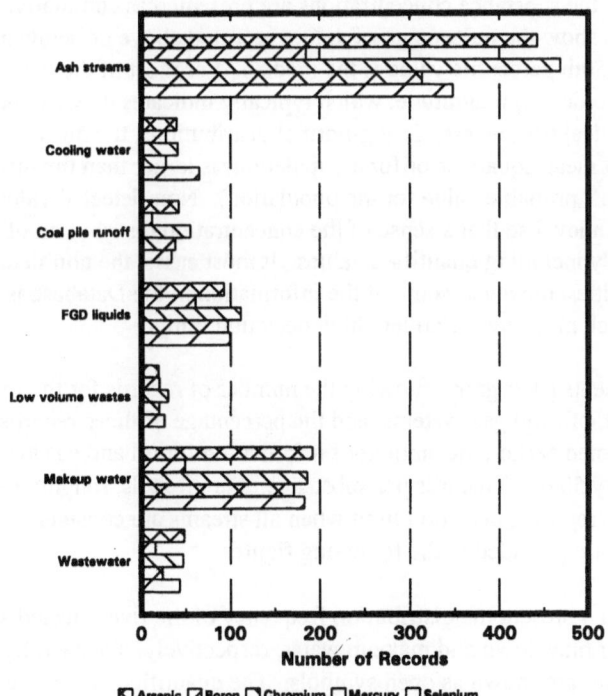

Fig. 1. Liquid Streams/Records in the PISCES Database
(Coal-Fired Power Plants

3.0 Stream Composition Distributions

The remainder of this paper presents PISCES Database information on several streams present at coal-fired power plants. Five elements have been selected for review: arsenic, boron, chromium, mercury, and selenium. In the review of reports for data entry, no culling of information occurs unless it is specifically identified as unacceptable for use by the Database quality ranking criteria. Many other substances have hundreds of records, including organic compounds and standard water quality parameters (pH, COD, BOD_5, alkalinity, etc.). These five substances were selected to provide an overview of the types of data present.

One of the frustrations of dealing with trace substances is the limitation of analytical methods to quantify low concentration levels. All analytical procedures have a lower limit where the presence of a substance cannot be positively confirmed, commonly called the detection limit. Values at or below the detection limit can provide valuable information but are also difficult to represent in summary statistics (i.e., what quantified value does one use to calculate the mean and standard deviation of nondetected values?).

In this paper, the substance concentrations are presented as cumulative frequency plots. These graphs show the percentage of values reported above or below a specific concentration. The 50th percentile value is the median. For most of the data sets, the values span several orders of magnitude, which typically indicates they can be described by log-normal statistical parameters. In log-normal distributions, the mean value (the statistic with the least mean square error for a population) is larger than the median (the statistic that is the most probable value for the population). Non-detected values are shown at their detection level so that a sense of the concentration distribution of a substance is not biased by only including quantified values. It most cases, the non-detected levels are the lowest quantities, however, some of the information in the Database is from less sensitive methods which may have relatively high detection limits.

Figure 2 presents a bar graph showing the number of records for the selected substances found in coal, oil, and gas systems, and the percentage of those records that are quantified. As denoted earlier, the amount of information for oil and gas systems is limited, approximately 50 to 60 records per substance. For all fuels, roughly half of the values are above the reported detection limit when all streams are considered. Individual stream distributions are provided in the following figures.

Figures 3 and 4 present the distribution frequency of the five selected substances in cooling tower blowdown and makeup water, respectively. On each figure, the non-detected values are shown as open symbols. The quantified values are shown in filled symbols. The cooling tower blowdown has mostly detected concentrations whereas there are many non-detected values for makeup water. Although there is no specific relationship between the two data sets (i.e., the makeup water samples may not correspond with any of the blowdown samples), the order of concentration ranges (boron > chromium > arsenic > selenium > mercury) are the same and the nominal magnitude differences

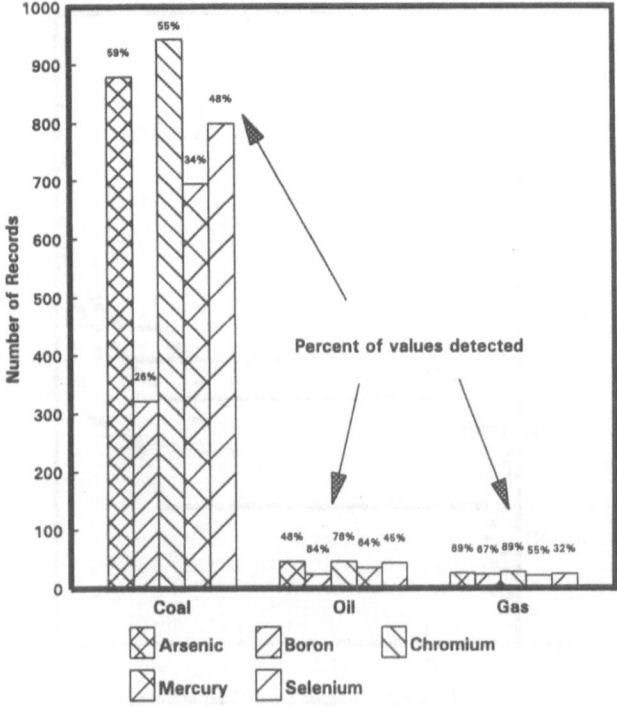

Fig 2. Liquid Streams Records in the PISCES Database

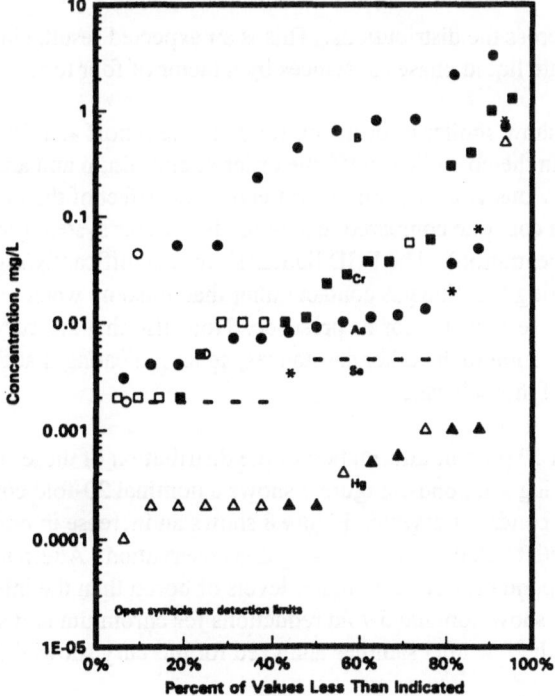

Fig. 3. Distribution of Substance Concentrations in Cooling Tower blowdown

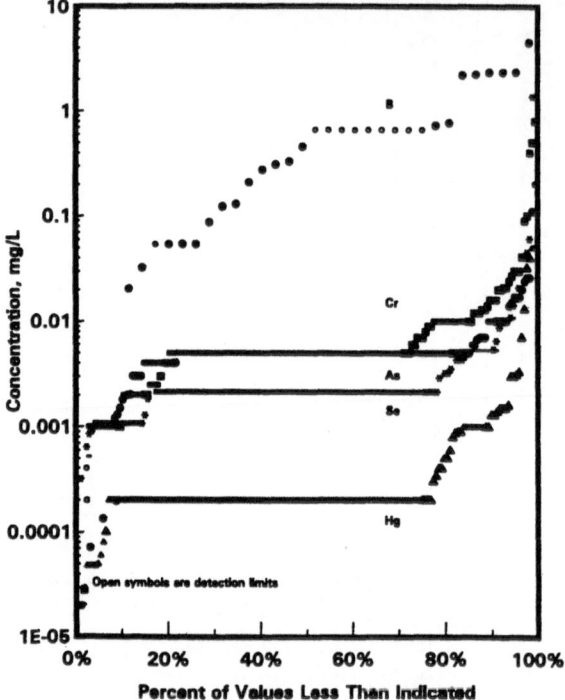

Fig. 4.   Distribution of Substance Concentrations in Makeup Water

appear similar across the distributions. This is an expected result, since most cooling towers concentrate liquid phase substances by a factor of four to ten.

Figures 5 and 6 show similar information for coal pile runoff and flue gas desulfurization (FGD) liquids. In the coal pile runoff, the order of chromium and arsenic are reversed from the two previous graphs, probably reflecting the effect of the different chemical environment in a coal pile compared to a circulating water system (acidic leaching versus evaporative concentration). The FGD liquids show a significantly higher ranking of selenium, indicating that flue gas contact rather than makeup water evaporative concentration effects are responsible for its presence. Note also that mercury concentrations are closer in concentration to the other substances, again indicating that the flue gas is a primary source of this substance.

Figure 7 through 10 present information of the distribution of these substances in streams entering and exiting ash ponds. Figure 7 shows a nominal 20-fold concentration reduction across ponds for arsenic. Figure 8 shows an increase in boron levels. The relatively few outlet values may be biasing this observation. Alternately, other streams entering the ash pond may contain higher levels of boron than the inlet sluice stream. Figures 9 and 10 show nominal 5-fold reductions for chromium and selenium across ash ponds. None of the few inlet samples analyzed for mercury showed quantified results.

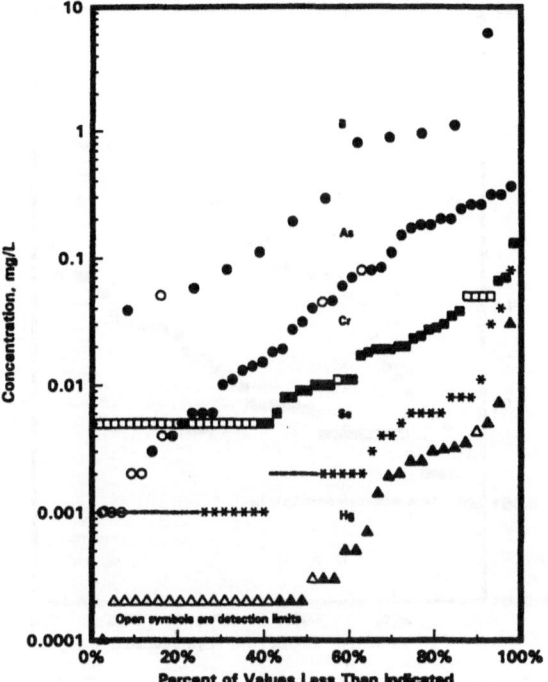

Fig. 5. Concentration Distributions in Coal Pile Runoff

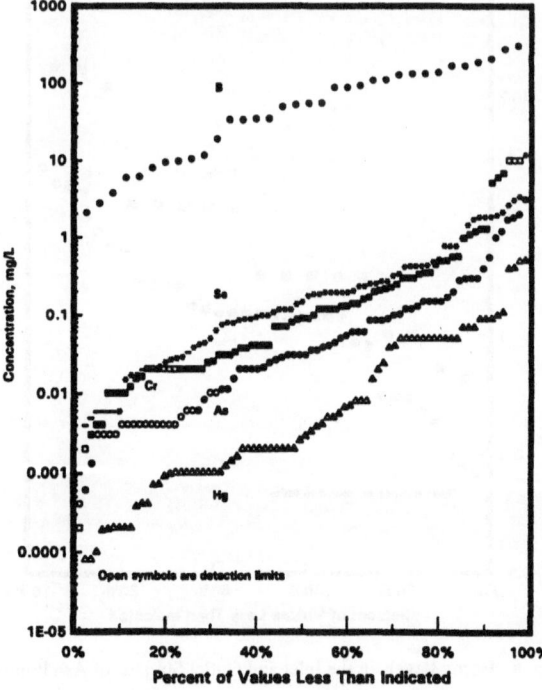

Fig. 6. Distribution of Concentrations in FGD Liquid

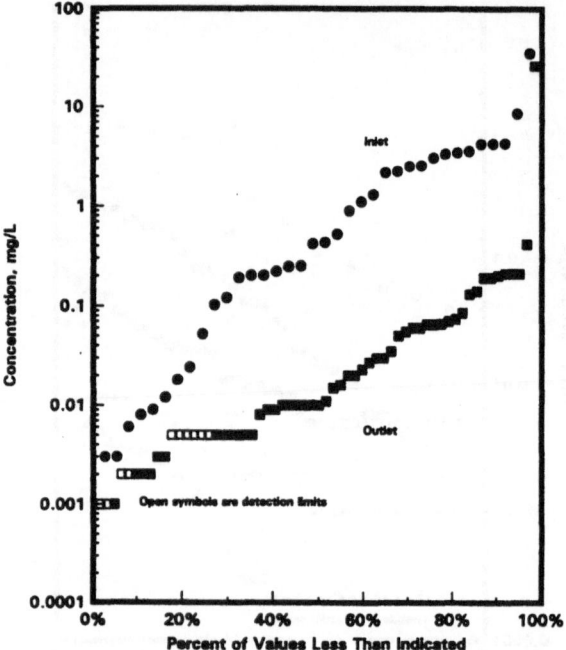

Fig. 7. Arsenic Levels in the Inlet and Outlet Streams of Ash Ponds

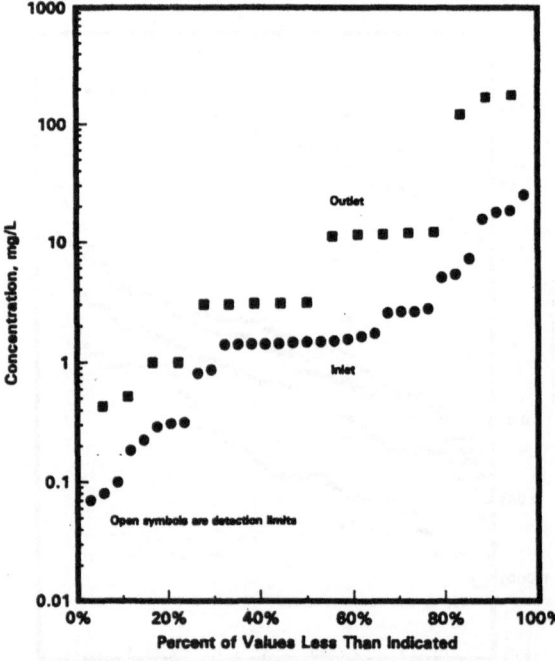

Fig. 8. Boron Levels in the Inlet and Outlet Streams of Ash Ponds

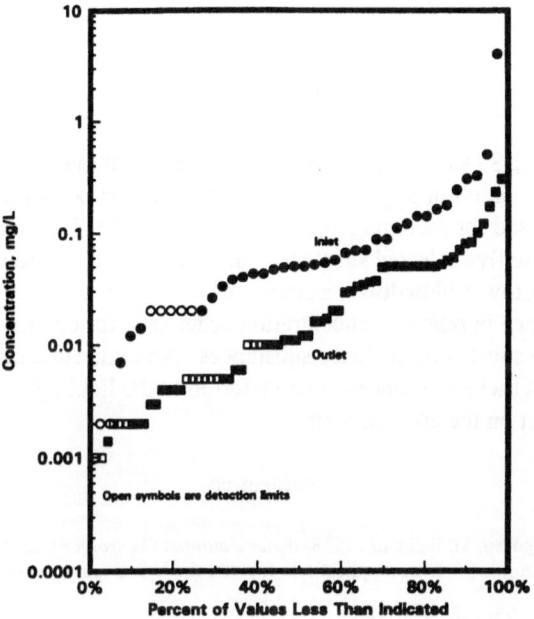

Fig. 9. Chromium Levels in the Inlet and Outlet Streams of Ash Ponds

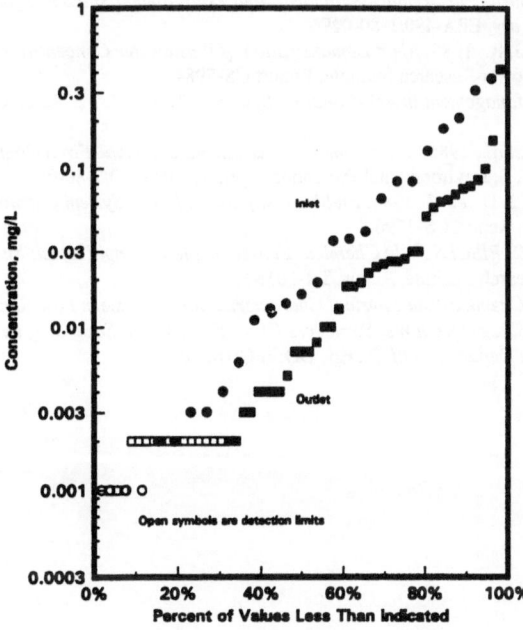

Fig. 10. Selenium Levels in the Inlet and Outlet Streams of Ash Ponds

Quantified effluent levels vary from 0.0001 to 0.01 mg/L, with a median level of 0.002 mg/L.

4.0 Conclusion

The use of the PISCES Database permits quick screening of massive amounts of information. Examination of large quantities of information permit identification of trends that can be used for focussing additional research. For instance, the relative concentrations of the five selected substances in makeup water to power plants is reflected in cooling tower blowdown, a stream which is primarily concentrated makeup water. The difference in relative concentration order seen for other liquid streams can be used to identify the sources of the these substances. An understanding of the chemical principles by which these substances are transferred to the liquid phase is the first step in reducing their effect on the environment.

References

Anderson. Orson L, Rogozen, M. B., et. al.: 1978, Water Pollution Control for Coal Slurry Pipelines, University of California ; Science Applications, Inc., for the U.S. Department of Energy, Report SAI-067-79-516.

Cox, D. B., Chu, T. J., and Ruane, R. J.: 1979, Characterization of Coal Pile Drainage, Tennessee Valley Authority for the U.S. Environmental Protection Agency, EPA-600/7-79-057.

Evans, J. C., Abel, K. H., et al.: 1985, Characterization of Trace Constituents at Canadian Coal-Fired Power Plants, Final Report, Batelle Pacific Northwest Laboratories for Canadian Electrical Association, Contract No. 001G194

Environmental Protection Agency: 1980, Development Document for Proposed Effluent Limitations Guidelines, New Source Performance Standards, and Pretreatment Standards for the Steam Electric Point Source Category, EPA-490/1-80-0296.

Hess, M. B., and Jones, G. R., 1988, Field Demonstration of Wastewater Concentration by Seeded Reverse Osmosis, Electric Power Research Institute, Report CS-5984.

Laing, L., 1987, Water Management in Ash-Handling Systems, Electric Power Research Institute, Report CS-5369.

Leavitt, C., Arledge, K., et al., 1985, Environmental Assessment of a Coal-Fired Controlled Utility Boiler, TRW Inc. for the U. S. Environmental Protection Agency, EPA-600/7-80-086.

Litherland, S. T., Colley, J. D., et al., 1984, Field Investigation of FGD System Chemistry, 1984, Electric Power Research Institute, Report CS-3796.

Radian Corporation, 1992, PISCES Field Chemical Emissions Monitoring Project: Site15 Emissions Report, Electric Power Research Institute, Report TR-105620.

Thompson, C.M., 1982, Chemical and Physical Characterization of Western Low-Rank Coal Waste Materials, Part 3: Sludges from Wet Scrubbers Using Fly Ash as a Scrubbing Reagent, Radian Corporation for the Department of Energy, DOE/FC/10200-T2

RIVRISK: A MODEL TO ASSESS POTENTIAL HUMAN HEALTH AND ECOLOGICAL RISKS FROM CHEMICAL AND THERMAL RELEASES INTO RIVERS

C.S. LEW, W.B. MILLS, K.J. WILKINSON, and S.A. GHERINI

Tetra Tech, Inc., 3746 Mt. Diablo Blvd., Suite 300, Lafayette, CA 94549

Abstract. RIVRISK, an interactive Windows™-based model, predicts ambient chemical concentrations, human health and ecological risks, and water temperatures due to chemical and thermal releases into rivers. RIVRISK simulates chemicals that enter rivers from pipes, through submerged diffusers, from groundwater seepage, and atmospheric plume deposition. Human health risks are calculated for up to 16 pathways via exposure routes of dermal contact, ingestion, and inhalation. Ecological risks are evaluated for aquatic organisms by comparing the river water concentration to a water quality criterion for freshwater organisms. Monte Carlo analyses of ambient concentrations and temperatures can also be performed based on uniform distributions or user-defined histograms of varied parameters. To illustrate the various features of the model, RIVRISK was applied to a case study based on the Glen Lyn Power Plant located along the New River in Virginia.

Key Words: human health risk; ecological risk; rivers; discharges

1. Introduction

Due to the fact that throughout the U.S there are hundreds of power plants and chemical plants that discharge to rivers, evaluation of impacts of these plants on rivers is important. Plant releases to a river can occur from direct discharges, atmospheric deposition from stack emissions, seepage from contaminated groundwater, surface runoff, and the cycling of river water for cooling purposes. To evaluate the impacts of these discharges, RIVRISK has been developed. RIVRISK is a two-dimensional analytical river model that predicts chemical concentrations, water temperatures, and both human health and ecological risks from exposure to the river water.

To illustrate how RIVRISK evaluates human health risks, the model is applied to a case study: the Glen Lyn Power Plant located along the New River in Virginia. The case study depicts the tasks and resources involved in data acquisition, illustrates the use of RIVRISK, and portrays conclusions that can be drawn from the model results. Furthermore, the risk analysis for the case study is performed for three individuals, a reasonably exposed individual (using 50th percentile data), a maximally exposed individual (using 95th percentile data), and a standard individual (using U.S. Environmental Protection Agency [U.S. EPA] defaults). This risk analysis is performed to illustrate the differences in the risk predictions that can result from using alternative exposure parameters.

2. The RIVRISK Model

RIVRISK evaluates the effects of point discharges, atmospheric deposition from stack emissions, groundwater seepage, and thermal discharges to a river. Figure 1 presents

Water, Air and Soil Pollution **90**: 123-132, 1996.
© 1996 *Kluwer Academic Publishers.*

a graphical illustration of the processes that are simulated in RIVRISK. In addition to performing the transport calculations for the exposure concentrations, RIVRISK also performs the subsequent calculations of risks to humans, based on the possible exposure routes from contaminated river water, and to aquatic organisms.

3. Concentration and Water Temperature Predictions

RIVRISK predicts the concentrations of both organic and inorganic chemicals in a river downstream from a power plant, chemical plant, or other source of chemical or thermal releases. Chemicals can enter the river from point discharges, through submerged diffusers, from groundwater seepage, and via atmospheric plume deposition. Each of these pathways has its own individual algorithm, which is solved separately, and the results are superimposed to obtain the total chemical concentration.

The dissolved chemical concentration in rivers due to diffusers is determined by a two-dimensional steady-state analytical solution to the advection-dispersion equation with modifications for multiple discharges. Groundwater seepage is represented by seepage along the shoreline next to the power plant. A method of images solution to the advection-dispersion equation is used to simulate this. Atmospheric deposition is represented by a square area of deposition. Up to five such areas can be simulated to mimic more general patterns of deposition. A simplified version of the advection-dispersion equation, which assumes dominance of deposition and longitudinal advection, is used to simulate atmospheric deposition.

RIVRISK contains an environmental chemistry module that can be used to evaluate chemical processes in the aquatic environment. Rate constants and half-lives for the processes of hydrolysis, photolysis, and volatilization can be calculated. These results can help the user to determine the relative importance of these different processes in their particular simulation and to estimate the composite dissolved and absorbed transformation rates, which are required inputs to the transport model. The linear equilibrium partition coefficient, also a required input to the transport model, can be estimated in this module. The environmental chemistry module also provides a component that allows verification of whether acid or base speciation is insignificant given that RIVRISK does not presently simulate these processes. In addition, speciation of 15 different metals can be evaluated.

For temperature predictions, RIVRISK evaluates up to two thermal discharges to the river. Downstream temperatures are simulated using a steady-state analytical solution to the energy balance equation.

Uncertainty analysis is important in modeling because of the uncertainty inherent in the input variables. RIVRISK uses the Monte Carlo technique as a method of analyzing the uncertainty in RIVRISK's predictions due to the uncertainty in the input data. In representing these random variables by distributions and performing Monte

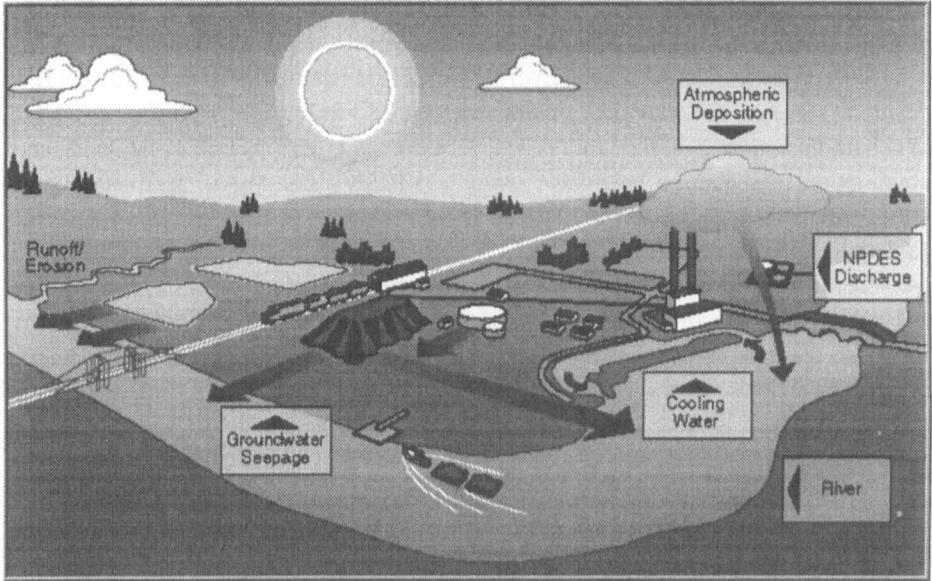

Fig. 1.   Conceptual illustration of the RIVRISK model.

Carlo analysis, the dependence of the solution on the parameters becomes more apparent.

In RIVRISK, Monte Carlo analysis of chemical concentrations and river water temperatures can be performed. Varied river, chemical, discharge, and temperature parameters can be given a uniform distribution or can be specified by a histogram. Cumulative distribution functions (CDFs) of concentration and temperature are generated at a specified location in the river. The mean, standard deviation, maximum, and minimum for these CDFs are also reported.

4. Human Health and Ecological Risks

Human health risks are calculated from the chemical concentrations using the method outlined in the U.S. EPA's Risk Assessment Guidance for Superfund (U.S. EPA, 1989). Both the lifetime carcinogenic risk and the noncarcinogenic risk (hazard index) are calculated for 16 exposure pathways: inhalation of volatiles in shower water, ingestion of water while showering, incidental ingestion of water while swimming, ingestion of drinking water, ingestion of soil, ingestion of leafy vegetables, ingestion of root vegetables, ingestion of finfish, ingestion of shellfish, ingestion of meat, ingestion of milk, dermal contact with water while showering, dermal contact with water while swimming, dermal contact with soil, inhalation of ambient air, and ingestion of mother's milk. The risk for each pathway as well as the total risk are calculated.

The carcinogenic risk versus time is also calculated when simulating multiple hydrologic periods, so that cumulative risks for exposure durations of up to a lifetime can be determined. For multiple hydrologic period simulations, river, discharge, deposition, seepage, and chemical parameters can be varied.

The human health risks for other media, such as groundwater and soil contamination, can also be calculated using RIVRISK's stand-alone risk calculation module. Media concentrations are specified by the user and the risks are calculated for a single exposure point.

A screening level approach is used to evaluate ecological risks, where surface water chemical concentrations are compared with chemical-specific water quality criteria for freshwater organisms to generate ecological hazard quotients. Hazard quotients are calculated for all emitted chemicals to evaluate which chemicals are dominating ecological risks.

5. Integrated Modeling Environment

RIVRISK is a Windows™-based model that has an integrated modeling environment, which allows the user to quickly evaluate a problem. The RIVRISK model integrates an interactive user interface for inputting data, the computational modules, and graphical and tabular output into a complete package that is easy to use and understand.

RIVRISK's graphical user interface allows quick access to input data. The input data is organized by process, each of which is represented by a button on RIVRISK's main screen (see Figure 1). Data is stored in the input screens to allow the user to quickly change a parameter and make another run. Each input screen also has context-sensitive help. Additionally, the input data can easily be viewed and printed for inclusion into a report.

Results from the RIVRISK model are presented in the form of graphs. The plot features, such as the line properties and size, are easily changed on the plot screen, and the plot can be sent directly to a printer. The corresponding data are also presented for viewing or copying to another program. Output available as plots versus distance downstream are human health risks (carcinogenic or noncarcinogenic), ecological risks (hazard quotient), concentration, water temperature, sediment concentration, and fish concentration. Other available plots are cross sections of concentration, CDFs of concentration and water temperature, contours of temperature, and risk versus time. Water quality criteria are plotted along with the predictions for easy comparison.

RIVRISK's chemical database contains chemical and risk parameters for 122 chemicals common to power plant and industrial releases. The risk parameters are based on published U.S. EPA data for risk analysis. The chemical data are automatically provided when the user selects a chemical and can be modified to suit a particular site.

6. Case Study Description

RIVRISK was applied to a case study involving Appalachian Power Company's Glen Lyn Power Plant. The objectives of applying the RIVRISK model to this case study are twofold. First, the case study illustrates the tasks involved in applying the model: data acquisition, use of the model, and drawing conclusions from the model results. Second, the human health risks from the plant releases for three different individuals are compared to illustrate that the data used for a risk assessment can make a difference in the resulting calculated risks.

The Glen Lyn Power Plant is located on the New River in Virginia. The New River is one of the headwaters of the Ohio River. It begins in North Carolina and flows northwest through Virginia to merge with the Greenbriar River in Hinton, West Virginia, to form the Kanawha River. In the vicinity of the Glen Lyn Power Plant, the river is approximately 140 meters wide (Dickson *et al.*, 1976). The 17-day consecutive low flow at Glen Lyn was approximately 32 m³/s for records from 1915 to 1965 (Dickson *et al.*, 1976). The river was assumed to be 1.5 m deep.

The Glen Lyn Power Plant is located at New River kilometer 153. The fossil-fuel plant was built in 1919 and has a current operating capacity of 350 megawatts (MW) (Dickson *et al.*, 1976).

Information regarding the discharge from the Glen Lyn Power Plant was taken from EPRI's PISCES database (Chow *et al.*, 1991) for a 350-MW power plant. A once-through cooling water discharge was assumed, with a wastewater discharge flow rate of 0.2 cfs; the flow rate of the cooling water discharge is higher. Chemical concentrations in the discharge were assumed to be mean concentrations and are presented in Figure 2. The data for atmospheric deposition from stack emissions are typical values extracted from two sources (Chu *et al.*, 1993 and Miller *et al.*, 1994). The deposition rates for each chemical are presented in Figure 2. Groundwater seepage data were extracted from the PISCES database for coal pile runoff and are also presented in Figure 2.

Human health risks were calculated for this case study for 14 of the 16 exposure pathways simulated by RIVRISK. The ingestion of mother's milk pathway was not applicable for this case since it applies only to chemicals with long half-lives. The inhalation of ambient air pathway was also not simulated. For calculation of risks, three individual receptors were defined: the reasonably exposed individual, the maximally exposed individual, and the standard individual. The risks to the reasonably exposed individual were calculated using the 50th percentile exposure data; the risks to the maximally exposed individual were calculated using the 95th percentile exposure data; and the risks to the standard individual were calculated using the current U.S. EPA default data. Fiftieth and 95th percentile data were extricated from exposure factor distributions available in the literature. Distributions are only readily available for some of the exposure parameters; thus, the standard U.S. EPA defaults were used for all three individuals when distributions were not available. Due to the large amount of data, the standard U.S. EPA defaults for

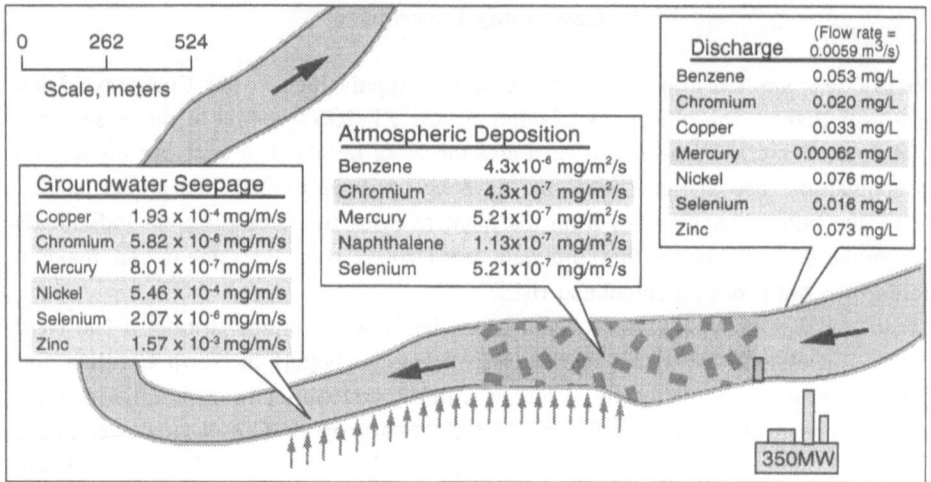

Fig. 2.   Releases from Glen Lyn Power Plant.

parameters without distributions are not presented; only references are provided (U.S. EPA, 1989; U.S. EPA, 1991; U.S. EPA, 1992; McKone and Knezovich, 1991; McKone, 1987; U.S. EPA, 1990; Travis and Arms, 1988).   The 50th and 95th percentile data for parameters with distributions along with the corresponding U.S. EPA defaults are presented in Table I for adults and children.

Both the carcinogenic and noncarcinogenic risks were calculated for the three exposure scenarios.   Benzene and chromium are the only chemicals contributing to the carcinogenic risks (chromium only contributes to the inhalation pathway) whereas all of the chemicals except benzene cause noncarcinogenic risks.   For this case study, a common toxic endpoint for the seven noncarcinogenic chemicals was assumed in calculating the hazard index.   Therefore, the noncarcinogenic risks from each chemical for each exposure pathway were simply added to produce a total hazard index for that pathway.   References for the following chemical properties used in the RIVRISK simulation of the case study are as follows: slope factors and reference doses (U.S. EPA, 1994a; U.S. EPA, 1994b, U.S. EPA, 1994c), cattle and milk transfer factors (MEPAS, 1992), soil to plant concentration factors (Baes *et al.*, 1984), skin permeability constants (U.S. EPA, 1992), chemical transfer efficiencies (McKone, 1987), absorption factors (Cal. EPA, 1984), and bioconcentration factors (U.S. EPA, 1986).

7. Results of Case Study

The maximum carcinogenic and noncarcinogenic risks for each of the three exposed individuals are presented in Tables II and III, respectively.   Note that the total maximum risk is not the sum of the maximum risks for each pathway because the maximum risks by pathway may occur at different receptor locations along the length

TABLE I

Risk exposure factors: 50th percentile, 95th percentile, and default data

Parameter	Child			Adult		
	50th Percentile	95th Percentile	U.S. EPA Default	50th Percentile	95th Percentile	U.S. EPA Default
Body weight (kg)[1]	13.1	19.1	15	71	100	70
Exposure duration (yrs)[2]	NA[10]	NA	NA	9	33	24
Tap water ingestion rate (L/day)[3]	0.62	1.43	1	1.1	2.5	2
Skin surface area (cm^2)[4]	5,780	7,660	7,600	18,300	22,400	23,000
Shower duration (min)[5]	6.8	14.6	15	6.8	14.6	15
Fish consumption (g/d)[6]	0	2.0	6.5	0	5.5	6.5
Shellfish consumption (g/d)[6]	0	2.8	6.5	0	8.2	6.5
Soil ingestion rate (mg/d)[7]	4.9	84	200	NA	NA	NA
Skin surface area for exposure to soil (cm^2)[8]	1,445	1,915	2,100	4,575	5,600	5,800
Soil adherence factor (mg/cm^2)[9]	0.16	1.2	1.0	0.16	1.2	1.0

[1]NHANES II; Brainard and Burmaster, 1992.
[2]Price et al., 1992.
[3]Roseberry and Burmaster, 1993.
[4]Child: NHANES II, and Costeff, 1966 (calculated from body weight); Adult: NHANES II and Murray and Burmaster, 1992 (calculated from body weight, height).
[5]James and Knuiman, 1987.
[6]Rupp et al., 1980.
[7]Calabrese et al., 1989.
[8]U.S. EPA, 1992 (25% of total skin surface area).
[9]Roels et al., 1980; Charney et al., 1980; Gallacher et al., 1984; Duggan et al., 1985; Que Hee et al., 1985; Driver et al., 1989.
[10]NA indicates that the parameter was not varied.

of the river. The risks are below the levels of concern for all three individuals; the total hazard index is below 1, and the total carcinogenic risk is below 10^{-6}.

For the carcinogenic risks (Table II), the results for the three exposed individuals are close; the risks to the standard individual are only slightly higher than the risks to the maximally exposed individual and are about 2.5 times higher than the risks to the reasonably exposed individual. The standard individual had the highest risks for all but three pathways; the maximally exposed individual had higher risks for the ingestion of drinking water pathway whereas risks from the ingestion of meat and milk pathways were zero for all three individuals. The ingestion of drinking water pathway shows higher risks since the ingestion rate of drinking water is higher for the maximally exposed individual than for the standard individual. The carcinogenic risks for the ingestion of meat and milk pathways are zero since the benzene cattle and milk transfer factors are zero.

For the total noncarcinogenic risks (Table III), the risks to the standard individual are the highest, about three times higher than the total risk to the maximally exposed individual and 10 times higher than the total risk to the reasonably exposed individual. The standard individual had the maximum risks for seven pathways: inhalation while showering; ingestion of soil, fish, and shellfish; and dermal contact

TABLE II

Maximum carcinogenic risk for each exposed individual by pathway

Pathway	Reasonably Exposed Individual	Maximally Exposed Individual	Standard Individual
Ingestion of water while showering	1.6×10^{-9}	1.8×10^{-9}	2.0×10^{-9}
Ingestion of water while swimming	6.8×10^{-10}	7.5×10^{-10}	8.7×10^{-10}
Ingestion of drinking water	2.3×10^{-8}	6.9×10^{-8}	5.9×10^{-8}
Ingestion of soil	6.7×10^{-13}	2.7×10^{-12}	5.2×10^{-12}
Ingestion of leafy vegetables	9.8×10^{-9}	1.7×10^{-8}	1.9×10^{-8}
Ingestion of root vegetables	2.4×10^{-8}	3.6×10^{-8}	4.0×10^{-8}
Ingestion of finfish	0	6.9×10^{-10}	1.4×10^{-9}
Ingestion of shellfish	0	1.0×10^{-9}	1.4×10^{-9}
Ingestion of meat	0	0	0
Ingestion of milk	0	0	0
Dermal contact while showering	3.3×10^{-9}	1.4×10^{-8}	1.6×10^{-8}
Dermal contact while swimming	1.3×10^{-8}	2.5×10^{-8}	2.8×10^{-8}
Dermal contact with soil	0	0	0
Total	1.9×10^{-7}	4.4×10^{-7}	5.0×10^{-7}

while showering and swimming and with soil. The maximally exposed individual had the highest risks for only one pathway, ingestion of drinking water. The reasonably exposed individual showed the highest noncarcinogenic risks for six pathways: ingestion of water while showering and swimming, ingestion of leafy and root vegetables, and ingestion of meat and milk. The reasonably exposed individual incurs higher noncarcinogenic risks for some pathways because the risks are calculated by dividing by the body weight of the individual, and the body weight is lowest for the reasonably exposed individual. This result applies only for the noncarcinogenic risks because the risks are calculated by comparing a daily dose to the reference dose; the risk is not calculated by multiplying by the exposure duration, which is much lower for the reasonably exposed individual, as it is for the carcinogenic risks. The total noncarcinogenic risks are highest for the standard individual because the primary pathways contributing to the risks for this individual are ingestion of finfish and shellfish. For the reasonably exposed individual, the risks due to these pathways are zero since the 50th percentile ingestion rates of finfish and shellfish are zero.

8. Summary and Conclusions

The RIVRISK model evaluates human health and ecological risks due to chemical and thermal releases to rivers. Both organic and inorganic chemicals can enter the river via direct discharges, atmospheric deposition, groundwater seepage, and surface

TABLE III
Maximum noncarcinogenic risks for each exposed individual by pathway

Pathway	Reasonably Exposed Individual	Maximally Exposed Individual	Standard Individual
Inhalation while showering	3.6×10^{-6}	5.5×10^{-6}	7.2×10^{-6}
Ingestion of water while showering	2.0×10^{-4}	1.4×10^{-4}	1.8×10^{-4}
Ingestion of water while swimming	2.0×10^{-4}	1.4×10^{-4}	1.8×10^{-4}
Ingestion of drinking water	2.5×10^{-3}	3.9×10^{-3}	3.5×10^{-3}
Ingestion of soil	6.9×10^{-6}	2.2×10^{-5}	6.6×10^{-5}
Ingestion of leafy vegetables	6.9×10^{-4}	4.7×10^{-4}	6.0×10^{-4}
Ingestion of root vegetables	2.3×10^{-3}	1.6×10^{-3}	2.0×10^{-3}
Ingestion of finfish	0	7.1×10^{-3}	0.029
Ingestion of shellfish	0	9.9×10^{-3}	0.029
Ingestion of meat	8.0×10^{-4}	5.5×10^{-4}	7.0×10^{-4}
Ingestion of milk	1.3×10^{-4}	8.9×10^{-5}	1.1×10^{-4}
Dermal contact while showering	5.5×10^{-8}	1.1×10^{-7}	1.4×10^{-7}
Dermal contact while swimming	5.0×10^{-7}	4.6×10^{-7}	5.8×10^{-7}
Dermal contact with soil	8.7×10^{-7}	5.9×10^{-6}	6.9×10^{-6}
Total	0.0068	0.024	0.066

runoff. Human health risks are calculated for up to 16 pathways via exposure routes of dermal contact, ingestion, and inhalation. Uncertainty analysis also can be performed via Monte Carlo.

The human health risks were evaluated for human exposure to the New River downstream of chemical releases from the Glen Lyn Power Plant using the RIVRISK model. Both the carcinogenic and noncarcinogenic risks were calculated for three individuals: a reasonably exposed individual, a maximally exposed individual, and a standard individual. Although the calculated maximum total carcinogenic and noncarcinogenic risks were highest for the standard individual, a breakdown of the risks by pathway indicated the following:

- For carcinogenic risks, the standard individual had the highest maximum risks for all pathways except ingestion of drinking water, for which the maximally exposed individual incurred slightly higher risks.
- For noncarcinogenic risks, either the standard individual or the reasonably exposed individual had the highest maximum risk for all pathways except ingestion of drinking water, for which the maximally exposed individual incurred slightly higher risks.

Acknowledgments

This model was developed with funding from the Electric Power Research Institute (EPRI) under the direction of Dr. Robert A. Goldstein. The authors would like to express their appreciation to Patti Heath, Gloria Sellers, Mary Porcella, Jennifer Kimberley, and Phoebe McClure for their work in producing and editing this paper.

References

Baes III, C. F., Sharp, R. D., Sjoreen, A. L., Shor, R. W.: 1984, *A Review and Analysis of Parameters for Assessing Transport of Environmentally Released Radionuclides through Agriculture*, DE85-000287.

Brainard, J. and Burmaster, D. E.: 1992, *Risk Analysis* **12(2)**, 267-275.

Calabrese, E. J., Barnes, R., Stanek III, E. J., Pastides, H., Gilbert, C. E., Veneman, P., Wang, X, Lasztity, A., Kostecki, P. T.: 1989, *Regulatory Toxicology and Pharmacology* **10**, 123-137.

California Environmental Protection Agency, Department of Toxic Substances Control: 1984, *Preliminary Endangerment Assessment Guidance Manual*.

Charney, E., Sayre, J. W., Coulter, M.: 1980, *Pediatrics* **65**, 226-231.

Chow, W. *et al.*: 1991, *Managing Air Toxics: Status of EPRI's PISCES Project*. Proceedings of the Conference on Managing Air Toxics: State of the Art, Washington, D.C.

Chu, P., Nott, B., and Chow, W.: 1993, *Results and Issues from the PISCES Field Tests*. Second International Conference on Managing Hazardous Air Pollutants, Washington, D.C.

Costeff, H.: 1966, *Archives of Disease in Childhood* **41**, 681-683.

Dickson, K. L., Cairns, Jr., J., Cherry, D. S., Stauffer, J.R.: 1976, *Thermal Ecology II*. Proceedings of Technical Information Center, Energy Research and Development Administration Symposium Held at Augusta, GA.

Driver, J. H., Konz, J. J., Whitmyre, G. K.: 1989, *Bull. Env. Cont. Tox.* **43**, 814-820.

Duggan, M. J., Inskip, M. J., Rundle, S. A., Moorcroft, J. S.: 1985, *Sci. Tot. Env.* **44**, 65-79.

Gallacher, J. E., Elwood, P. C., Phillips, K. M., Davis, B. E., Jones, D. T.: 1984, *Arch. Dis. Child.* **59**, 40-44.

James, J. R. and Knuiman, M. W.: 1987, *Am. Stat. Ass.* **82(399)**, 705-711.

McKone, T. E.: 1987, *Environmental Science and Technology* **21**, 1194-1201.

McKone, T. E. and Knezovich, J. P.: 1991, *Journal of Air and Waste Management Association* **40**, 282-286.

Miller, A. C., Srivastava, R. K., Ryan, J. V.: 1994, *Environmental Science and Technology* **28**, 1150-1158.

Murray, D. M. and Burmaster, D. E.: 1992, *J. Exp. Anal. Env. Epidem.* **2(4)**, 451-461.

National Health and Nutrition Examination Survey (NHANES II): 1976-1980, February.

Pacific Northwest Laboratories: 1992, *Multimedia Environmental Pollution Assessment System (MEPAS)*. Prepared for Battelle Memorial Institute, Version 3.x.

Price, P. S., Sample, J., Strieter, R.: 1992, *Risk Analysis* **12(3)**, 367-382.

Que Hee, S. S., Peace, B., Scott, C. S., Boyle, J. R., Bornschein, R. L., Hammond, P. B.: 1985, *Environmental Research* **38**, 77-95.

Roels, H. A., Buchett, J. P., Lauwerys, R. R., Bruaux, P., Claeys-Thoreau, F., Lafontaine, A. and Verduyn, G.: 1980, *Environmental Research* **22**, 81-99.

Roseberry, A. M. and Burmaster, D. E.: 1993, *Risk Analysis* **11(2)**, 339-342.

Rupp, E. M., Miller, F. L., Baes, C. F.: 1980, *Health Physiology* **39**, 165-175.

Travis, C. C. and Arms, A. D.: 1988, *Environmental Science and Technology* **22**, 271-274.

U.S. EPA: 1986, *Superfund Public Health Evaluation Manual*. EPA/540/1-86/060.

U.S. EPA: 1989, *Risk Assessment Guidance for Superfund, Human Health Evaluation Manual Part A*. EPA/540/1-89/002.

U.S. EPA: 1990, *Methodology for Assessing Health Risks Associated with Indirect Exposure to Combustor Emissions*. Cincinnati, OH, EPA/600/6-90/003.

U.S. EPA: 1991, *Human Health Evaluation Manual, Supplemental Guidance: Standard Default Exposure Factors*. Washington, D.C., OSWER Directive 9285.6-03.

U.S. EPA: 1992, *Dermal Exposure Assessment: Principles and Applications*. Washington, D.C., EPA/600/8-91/011B.

U.S. EPA: 1994a, *Integrated Risk Information System (IRIS)*. Office of Research and Development, Environmental Criteria and Assessment Office, Cincinnati, OH.

U.S. EPA: 1994b, *Health Effects Assessment Summary Tables (HEAST)*. Washington, D.C., EPA/540/R-94/020.

U.S. EPA: 1994c, *Region IX Preliminary Remediation Goals (PRGs)*. San Francisco, CA.

RISKS DUE TO GROUNDWATER CONTAMINATION AT A PLUTONIUM PROCESSING FACILITY

I. DATSKOU[1], and K. NORTH[2]

[1]Center for Risk Management, Oak Ridge, TN 37830*, [2]University of Tennessee, Knoxville, TN 37996

*Managed by Lockheed Martin Energy Research Corp., under contract no. DE-AC05-96OR22464 with the U.S. Department of Energy.

Abstract. The specter of contaminated groundwater looms over industrialized, suburban, and rural areas. The sources of groundwater contamination are many and the contaminants numerous. Historical waste disposal practices pose the greatest threat to groundwater in the United States. Common solvents such as trichloroethylene, tetrachloroethane, benzene, and carbon tetrachloride have been found in widespread areas. For this study, a risk assessment is performed using a multimedia environmental transport model to estimate public risk from a contaminated groundwater plume. From 1955 to 1973, a crib above the groundwater plume was used for the disposal of plutonium processing waste. The plume contains chemical and radioactive wastes that could pose a health threat to people living in the vicinity of the site. Remedial designs are selected for remediating the different contaminants in the groundwater plume through a multiple phase treatment process. This study evaluates: baseline health risks to the public, health risk reduction to the public as a result of the remedial activities, health risk to the workers directly involved in cleaning up the site, and costs associated with each remedial activity.
Keywords: risk assessment, remediation, groundwater plume, worker risk, cost-benefit analysis, cleanup objectives.

1. Introduction

The specter of contaminated groundwater looms over industrialized, suburban, and rural areas. The sources of groundwater contamination are many and the contaminants numerous. Historical waste disposal practices pose the greatest threat to groundwater in the United States.

Because the contamination of groundwater is a vital issue on the national agenda, risk assessments must be performed to analyze and optimize, the risk reduction, the cost, and the technology effectiveness for cleanup of contaminated sites.

In this paper, a risk assessment is performed on a groundwater release site (i.e., specific location where contaminant releases may occur) to show the risks, costs, and technologies associated with cleanup. The site will be kept anonymous for this paper since the work is in its draft stages. The ultimate goal is evaluating the most efficient and cost-effective methods while minimizing health risks to receptors. The study examines three representative remedial options   complete   removal/treatment,   containment/control,   combination treatment/containment—and evaluates technologies typically associated with these options. These three options are compared to the no action option.

2. Description of Site

The release site is a contaminated groundwater plume located beneath a crib, which is a subsurface liquid waste disposal unit designed to dispose of liquid wastes directly into the soil column. This plume contains chemical and radioactive wastes that could pose a health threat to people living in the vicinity of the site. From 1955 to 1973, the crib above the

Water, Air and Soil Pollution **90**: 133-141, 1996.
© 1996 *Kluwer Academic Publishers.*

plume was used for the disposal of plutonium processing waste. Extensive discharging of liquid waste in the area has driven the contamination plume into the aquifer below, where up to 22 km² are contaminated (ORNL, 1995a).

3. Contaminants of Concern

Source term data for the contaminated groundwater, taken from 1990 field studies, are listed in Table I and Table II. The contaminants were broken down into two groups based on the extent of the contaminated area. Average areas of 1.6 km² for Group I and 15 km² for Group II were assumed (ORNL, 1995a).

The concentrations of the contaminated groundwater constituents come from a network of sampling wells. The deepest well used for sampling is 16.5 m into the aquifer; therefore, the depth of the contamination is somewhere between 16.5 m and the 76 m depth of the aquifer. For our conservative model it is assumed that the plume extends throughout the 76 m of the aquifer (ORNL, 1995a). Modeling assumptions can be found in ORNL, 1995a,b,c.

The source term concentrations for ^{234}U, ^{235}U, and ^{238}U were given as total concentrations for ^nU. It was assumed that these isotopes are present in the ratios of their natural abundance. Unknown amounts of plutonium, ^{90}Sr, and ^{241}Am have also been detected in the groundwater. Since the contaminated areas and constituent inventories are unknown, these contaminants were not modeled (ORNL, 1995a).

The contaminants present in the groundwater plume came from aqueous and organic wastes that were released into the overlying trenches and cribs. Since the contamination of interest is only the contaminated groundwater, it is assumed that all of the contamination is already in the groundwater and that no additional release from the vadose zones occurs (ORNL, 1995a).

The contaminants evaluated for potential receptors are carcinogens, noncarcinogens, and radionuclides. Health effects resulting from exposure to carcinogens are reported as cancer incidence. Health effects resulting from exposure to noncarcinogens are reported as hazard indices. The health effects resulting from exposure to radionuclides are reported as cancer incidence, cancer fatalities, and dose. In addition to these exposure-related risks and hazards, remedial workers are assessed for fatalities, injuries, and illnesses resulting from construction-related activities and heat stress associated with wearing protective equipment (ORNL, 1995b; ORNL, 1995c).

4. Results

The three remedial options (complete removal/treatment, containment/control, and combination treatment/containment) analyzed for this study are compared to the no action option. This section provides an integrated analysis of the following variables:
- public risk
- remediation worker risk
- remediation cost
- postremediation risk

TABLE I

Source Term Data for Group I Contaminants

Constituent	Surface Area (cm²)	Concentration (mg/L or pCi/L)	Calculated Inventory (g or Ci)
Chloroform	1.51E10	1.08E–1	2.49E6
Chromium VI	1.60E10	9.14E–1	2.23E6
Cyanide	9.48E9	1.9E–2	2.74E5
Fluoride	1.43E10	2.47	5.39E7
Trichloroethylene	1.18E10	1.29E–2	2.32E5
Uranium-234	3.21E10	2.81E1	1.37
Uranium-235	3.21E10	1.20	5.85E–2
Uranium-238	3.21E10	2.60E1	1.27

TABLE II

Source Term Data for Group II Contaminants

Constituent	Surface Area (cm²)	Concentration (mg/L or pCi/L)	Calculated Inventory (g or Ci)
Carbon Tetrachloride	8.21E10	7.78E–1	9.74E7
Iodine-129	1.29E11	6.17	1.21
Nitrate	2.12E11	1.96E2	6.33E10
Technetium-99	1.07E11	6.02E3	9.80E2

The study analyzes three receptor scenarios, which are described briefly as follows.

• Offsite receptor. The offsite scenario is chosen to provide an estimate of risk to actual populations living around the installation. Offsite receptors include people living within an 80-km (50-mi) radius of the site. The only potential pathway through which the public could be exposed is transport of the contaminants through the aquifer to a river flowing 10 km from the site. Six different water intake locations are located in the river adjacent to the site. The closest intake is 6,000 m downstream. These intakes supply drinking water to 525,000 people and irrigation water and livestock water to farmers (ORNL, 1995a). In the results section this receptor is presented as offsite MEI (maximum individual exposure) as well as offsite public that represents the population risk.

• Remediation worker. This scenario is evaluated to provide an estimate of risk due to occupational exposure faced by workers during remediation. Remediation workers are assumed to be employees involved with remediation who work at the site 8 hours/d, 250 d/year for a maximum of 7 years. Technology specific assumptions can be found in ORNL, 1995c.

● Hypothetical homesteader receptor. The homesteader scenario is evaluated to give an upper bound estimate of onsite risks. Under this most conservative of risk estimates, the site is under no institutional controls. In this case, exposure via contaminated groundwater is possible in addition to exposure to the surface water body. Additional assumptions can be found in ORNL, 1995b and EPA, 1989. In the results section this receptor is presented as onsite receptor (the numbers reflect MEI and not population risk for this case).

4.1 REMEDIAL OPTIONS

The four options that this study analyzes are briefly described in the following sections.

4.1.1 No action
The no action option assumes that no remediation of the site is attempted. This option addresses risk that could result for two separate receptor exposure scenarios (i.e., hypothetical onsite and offsite receptor).

4.1.2 Containment/control option
This option utilizes containment and/or control technologies in an attempt to reduce risks to below 1E-6. The scope of this remedial option will involve maintaining the controls. The long term groundwater monitoring system consists of eight groundwater monitoring wells and the monitoring of those wells annually for volatile organic compounds, dissolved metals, nitrates, fluoride and radionuclides. Table III presents the technologies employed for this option, the number and classification of workers involved with the cleanup, the duration of each technology, and the costs associated with each technology (LBA, 1995).

TABLE III

Estimated Labor and Cost Summary for the Containment/Control Option

Technology	Number of Workers*	Duration (years)	Total Cost (2005 $)*
Install monitoring wells(prep 82 m deep)	1S,1O/E,1L	0.004	1,500
Install 8 monitoring wells (capital)	1S,1O/E,1L	0.242	711,000
Annual			
monitor wells (O&M)		215.385	460,000
		subtotal 1	1,172,500
Overhead & profit (60 % of subtotal)			703,500
Other costs (health & safety)			117,000
		subtotal 2	1,993,000
Contingency (50 % of subtotal 2)			996,500
TOTAL COST (approximately)			3,000,000

* S=Supervisor, O/E=Operator/Engineer, and L=Laborer.

* Total cost is calculated by escalating to year 2005 dollars and by adjusting the cost for minimum labor requirements and increased levels of personal protection.

4.1.3 Complete removal/treatment option

A schematic of the proposed remedial action is provided in Figure 1. The goal of the complete removal/treatment option is to reduce all potential risks to <1E–6 and contaminant concentrations to below the current MCLs. Remediating this extensive groundwater plume under this scenario involves a groundwater extraction system followed by multiple phase ex-situ treatment system. The contamination exists in two overlapping plumes, one consisting of VOCs and the other comprising a mixture of heavy metals, nitrates, and radionuclides. The remediation strategy for this scenario calls for the pumping of seven pore volumes of the inorganic plume and 10 pore volumes of the VOC plume. Several pore volumes must be pumped because of adsorption of VOC.

The water from the VOC plume is treated with air stripping. Water from both plumes is then treated to remove the other inorganic contaminants present in the groundwater. Fluoride and cyanide are removed via a coagulation and flocculation process with gravity separation to remove the suspended solids. Next, an aerobic biological process is used to remove the nitrates present in the water stream. The remaining radionuclides and chromium are removed as best as possible with an ion exchange process in the last stage of the remediation. The treated effluent stream is then returned to the aquifer through injection wells. The ultimate condition of the site will consist of an open vegetative field. While it is the intent of this alternative to achieve a post-remedial risk less than or equal to threshold levels, this goal cannot be achieved using presently available removal/treatment techniques (LBA, 1994).

Table IV lists the technologies employed for this option, the number and classification of workers involved with the cleanup, the duration of each technology, and the costs

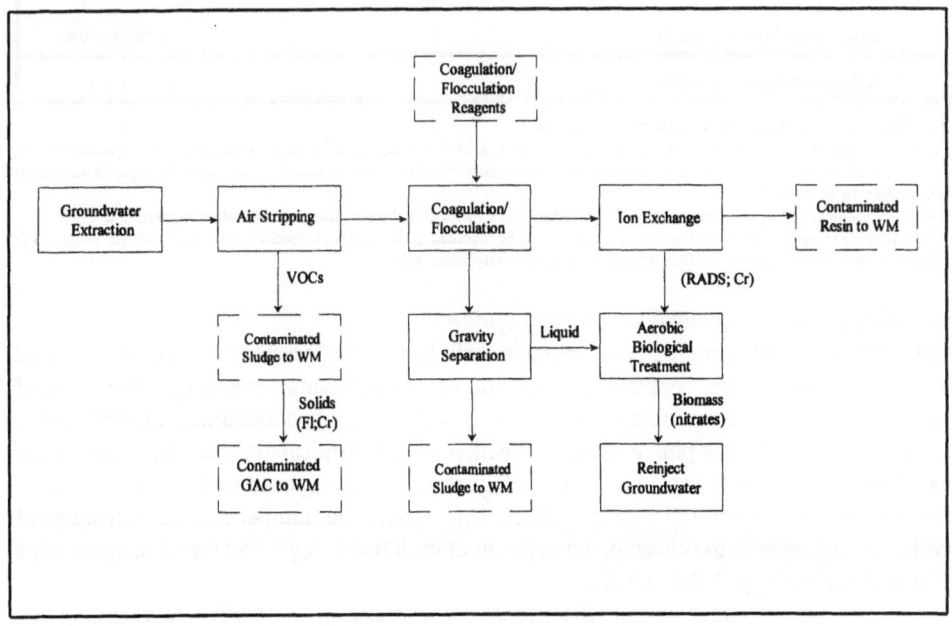

Fig. 1. Treatment Train for the Complete Removal and/or Treatment Option.

associated with each technology (ORNL, 1995a).

TABLE IV

Estimated Labor and Cost Summary for the Complete Removal/Treatment Option

Technology	Number of Workers[a]	Duration (years)[b] capital/operation	Total Cost ($)[c]
Extraction well; Water Recovery	1S,2O/E,1L	0.909/27	3,400,000,000
Air Stripping	1S,2O/E,1L	0.812/27	377,000,000
Thermal Oxidation	1S,2O/E,1L	0.018/27	275,000,000
Biological Treatment	1S,2O/E,1L	0.648/27	277,000,000
Coagulation/Flocculation	1S,2O/E,1L	0.037/27	644,000,000
Gravity Separation	1S,2O/E,1L	0.037/27	108,000,000
Ion Exchange	1S,2O/E,1L	0.485/27	127,000,000
Groundwater Reinjection	1S,2O/E,1L	0.205/27	302,000,000
Package Contaminated Materials[*]	1S,3O/E,3L	0.019/13	
Transportation of Waste	1S,1O/E,1L	13.5	76,000,000
		subtotal 1	5,586,000,000
Overhead/Profit (60 % of subtotal 1)			3,351,000,000
Other costs (health & safety)			560,000,000
		subtotal 2	9,497,800,000
Contingency (50 % subtotal 2)			4,748,800,000
TOTAL COST (approximately)			$14,300,000,000

[a] S = Supervisor,   O/E = Operator/Engineer,   L = Laborer
[b] For groundwater extraction and treatment operations, durations are not directly related to labor requirements (i.e., person-hours); rather, durations are computed based on the number of wells in operation, the estimated pumping rates, and the required number of pore volumes to be extracted.
[c] Total cost is calculated by escalating to year 2005 dollars and by adjusting the cost for minimum labor requirements.
[*] Costs for packaging contaminated material are assumed to be included in the costs of remedial technologies which produce the material and have been included for the purpose of labor risk assessment only.

4.1.4 Combination treatment/containment option

Under this option, the source of contamination will be removed/treated and/or controlled and contained to reduce and control the migration of contaminants remaining. The selected remedial design for this option is to remove and treat the highly contaminated portions (hot spots) of the groundwater plume using treatment methods identical to those employed for the complete removal/treatment option (check Figure 1 for the proposed remediation process). Table V lists the technologies employed for this option, the number and classification of workers involved with the cleanup, the duration of each technology, and the costs associated with each technology (LBA, 1994).

As for the previous option, the process is essentially ineffective at reducing the concentrations of contaminants of concern within the aquifer and so this site must remain restricted to the extent that the groundwater may not be utilized.

TABLE V

Estimated Labor and Cost Summary for the Combination Treatment/Containment Option

Technology	Number of Workers[a]	Duration (years)[b] capital/operation	Total Cost ($)[c]
Extraction well; Water Recovery	1S,2O/E,1L	0.509/20	1,400,000,000
Air Stripping	1S,2O/E,1L	0.606/20	45,000,000
Thermal Oxidation	1S,2O/E,1L	0.018/20	16,000,000
Biological Treatment	1S,2O/E,1L	0.648/20	121,000,000
Coagulation/Flocculation	1S,2O/E,1L	0.037/20	274,000,000
Gravity Separation	1S,2O/E,1L	0.037/20	47,000,000
Ion Exchange	1S,2O/E,1L	0.485/20	62,000,000
Groundwater Reinjection	1S,2O/E,1L	0.205/20	149,000,000
Transportation of Waste	1S,1O/E,1L	10	32,000,000
		subtotal 1	2,146,000,000
Technology Specific Costs			1,170,000
		subtotal 2	2,147,170,000
Overhead/Profit (60 % of subtotal 2)			1,288,302,000
Other costs (health & safety, site supervision)			218,000,000
		subtotal 3	3,623,472,000
Contingency (50 % subtotal 2)			1,826,736,000
TOTAL COST (approximately)			5,500,000,000

[a] S = Supervisor, O/E = Operator/Engineer, L = Laborer
[b] For groundwater extraction and treatment operations, durations are not directly related to labor requirements (i.e., person-hours); rather, durations are computed based on the number of wells in operation, the estimated pumping rates, and the required number of pore volumes to be extracted.
[c] Total cost is calculated by escalating to year 2005 dollars and by adjusting the cost for minimum labor requirements. Costs for packaging contaminated material are assumed to be included in the costs of remedial technologies which produce the material and have been included for the purpose of labor risk assessment only.

4.2 RISK ESTIMATES

Public risks generated for pre-remediation and post-remediation site conditions and remediation worker risks posed by remedial activities at the groundwater as source site are summarized in Table VI. Cancer fatalities, cancer incidence, and dose are presented in each box of the table. Cancer fatality risk is the first number, cancer incidence is the second number, and dose is the third. If only one number appears in a box of the table, it indicates that all risks are represented by that one number (e.g., all risks may be less than 1E-6). For worker risks, risks are reported as exposure-related risks and physical hazards. Physical hazards are summed and presented as fatalities and injuries/illnesses incurred by both construction and heat stress. Fatalities are represented by the top number in the box, and illnesses or injuries are represented by the bottom number in each box.

The contaminants that pose the greatest risks and hazards are carbon tetrachloride, nitrate (HI >1), and chromium VI (HI >1) and the primary exposure route is the direct ingestion of the contaminated groundwater. For offsite receptors, nitrate and carbon tetrachloride produce the highest risks (and hazards for the noncarcinogens) via ingestion. Ingestion of hexavalent chromium also poses a potential threat to the hypothetical onsite receptor. Exposure routes of concern during remediation are inhalation, ingestion, and direct radiation. All workers are assumed to wear level D PPE, except during the extraction well mobilization and package contaminated materials operation phases where the workers wear level C PPE.

5. Conclusions

Due to the extent of the contamination and limitations of the extraction and remedial technologies, a considerable amount of residual contamination remains after remediation. This residual contamination comprises components that cannot be extracted from the plume with the existing extraction technologies and those in the effluent stream that cannot be removed with the remedial processes. According to EPA action on a site would take place if the no action risk is greater than 1E-6. For the plume that was studied here the risk levels were unacceptable and so several remedial options were analyzed. Under the treatment option the cost is prohibitive and the risk reduction wasn't acceptable. During a cost benefit analysis the only logical option would be the containment and control option, since the cost is reasonable and the risk reduction for the public MEI is acceptable. Figure 2 presents a comparison among remedial options of cost, post-remedial risk and worker hazard.

TABLE VI

Health Risk[a] Summary for the Contaminated Groundwater

Remedial Options	Offsite MEI[a]	Offsite Public[a]	Onsite Receptor[a]	Remedial Worker[b]	Remedial Worker[b]	Offsite (during)[a,c]	Cost (2005 $)
No Action	<1E-6 <1E-6 <1E-6	2E-1 1.4 4E2	1.1E-3 7.5E-3 2.3	N/A N/A N/A	N/A N/A	N/A N/A N/A	N/A
Treatment	<1E-6 <1E-6 <1E-6	1E-1 8E-1 4E2	1.1E-3 3.8E-3 2.2	<1E-6 <1E-6 <1E-6	7.3E-2 2E1	<1E-6 3.0E-3 <1E-6	14 billion
Containment/ Control	<1E-6 <1E-6 <1E-6	3E-2 6E-3 3E1	2E-4 3E-5 3E-1	<1E-6 <1E-6 <1E-6	2.6 E-4 7.1E-2	<1E-6 3.0E-6 <1E-6	3 million
Combination Treatment/ Containment	<1E-6 <1E-6 <1E-6	2E-3 9E-4 2	4E-5 6E-6 2E-2	<1E-6 <1E-6 <1E-6	7.1E-2 2.0E1	<1E-6 7.2E-3 <1E-6	5.5 billion

[a]Risks for each receptor are presented as cancer fatalities, cancer incidence, and dose respectively for each option. Public refers to population risk.
[b]Hazards for this receptor are presented as fatalities, and injuries respectively for each option.
[c]Risks during remediation to the offsite public

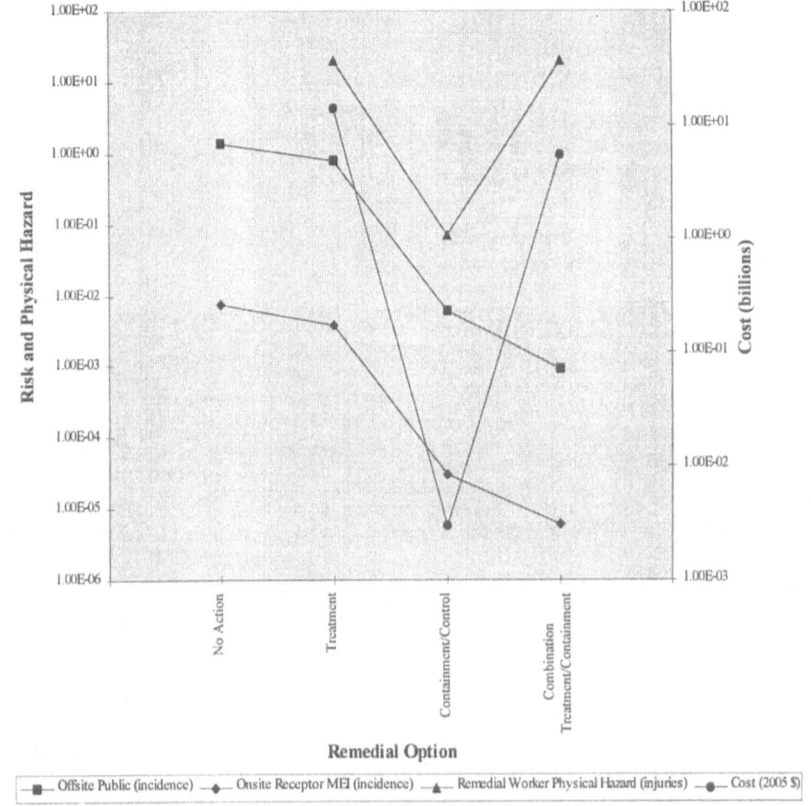

Fig 2. Comparison of Remedial Options Using Cost, Physical Hazard and Risk.

References

Environmental Protection Agency - Office of Emergency and Remedial Response (EPA): 1989, *Risk Assessment Guidance for Superfund (Vol. 1) Human Health Evaluation Manual (Part A), Interim Final.* EPA/540/1-89/002.

Louis, Berger & Associates (LBA): 1994, *Engineering Analysis for PEIS Example Cases.*

Oak Ridge National Laboratory (ORNL): 1995a, *Analysis of the 10 Example Cases In Support of the Environmental Restoration Programmatic Alternatives Assessment: Volume IV Public and Worker Risk Assessment Results.* ORNL-6866.

Oak Ridge National Laboratory (ORNL): 1995b, *Environmental Restoration Programmatic Environmental Impact Statement Methodology for Estimating Human Health Risks.* ORNL-6865.

Oak Ridge National Laboratory (ORNL): 1995c, *DOE Worker Health Risk Evaluation Methodology for Assessing Risks Associated with Environmental Restoration and Waste Management.* ORNL-6833.

TVA-EPRI RIVER RESOURCE AID (TERRA)
Reservoir and Power Operations Decision Support System

K. Lindquist, M. McGee, and L. Cole

*Tennessee Valley Authority Engineering Laboratory, Post Office Drawer E
Norris, Tennessee 37828*

Abstract. The Tennessee Valley Authority (TVA) and the Electric Power Research Institute (EPRI) have developed software to assist utilities in coordinating power and reservoir operations while balancing multiple objectives. The INTEGRAL project consists of a Decision Support System (DSS) and a Power and Reservoir System Scheduling Model (PRSYM). The DSS, as tailored to the TVA power and reservoir system, is the TVA-EPRI River Resource Aid (TERRA). TERRA is a decision support system that integrates tracking, display, and modeling tools. TERRA provides historical, current, and forecast data to users at several locations to help resolve problems in operation, forecasting, and planning. Graphical displays are featured for individual power plants and reservoirs and on a system-wide basis. TERRA tracks constraints that affect operations and flags non-compliance. Constraints consist of flow, dissolved oxygen, and temperature limits for regulatory compliance, recreation, lake improvement, and special operations. Numerical models may be run from TERRA to predict water temperatures and flows at selected power plants to aid in reservoir operations to comply with thermal limits and minimum flow requirements. TERRA uses a geographic information system (GIS) background map of the Tennessee River Valley. The TERRA software is adaptable to other reservoir systems by changing the GIS background map and reconfiguring other system features.

Key Words: decision support system, reservoir operations, water resource management, numerical modeling, thermal compliance, geographic information system, water quality, dissolved oxygen

1. Introduction

Managing large reservoir and river systems is becoming increasingly complex due to the need to simultaneously meet many conflicting objectives. Managing these systems requires increasingly sophisticated tools. Developing improved tools for coordinating scheduling and planning of power generation and reservoir operations is the purpose of INTEGRAL, a tailored-cooperation project of the Tennessee Valley Authority (TVA) and the Electric Power Research Institute (EPRI). The INTEGRAL project consists of a Decision Support System (DSS) and a Power and Reservoir System Scheduling Model (PRSYM). The objective in developing the decision support system is to provide software to help utilities coordinate power and reservoir system operations while protecting water quality, meeting water supply needs, and managing reservoir levels. TVA worked with the Center for Advanced Decision Support for Water and Environmental Systems (CADSWES) to develop the system (CADSWES, 1993a,b).

The DSS, as tailored to the TVA power and reservoir system, is the TVA-EPRI River Resource Aid (TERRA). TERRA runs on two Unix servers in Knoxville and Chattanooga, Tennessee. Users at widely dispersed decision-making locations access these servers via Unix workstations and IBM-compatible personal computers, which are

Water, Air and Soil Pollution **90**: 143-150, 1996.
© 1996 *Kluwer Academic Publishers.*

connected by a corporate wide-area network. The system uses a Geographic Information System (GIS) background map of the Tennessee River Valley, which includes portions of seven states. The user accesses information by clicking on icons for meteorological stations, dams, or hydroelectric, nuclear, and fossil power plants on the map. Other map layers include reservoirs, state boundaries, rivers, streams, railroads, and roads. See Figure 1.

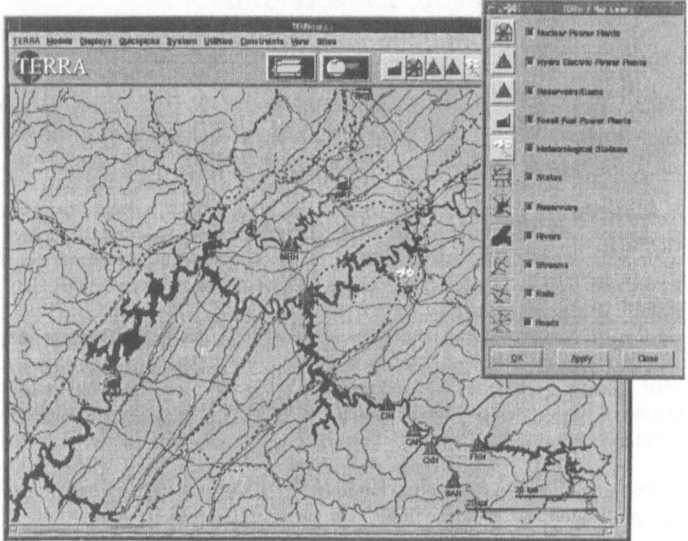

Fig. 1. TERRA GIS Background Map, Enlargement of Knoxville, Tennessee Area

TERRA serves five primary functions (Miller et al., 1993):

1. **Manage Current and Historical System Status Data** - Current year, previous year, near real time, and forecast data are automatically updated, stored, and integrated into TERRA displays. Summaries of historical data are also included.

2. **Manage System Constraints** - Important power and reservoir system constraints, commitments, and guidelines that affect operations are updated, stored, and displayed.

3. **Track System Compliance** - Current system status is checked for compliance with regulations, commitments, and guidelines. Measured data, scheduled operation, and forecasted operation are checked against these constraints. A warning system indicates when thresholds and limits are broken.

4. **Aid Power and Reservoir System Scheduling** - Data and numerical models used in daily power and reservoir scheduling are updated, stored, and integrated into the system.

5. **Integrate Forecasted and Operational Planning Information** - Data and numerical models used in power and reservoir system forecasting and planning are integrated into TERRA to anticipate problems and plan resource utilization.

2. System Data

TERRA provides a common set of historical, current, and forecast data to assist in rapidly resolving problems in the operation, forecasting, and planning of the power and reservoir systems. Much of the data is uploaded directly from automatic data acquisition systems. TERRA manages system status data such as pool levels, turbine discharges, water temperatures, and dissolved oxygen levels. Site-specific or system-wide displays are available for reservoir, power system, and environmental parameters. For the hydropower system, TERRA tracks discharges, power generation, and elevations at each hydroelectric plant, as shown in Figure 2.

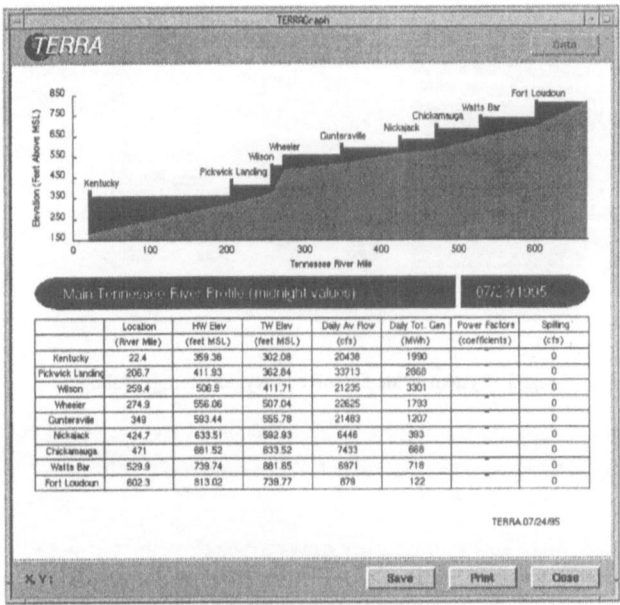

Fig. 2. Hydropower System Profile

Actual and forecasted discharge temperatures and dissolved oxygen may be displayed for each dam. Current tributary storage is available along with the minimum operating guide. Storage volumes are exhibited as bar charts showing the quantity of water in the reservoir of particular temperature and dissolved oxygen properties for the current and previous year, as shown in Figure 3.

In addition, elevation versus volume curves are available on-line for each reservoir. Meteorology from multiple stations is available on the system to display or use as input for a reservoir water temperature model. Meteorological data are uploaded three times daily. The current weather data are available, along with the 3 - 5 day forecast and the 6 - 10 day forecast for various areas in the reservoir system. Hourly forecasts of air temperature, dew point, humidity, wetbulb temperature, wind, and cloud cover are on-line.

Schedules for plant outages are quickly available to facilitate planning of seasonal activities to take advantage of down times. The annual maintenance schedule for the twelve fossil power plants in the TVA system is shown in Figure 4.

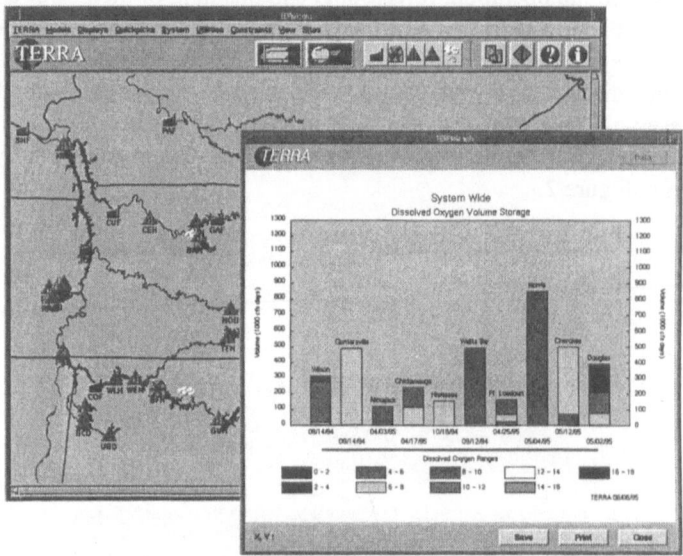

Fig. 3. System-Wide Dissolved Oxygen Storage Volume

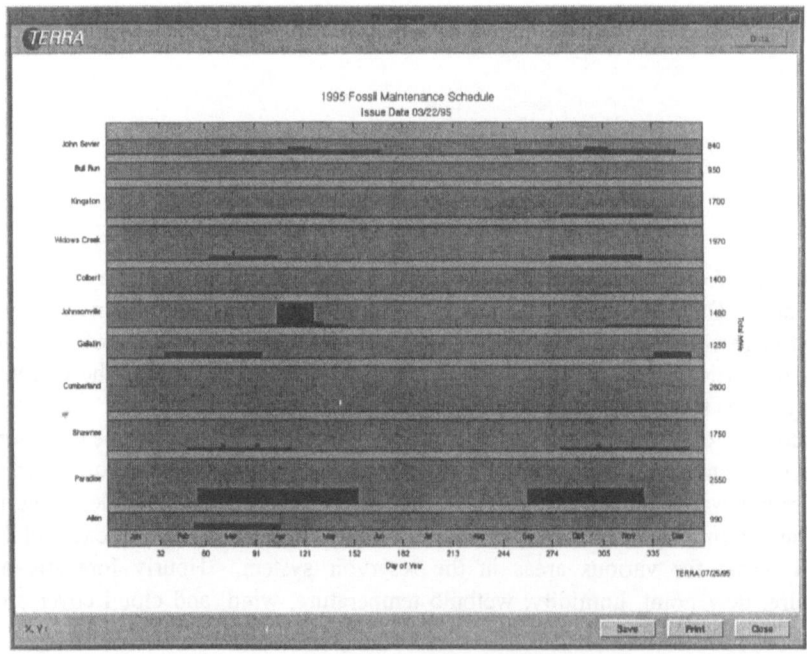

Fig. 4. Fossil Plant Maintenance Schedule

3. Managing Constraints and Tracking Current System Status

A database is maintained by TERRA containing current power and reservoir system constraints, commitments, and guidelines that guide daily reservoir operations. The database is updated as changes are made during daily scheduling activities or as new operating policies are adopted (Shane et al., 1993). Guidelines cover such issues as dissolved oxygen, minimum flow and temperature limits for endangered species, permit compliance, safety, recreation, lake improvement, unit outages, and special releases. Special releases are scheduled for spill mitigation, fishing opportunities, mussel relocation, Olympic whitewater trials, and other purposes. System health summaries are displayed. A warning system alerts users to potential problems and actual non-compliances. For example, TERRA tracks compliance with National Pollutant Discharge Elimination System thermal limits at each fossil fuel power plant and nuclear power plant in the power system and compliance with Nuclear Regulatory Commission safety standards for intake water temperatures at the nuclear plants. A sample weekly display of compliance with thermal limits at a nuclear power plant is shown in Figure 5. The user can also track and project seasonal cold water availability to mitigate thermal power plant discharges (Ostrowski et al., 1991).

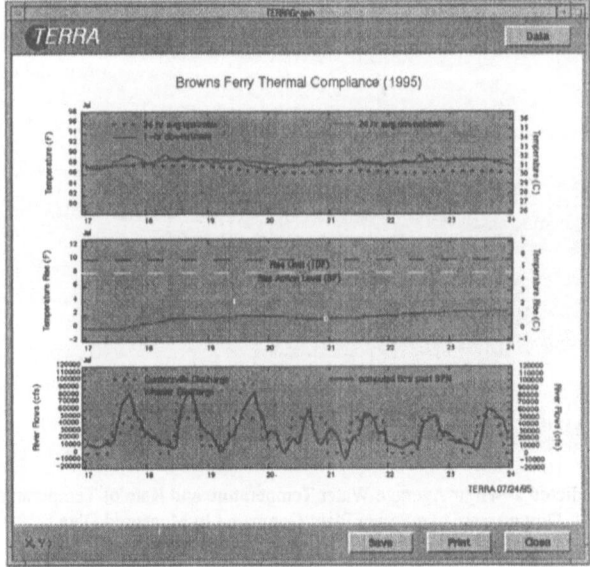

Fig. 5. Thermal Compliance Display for Browns Ferry Nuclear Plant

4. Integrated Modeling to Aid Scheduling and Protect Water Quality

An integrated modeling system has been developed to support the scheduling, forecasting, and planning functions of the INTEGRAL system. To address a spectrum of modeling needs, the modeling system includes resident models and non-resident models.

4.1. RESIDENT MODELS

TERRA allows operators to run resident site-specific hydrothermal scheduling models to predict intake and downstream water temperatures at power plants (Ostrowski et al., 1988). TERRA manages a database of privileges that allow only authorized users to run model simulations impacting daily operations. The flow models utilize the prescheduled hourly dam releases, power generation, and reservoir geometry in the TERRA database as input. The results are graphically displayed to aid decision-makers in daily reservoir operations. TERRA compares predictions from site-specific water temperature models with measured temperatures as shown in Figure 6. The system displays results for decision-makers responsible for daily reservoir and power operations so they can effectively resolve conflicts and interact with those responsible for environmental compliance.

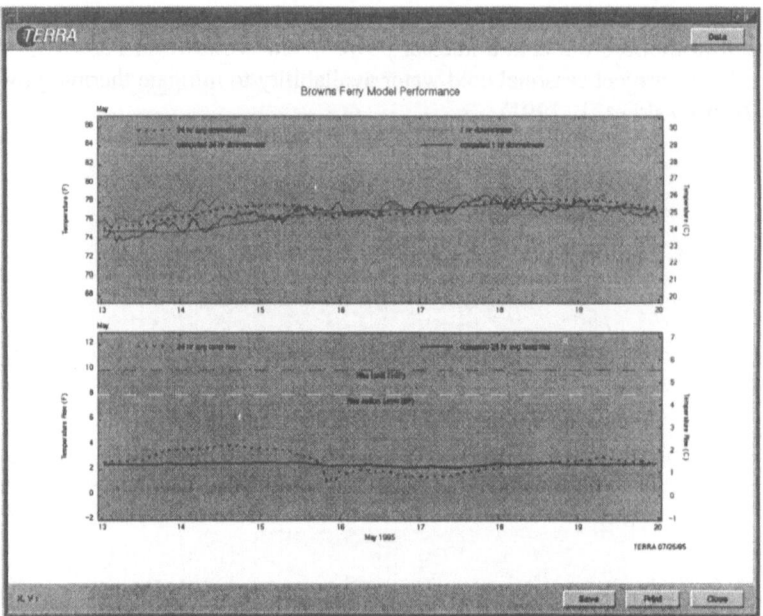

Fig. 6. Predicted 24-Hour Average Water Temperature and Rate of Temperature Rise
Downstream of a Power Plant Compared to Measured Data

Other site-specific models could include water quality at critical river reaches or detailed routing models at particular locations. TERRA will allow for the incorporation of system-wide or site-specific models relevant to other utilities or water resource agencies.

4.2. NON-RESIDENT MODELS

TERRA allows the user to apply existing utility-specific models in conjunction with the decision support system. Examples include hydrologic and stream flow forecasting models, reservoir and power system operational and planning models, and detailed water quality models. Interfaces have been developed to share data between TERRA and non-resident models.

An example of a non-resident model used in the TVA application is the system-wide temperature model (Ostrowski et al., 1991). Results are graphically displayed on TERRA showing the actual and forecasted intake temperature for specific sites, along with the historical mean, range of 10 to 90 percentiles, action levels, regulatory limits, and 90-day forecasted intake temperature for the power plant, as shown in Figure 7. This model assists mid-range reservoir system planning.

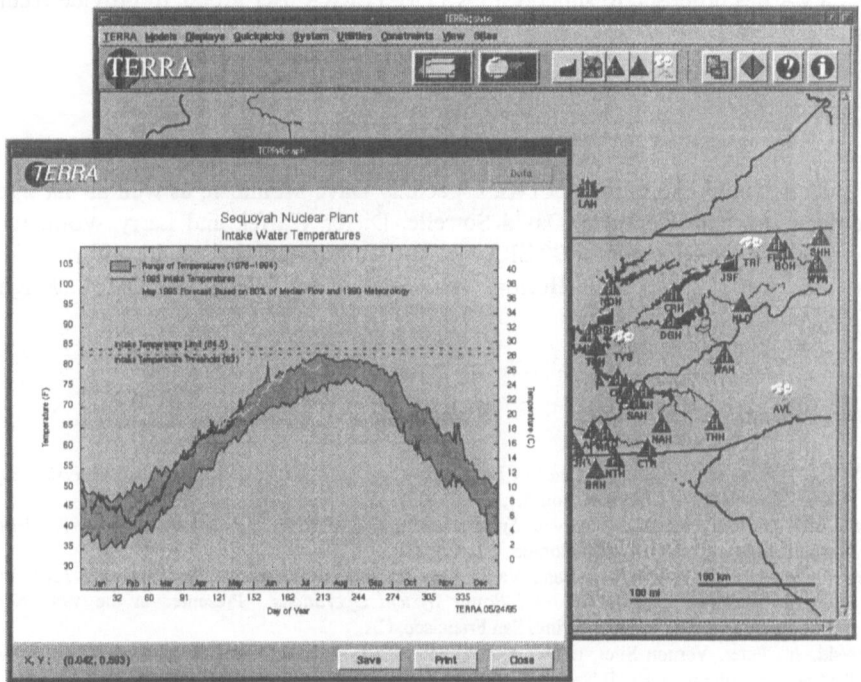

Fig. 7. Year-to-Date Measured Intake Temperature and 90-Day Forecast

5. Portability

Special care was taken in the development of TERRA to allow tailoring of the decision support system framework to the needs of any power utility or water resource agency.

1. A standard wide area communication platform (TCP/IP) is used.
2. An X-based interface allows access from either UNIX or WINDOWS computing environments.

3. TERRA utilizes the widely accepted ORACLE® relational database structure.
4. The GIS background map is in the standard ARCINFO® format.

The system is highly adaptable, and it will continue to evolve in response to users' needs.

6. Project Schedule and Future Work

The INTEGRAL Project was initiated in 1992. TERRA is currently operational and documentation has been completed. TVA and CADSWES will work with a second power utility to demonstrate the generic nature of the DSS framework. An internet hypertext interface is currently under development. Possible future enhancements include added flexibility to customize data visualization and analysis; portray multi-objective tradeoffs, risk, and uncertainty analysis; and provide general statistical, graphics, and reporting tools. TVA has proposed to support a PRSYM-TERRA user group, to provide technical assistance to other users.

Acknowledgements

The authors would like to thank EPRI, especially Dave McIntosh, as well as our internal customers: Morgan Goranflo, David Sorrelle, Dieter Waffel, and Larry Wolfe for co-sponsoring the development of TERRA. The authors appreciate the efforts of Amy Dickinson, Adam Foust, Dave Herron, Vern Siler, Cindy Webb, and Darryl Wright on this project.

References

CADSWES: 1993a, *TERRA Phase 2 Detailed Design for the TVA/EPRI INTEGRAL Project Decision Support System*. University of Colorado, Boulder, CO.

CADSWES: 1993b, *Power and Reservoir System Model (PRSYM) Design Document for 1993 Prototype Implementation*. University of Colorado, Boulder, CO.

Miller, Barbara, Peter Ostrowski, Jr., and Vahid Alavian: 1993, *Development of Enhanced Tools for the Integrated Analysis of Reservoir and Power System Operations*. Presented at the 1993 National Conference on Hydraulic Engineering, San Francisco, CA.

Ostrowski, Jr., Peter, Vernon Siler, and Darryl Wright: 1991, *Tracking Cold Water in a Large Reservoir System*. Proceedings of ASCE National Conference on Hydraulic Engineering, Nashville, TN.

Ostrowski, Jr., Peter, Barbara Miller, and Ming Shiao: 1988, *Meeting Environmental Commitments at Power Plants Using Real-Time Reservoir Scheduling*. Proceedings of ASCE Conference on Computerized Decision Support Systems for Water Managers, Fort Collins, CO.

Shane, Richard, Peter Ostrowski, Jr., and Morgan Goranflo: 1993, *The INTEGRAL Project--Providing Enhanced Tools for Coordinated Planning and Scheduling of Power Generation and Multipurpose Reservoir Operations*. Presented at the Conference on Environmental Commerce, Chattanooga, TN.

PART IV

CONTROL STRATEGIES

BEST MANAGEMENT PRACTICES FOR STORM WATER AT INDUSTRIAL FACILITIES

M. ZIMMERMAN[1] and C. MURPHY[1]

[1]Clayton Environmental Consultants, Inc., 1252 Quarry Lane, Pleasanton, California 94566

Abstract. Contaminated storm water runoff has been identified as a significant source of degradation to our nation's waterways. Industrial and construction activities are primary sources of this contamination. After identifying this problem, Congress passed the Clean Water Act Amendments of 1987 which required EPA to publish regulations to control storm water discharges from certain activities. As a result, industrial facilities subject to the program are required to obtain permits and implement controls referred to as Best Management Practices (BMPs) to reduce the pollutants in the storm water runoff from their sites. BMPs can be implemented to reduce contaminants from entering storm water (source controls) or to treat storm water after it has contacted industrial areas (treatment controls). In general, source controls are more effective in reducing pollutant levels in storm water runoff, and are therefore, preferred by EPA and other regulatory agencies. This paper provides an overview of the EPA Storm Water Program which also applies to most state programs and discusses methods to identify and implement BMPs.

Key Words: Best Management Practices (BMPs), Clean Water Act (CWA), discharge, EPA, general permit, housekeeping, National Pollutant Discharge Elimination System (NPDES), runoff, sheet-flow, source control, storm water pollution prevention plan (SWPPP), and treatment control.

1. Introduction

The National Pollutant Discharge Elimination System (NPDES) is a Federal program which requires a permit to discharge wastewater into waters of the United States from any point source. Since the program was passed in 1972, permitting and control efforts primarily focused on reducing pollutants from discharges of industrial process wastewater and municipal sewage (NARA, 1990). However, as the NPDES program was implemented and these industrial sources were controlled, it became apparent that other factors were contributing to the nation's water quality problem. Through careful evaluation, storm water runoff was identified as a significant pollutant source.

After recognizing the problem, the Clean Water Act (CWA) was modified in 1987 to establish the framework for addressing storm water discharges. Under these amendments, EPA issued regulations which require permits for storm water discharges associated with specific industrial activities and discharges from certain municipal separate storm sewer systems. "Discharges associated with industrial activity" were defined as *the discharge from any conveyance which is used for collecting and conveying storm water and which is directly related to manufacturing, processing or raw materials storage areas at an industrial plant* (NARA, 1995). Discharges associated with industrial activity include storm water runoff from industrial plant yards, immediate access roads and railroad sidings, drainage ponds, material handling/maintenance areas, residual treatment areas, and loading/unloading areas. EPA identified a wide variety of industrial facilities (listed in Title 40 of the Code of Federal Regulations Section 122.26 (40 CFR 122.26)) to be subject to the NPDES permit program which they developed. These initial requirements were known as Phase I of the storm water regulations.

Water, Air and Soil Pollution **90**: 153-162, 1996.
© 1996 *Kluwer Academic Publishers.*

In August 1995, USEPA finalized Phase II of the storm water regulations which generally covers storm water discharges from small facilities. Phase II does not affect large industrial facilities, such as utilities, covered in the original Phase I regulations. However, if any facilities were not required to develop storm water plans under the Phase I regulations, they should be required to do so under the Phase II regulations. Local water quality agencies set their own priorities to identify the timing for these facilities to comply with the Phase II regulations.

Under the NPDES permit program, dischargers have three options for applying for permit coverage: (1) submit an individual application, (2) participate in a group application, or (3) submit a notice to be covered by a general permit. If a facility had a NPDES permit which covered storm water at the time the regulations were promulgated, they were not required to obtain a permit until expiration of their existing permit. The EPA NPDES storm water permit program is now well underway and several states have developed their own programs. The state programs are typically very similar to the EPA program since they must be at a minimum as strict as the federal regulations.

In addition to requiring permits, regulations passed in September 1992, required some facilities subject to the program to comply with certain standards such as sampling, monitoring, and implementing BMPs to reduce the contaminants in storm water runoff. BMPs include a wide range of management procedures, schedules of activities, prohibitions on practices, and other management practices to prevent or reduce pollutants in storm water runoff.

2. BMPs, Waste Minimization for Storm Water

Any management practice which is designed to prevent or reduce contaminants in storm water qualifies as a BMP. BMPs in the NPDES program are analogous to waste minimization practices in EPA's solid and hazardous waste program. Similar to waste minimization strategies, BMPs may include operating procedures, training, inspections, maintenance, structural changes, and end-of-pipe treatment.

BMPs which are designed to prevent or minimize contact of pollutants with storm water (referred to as source control BMPs) are preferred by EPA to those which treat storm water contaminated with pollutants (referred to as treatment control BMPs). For example, two BMPs which could be applied to a truck unloading operation would be: (1) provide a cover over the area, or (2) provide a drain routed to an oil/water separator. The cover would be a better BMP choice since it would prevent the storm water from contacting any pollutants rather than treating the water after it has potentially been contaminated. Another advantage to source controls is that they are generally more cost effective than treatment controls.

3. How to Select Best Management Practices

Each industrial facility subject to EPA's storm water program is required to implement BMPs to reduce the amount of pollutants in the storm water runoff from the site. To identify the BMPs appropriate for a facility, it is important to first identify the activities onsite which may contribute to pollution of the storm water runoff. Most likely, each activity identified will have several different BMPs which may be viable. Each BMP should be evaluated and prioritized to determine the best option. (As previously stated, source controls are typically preferred to treatment options.) As a result of this evaluation, the BMP chosen for some areas may actually be a combination of several BMPs.

Several factors may be considered to determine the adequacy of BMPs chosen for a particular facility. First, EPA has defined the following minimum BMP criteria for general permits:

- Termination of all non-storm water discharges;
- Good housekeeping;
- Preventative maintenance;
- Spill prevention and response;
- Inspections;
- Employee training;
- Recordkeeping and internal reporting procedures; and
- Sediment and erosion control.

At a minimum, a facility's BMPs should encompass the above items. After implementing these BMPs, the facility should evaluate their effectiveness. This can be done through visual inspections or sampling of the storm water discharge points to detect any pollutants in the runoff. If it appears that the existing BMPs are not adequate or effluent limits are not being met, alternative BMPs should be considered.

To help facilities with the process of selecting BMPs, several agencies and consultants in California developed the "California Storm Water Best Management Practice Handbook" (Camp Dresser *et al.*, 1993). This book, although developed by California agencies, presents BMPs which are also applicable to industrial facilities anywhere, including facilities subject to the Federal storm water program. This handbook outlines a seven-step program for selecting BMPs, described below.

3.1 STEP 1: IDENTIFY ACTIVITIES AND POTENTIAL SOURCES OF POLLUTANTS

The activities which could contribute to storm water runoff pollution should be identified. EPA has identified six activities as being major potential sources of pollutants: (1) loading or unloading of bulk materials, (2) outdoor storage of raw materials, (3) outdoor process activities, (4) dust- or particulate-generating processes, (5) illicit connections or management practices, and (6) waste disposal practices (NARA, 1993). Other activities to identify as potential sources include vehicle fueling, maintenance, and equipment washing.

3.2 STEP 2: IDENTIFY AND EVALUATE BMPS IN PLACE

The facility should identify whether any BMPs may already be in place and evaluate their sufficiency. It is likely that most facilities subject to the storm water permit

program already have some BMPs in place. These may have been implemented as a
requirement for (1) another regulatory program, (2) good business practices such as
housekeeping, or (3) financial reasons. For example, a tank which holds hazardous
waste should already be equipped with secondary containment because of the standards
for tanks under EPA's Resource Conservation and Recovery Act (RCRA). Existing
structures, such as this one, would therefore qualify as BMPs under the storm water
program.

For each activity, any existing engineering controls or operational practices which
could qualify as a storm water BMP should be identified and evaluated as to whether
they are sufficient. The evaluation may include making visual observations of
discharge points or collecting samples of storm water discharges. The evaluation
should determine if some existing programs or structures may have to be modified or
enhanced to effectively control pollutants in storm water.

3.3   STEP 3: IDENTIFY AND ELIMINATE NON-STORM WATER DISCHARGES

The facility should identify and eliminate any non-storm water discharges to the storm
drain. Non-storm water discharges include any water used directly in the
manufacturing process (process wastewater), air conditioning condensate and coolant,
non-contact cooling water, cooling equipment condensate, vehicle and equipment wash
water, sink and drinking fountain wastewater, sanitary wastes, spills, or other
wastewaters. If non-storm water discharges are identified, they should be eliminated by
blocking and/or rerouting the drainage and finding an alternative disposal option for the
wastewater.

The facility may use several techniques to identify their non-storm water discharges
(EPA, 1992). Some examples are:

-   Visually inspect the discharge points during dry weather.

-   Review the piping schematics and as-built drawings to determine if there are any
    illicit connections to the storm water drainage system.

-   Perform a smoke test during dry weather by filling the storm water
    collection/drainage system with smoke. If smoke appears at the base of a toilet
    or other non-storm water inlet, an illicit connection may exist.

-   Perform a dye test by injecting a dye into the sanitary or process wastewater
    system. Examine the storm water collection/drainage system discharge points
    for discoloration.

3.4   STEP 4: IDENTIFY LOW- AND MODERATE-COST SOURCE CONTROL BMPs

Source control BMPs are those BMPs which control the pollutants at the source by
preventing contact between the storm water and the pollutant. The facility should
identify low- and moderate-cost source control BMPs. Since source control BMPs are
generally more effective than treatment control BMPs (those that treat polluted storm

water), they should be considered first. At a minimum, the low-cost BMPs which were identified by EPA as minimum requirements should be implemented. These minimum requirements or BMPs are listed previously in this paper in Section 3. How to Select Best Management Practices.

One example of a moderate-cost source control BMP would be to divert runoff water around the activity with berms. Another option would be to identify the most significant pollutant sources and segregate them from the less significant sources. Then, a BMP such as a cover could be used for the segregated area of most significant sources at a lower cost than covering a combined area for both types of sources.

3.5   STEP 5: IDENTIFY OTHER SOURCE CONTROL BMPs

If low- and moderate-cost source control BMPs are not adequate, the facility should identify other source control BMPs. The BMPs are not adequate if the contaminant load on the runoff is significant or if applicable effluent guidelines are not being met. For example, an equipment washing area may be impractical to cover and may require a costly containment system. In this case, the facility could look into segregating the area and disposing of the water offsite. To reduce the cost of implementing BMPs for these activities, the more significant sources could be segregated from the less significant ones.

3.6   STEP 6: CONSIDER TREATMENT CONTROL BMPs

The facility should consider treatment controls if source control BMPs are not adequate. Treatment control BMPs are those which treat storm water runoff, which may have already been contaminated, prior to discharge. Treatment control BMPs may be needed if:

- They are more cost-effective than the source controls identified in Steps 4 and 5;
- An effluent guideline cannot be met with source control BMPs;
- A particular pollutant of concern can only be controlled with a treatment control BMP; or
- They are required by an agency (such as a local water district).

To reduce the cost of treatment control BMPs, a facility should consider the following options:

- Reroute piping and drainage so that all activities to be controlled are routed into one control device, or grade the facility to a central accumulation point.

- Divert the sources to an existing wastewater treatment system.

- For activities covering a small area, ask the sewer district for permission to install a diversion pipe with a manual valve to the sanitary sewer. When the activity is in progress, the valve would be opened to divert process water directly into the sanitary sewer; at all other times, storm water would be diverted into the storm drain(s).

- Discuss the possibility with the sewer district of discharging only water from the initial part of a storm event (e.g. the first hour) to the sanitary sewer. After this initial period, the valve would be closed and storm water would be diverted into the storm drain(s). This may be sufficient to prevent contamination from entering the storm drains, since most of the pollutants will be washed away during the initial period.

- Consider leasing BMP equipment and using it onsite for a wet season. For example, if a facility has a problem with oil and grease in their discharges, they could lease an oil/water separator. The facility could evaluate whether the equipment will effectively remedy their problem and decide whether to make the capital investment prior to purchasing equipment. If the equipment is not adequate, another vendor or equipment type can be tried while minimizing costs.

3.7   STEP 7: PREPARE BMP LIST AND PRIORITIZE

The final step in selecting BMPs is to develop a list of all BMPs evaluated for each activity and determine which ones will be implemented. It is possible that some will be implemented immediately, while others will be phased in, due to cost or time constraints. Remember to first implement low-cost and moderate-cost source control BMPs. Finally, it is important to evaluate the BMPs chosen to ensure that they meet with the minimum requirements of the facility's permit, including numeric effluent limits when applicable.

4. Examples of Best Management Practices

In general, BMPs can be placed into two categories: (1) operational controls and (2) engineering controls. Operational controls include management practices which are implemented, such as housekeeping, training, inspections, and maintenance. Engineering controls include modifications to structures onsite, such as containment devices, settling basins, neutralization units, and oil/water separators. Operational controls tend to be more cost-effective and are therefore usually preferred by industry.

As mentioned earlier, BMPs can also be categorized as either source controls or treatment controls. Source controls are generally more effective than treatment controls and are therefore preferred by agencies.

4.1 OPERATIONAL CONTROL BMPs

Operational control BMPs are those which do not require structural modifications to the facility. Rather, they consist of management practices which are designed to reduce or eliminate the amount of pollutants in the storm water. Operational controls are generally less expensive to implement than engineering controls. Examples of operational control are presented below.

4.1.1 Eliminate Non-Storm Water Discharges

Non-storm water discharges to the storm drain should be identified and ceased. Step number 3 above in "How to Select Best Management Practices" discusses how to identify these discharges. Once identified, alternative disposal options should be used for the wastewater.

4.1.2 Eliminate Source

The most effective way to eliminate or reduce the pollutants in storm water is to eliminate the activity. This BMP, however, is not always feasible. Examples of BMPs to eliminate the sources include the following:

- Fuel vehicles at an offsite fueling station instead of fueling onsite.
- Take vehicles to a local car wash instead of washing them onsite.
- Substitute the use of a hazardous material for a nonhazardous material.
- Bring the activity indoors.
- Cover containers and waste piles.
- Eliminate or minimize the use of pesticides in landscaping activities.

4.1.3 Good Housekeeping

Another extremely effective BMP is to implement good housekeeping practices. Good housekeeping will significantly minimize, if not eliminate, the amount of pollutants which contact storm water. Examples of good housekeeping BMPs include the following:

- Sweep down areas periodically.
- Use drip pans at manifolds and bulk loading/unloading areas.
- Keep drip pans covered or clean when not in use.
- Use drip pans or containment pans for maintenance activities (put a dirty shovel in a bucket instead of on the ground).
- Store flexible hoses in a manner that material will not drip onto the ground (e.g., store them on a rack with caps on them).
- Keep residues off the exterior of equipment and containers.
- Clean up any spills immediately.

4.1.4 Spill Prevention and Response

Proper spill prevention and response procedures will help minimize pollutants exposed to storm water. Spill prevention measures can include engineering devices such as high-level alarms on tanks and secondary containment systems. They can also include operational procedures such as regular inspections and preventative maintenance of equipment.

Spill response procedures include programs to ensure that spills are quickly identified and remediated. They may also include internal reporting procedures, alarms, communication devices, and spill response kits. Procedures such as these may already be in place and may be required by other programs such as the Spill Prevention Control and Countermeasures (SPCC) Plan, the Business Plan or Hazardous Materials Management Plan (HMMP), the Emergency Response Plan, or the Hazardous Waste Contingency Plan (HWCP).

4.1.5 Inspections and Testing
Regular inspections of materials handling and storage areas should identify leaks, spills, or residues which could enter the storm drainage system. In addition, BMP control measures such as containment devices and erosion control measures may be inspected regularly to ensure that they are operating properly. Storm water discharge points (such as ditches) may also be inspected to identify any indications of contaminant discharge (such as discoloration of the soil). In addition, structures and equipment may be subjected to periodic testing to maintain their integrity (e.g., tanks receive periodic leak tests).

4.1.6 Regular Maintenance
Proper regular maintenance of structures and equipment will help prevent failure of the equipment which could result in a leak or spill ultimately discharging pollutants into the storm system. For example, pumps which are properly serviced are less likely to leak than those which are not maintained or serviced.

4.1.7 Inspection and/or Sampling of Storm Water Prior to Discharge
If storm water is contained prior to discharge, it could be inspected or sampled to ensure that it is not contaminated. If it contains pollutants, it could be treated prior to discharge or sent to an offsite disposal facility. If the inspection or sampling indicate it does not contain pollutants, the storm water can be discharged according to the facility's permit requirements.

4.1.8 Employee Training
Employees should be trained in proper implementation of the BMPs so that they are effectively implemented. Employees should also be trained in identifying storm water pollution problems and notify supervisors and management to correct the problems.

4.2   ENGINEERING CONTROLS

Engineering controls are those which use structural devices to control pollutants in storm water. Engineering controls tend to be more expensive to implement than operational controls. Also, they are generally not as effective as operational controls, since they focus on treating the storm water after it has contacted potential pollutants instead of controlling the source. Examples of engineering controls are presented below.

4.2.1 Reuse of the Storm Water

Storm water which will be discharged into the sanitary sewer system may be reused in a process onsite prior to discharge (for example, as cooling water if the potential contaminants are not harmful to the cooling equipment or process). Although this practice does not reduce the amount of pollutants in the storm water, it does reduce the amount of storm water which is discharged from the site. Prior to being implemented, this BMP may require modifications to the piping and drainage system.

4.2.2 Containment and Diversion Structures

Containment structures may be used to control storm water by:

1. Preventing spills or leaks from entering the storm water drainage system.

2. Containing storm water so that it can be inspected or treated prior to discharge or disposal offsite.

3. Reducing the amount of storm water runon into the process area.

4. Containing process waters, such as truck-wash water, which would otherwise be discharged to the storm drainage system.

If containment devices are equipped with discharge valves, they should be kept closed when not being used. Water should also be inspected or sampled prior to opening the valve and discharging to a storm drain. The inspection or sampling should ensure that the water meets the discharge effluent or water quality standards in the facility's permit.

4.2.3 Sediment and Erosion Control

Erodible surfaces could contribute significantly to pollutants in the storm water. Erodible surfaces could result from industrial activities which disturb the soil, such as grading and construction activities. Some methods to control erosion are:

- Preserve natural vegetation or revegetate bare areas.
- Pave the area.
- Divert storm water around the area.
- Utilize chemical stabilizing agents designed for erosion control.
- Utilize geomembranes or geotextiles.

4.2.4 End-of-Pipe Treatment

End-of-pipe treatment devices are often used to treat storm water which is already contaminated. Treatment of storm water prior to discharge may be necessary if other BMPs are not sufficient or if effluent guidelines are not being met. Examples of such devices include settling basins, neutralization units, filters, and oil/water separators. If end-of-pipe treatment units are used, they should be inspected regularly and subjected to a regular maintenance program to ensure that they are being effectively operated.

4.2.5 Air Pollution Control for Particulates

Particulates or dust originating from onsite activities may be a significant source of storm water pollution because of the material settling on the ground. These activities should be controlled to minimize the particulate air emissions. Controls such as bag houses or precipitators may be used on activities or units which are easily contained (such as a storage silo). Activities or units which take up a larger area, such as a coal pile, could be controlled with dust suppressants.

5. Conclusion

Facilities which are subject to an EPA or state storm water program must implement BMPs to reduce the contaminant load in storm water runoff from their facility. BMPs which are designed to prevent storm water contact with pollutants are referred to as source control BMPs. BMPs which treat storm water after it has become contaminated are referred to as treatment control BMPs. In general, source control BMPs are more effective in reducing pollutant levels in storm water runoff, and are therefore preferred by EPA and other regulatory agencies.

BMPs should be implemented in the following order: (1) use BMPs which are already in place at the facility; (2) eliminate non-storm water discharges; (3) use low- and moderate-cost source control BMPs, including those identified by EPA as the minimum requirement; (4) use other source control BMPs; and (5) when all other attempts are not viable or fail, use treatment control BMPs.

In selecting BMPs, facilities may implement operational controls or engineering controls. Operational controls are those which do not require structural modifications to the facility. Some examples of operational controls include housekeeping, spill prevention and response, inspections, and maintenance. Engineering controls are those which use structural devices to control pollutants and include items such as berms, air pollution control devices, and end-of-pipe treatment devices. In general, operational controls tend to be more cost effective than engineering controls.

References

Camp Dresser & McKee, Larry Walker Associates, Uribe and Associates, Resources Planning Associates: 1993, *California Storm Water Best Management Practices Handbook*, 1-1.

Environmental Protection Agency (EPA): 1992, *EPA Storm Water Management for Industrial Activities, Developing Pollution Prevention Plans and Best Management Practices*, September 1992 (EPA 832-R-92-006), 2-17.

National Archives and Records Administration. Office of the Federal Register (NARA): 1990 (November 16), *Federal Register*. Volume 55, Number 222, 47990.

National Archives and Records Administration. Office of the Federal Register (NARA): 1993 (November 19), *Federal Register*. Volume 58, Number 222, 61154.

National Archives and Records Administration. Office of the Federal Register (NARA): 1995, *Code of Federal Regulations 40, Parts 100-149*, Section 122.26.

COAL SORBENT SYSTEM FOR THE EXTRACTION AND DISPOSAL OF HEAVY METALS AND ORGANIC COMPOUNDS

R. F. Maddalone

TRW, One Space Park 01/2040, Redondo Beach, CA 90278

Abstract. The Coal Sorbent System (CSS) utilizes a newly developed coal-based sorbent, which has the combined characteristics of an activated carbon and ion exchange resin. The low cost and excellent combustion characteristics (including very low sulfur and ash) of the coal sorbent enabled the development of a system to extract metals and organics from aqueous streams, concentrate them on the sorbent and encapsulate the metals (destroy organics) utilizing developed incineration or vitrification processes. Coal Sorbent is produced by the chemical leaching of ordinary coal using the TRW Molten Caustic Leaching (MCL) process. The coal sorbent produced from the MCL process is much like activated carbon with a large internal surface area (up to 1000 m^2/g). In addition to its high surface area, the coal sorbent, unlike activated carbon, has inherent carboxyl groups much like ion exchange resins that can remove heavy metals from wastewater. Heavy metal capacities up to 70 mg/g (7%) of coal sorbent have been demonstrated. The coal sorbent has both a high surface area and a large concentration of carboxyl groups on the surface. These two features enable the Coal Sorbent to be used as a wastewater treating medium to remove both organics (high carbonaceous surface area) and heavy metals (carboxyl ion exchange groups). Once the heavy metals and organics are adsorbed on the virtually sulfur and ash-free coal sorbent, the spent coal sorbent can be processed by commercially available combustion/vitrification processes to encapsulate the metals and destroy the organics. Waste volume reductions as high as 420,000:1 have been demonstrated for uranium groundwater applications at an estimated cost <u>through disposal</u> of <$0.001/gallon water treated.

Keywords: Coal, activated carbon, ion exchange, uranium

1. Introduction

Of major concern with groundwater contamination is the potential for contaminating aquifers that supply drinking water or feed waterways. Permanent cleanup of contaminated groundwater requires that the hazardous materials be removed and the resulting waste products are stabilized or detoxified. The coal sorbent system (CSS) deals with complete end-to-end clean-up, unlike other processes that deal with simply the removal or disposal steps. For example, ion exchange, due to high resin cost, deals only with the removal of the metal from the waste stream, but does not produce a product that is readily disposed of without additional processing of a liquid waste. In contrast, the CSS process is designed to address the total problem of removing contaminants from aqueous wastes and stabilizing them in an environmentally safe manner.

The CSS process operates as follows: 1) coal is converted to the unique sorbent medium by treatment with caustic at high temperatures utilizing TRW's Molten Caustic Leaching (MCL) Process technology in our existing U.S. Department of Energy test circuit at San Juan Capistrano, California, 2) the product coal sorbent, which is the unique sorbent, is utilized in a portable, fixed-bed processing unit for treatment of contaminated water, reducing heavy metals and radioactive contaminants to parts per billion levels, 3) the loaded sorbent is as combustible as ordinary coal, and when burned in a DOE or commercial vitrifier/combustor along with a glass, such as borosilicate, can produce an encapsulated material suitable for storage. Overall waste volume reduction ratios are estimated at up to 420,000:1 while reducing heavy metals in the treated water to below

Water, Air and Soil Pollution **90**: 163-171, 1996.
© 1996 *Kluwer Academic Publishers.*

ppb levels. The basic operations of the CSS process have been reduced to practice under TRW IR&D funds and a patent has been issued (Maddalone, 1992).

2. Characteristics of the TRW Coal Sorbent

TRW's coal sorbent has carboxyl groups providing ion exchange sites for heavy metal adsorption and high surface area for the adsorption of organics. Single point nominal surface areas are in the range of 600 to 1,000 m^2/g. It is well established that brown coals and peat, which both have a high content of oxygenated compounds and humic acids, make excellent natural adsorbers of heavy metals. Figure 1 contains the results from surface and titration analyses of the coal sorbent. Electron Spectroscopy for Chemical Analysis (ESCA) analysis (Figure 1a) indicates that the coal surface consists of 87% carbon, 10% oxygen, with traces of nitrogen and sulfur. The ESCA scan suggests that at least three oxygen (carboxyl, hydroxyl and aldehydic) functionalities appear to be present on the surface of the coal. A sample of the untreated coal (Kentucky #13) was placed in a Fourier Transform Infrared Spectroscopy (FTIR) diffuse reflectance cell and its spectra collected. A sample of the acid and water washed Coal Sorbent made from that coal source was the analyzed in the same fashion. Using the digital processing associated with FTIR, the starting coal sample was subtracted from the Coal Sorbent spectra. The three spectra are shown in Figure 1b. If the samples were a perfect match the resulting difference spectra would be a straight line. The Figure 1b shows that the MCL processing of coal significantly changed the coal surface. First there is a decrease in the C-H bands indicating a removal of aliphatic hydrocarbons (confirmed in our analysis of the off-gas from the MCL process). Secondly and more importantly, the strong absorbance at 5.6 to 6.0 m indicate the presence the carboxylic acid groups. Finally, Figure 1c shows the results of a titration of a sample of Gravimelt processed coal starting in the acid form. The first derivative plot of the titration curve indicates that the inflection point (pK_a) is near 4.0. For reference benzoic acid has a pK_a of 4.18 again providing support to the supposition that carboxylic acids are present.

The weight of evidence from these physical and chemical tests support our contention that carboxylic acid groups are primary absorber in the Coal Sorbent. This variant of the Coal Sorbent produced from the MCL process is called Coal Sorbent-Weak Acid (CS-WA) for the weak acid (carboxylic acid) sorbent group.

TRW has conducted a significant number of laboratory tests of the proposed application using cadmium, lead, nickel and uranium as the test metals (Table I) and methylene blue (Table II) as the test organic compound. These removal efficiency tests were initially conducted as batch tests with the coal sorbent in excess. Both the coal sorbent (by titration) and the test solution were adjusted to the test pH (generally between 5 and 7) before they were allowed to contact. Test solutions containing approximately 7,000 mg/L $NaNO_3$ as the background were filtered and analyzed after pH adjustment to ensure that no precipitates were formed prior to contact with the coal sorbent. The filtered sample at the analyzed concentration was then contacted with a fixed amount of coal sorbent in a test tube. Table I shows that the coal sorbent is an effective adsorber for solutions of cadmium, lead, nickel, and uranium ions.

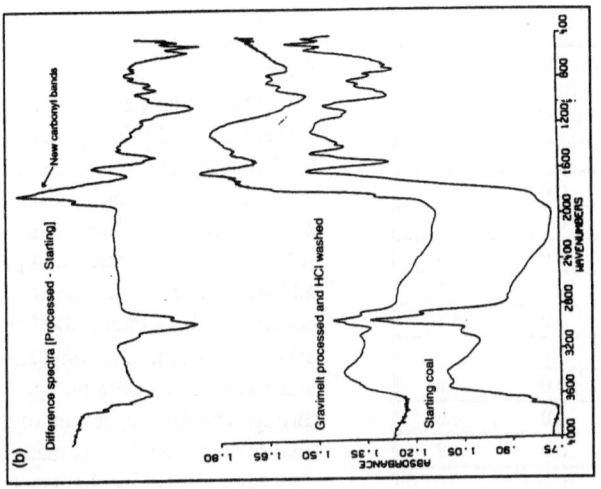

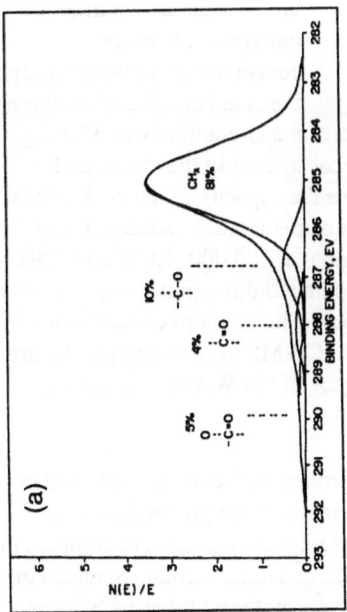

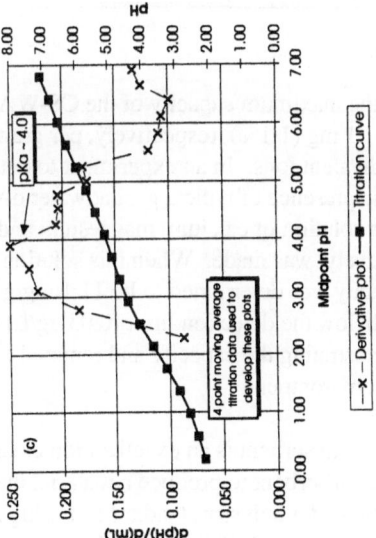

Figure 1. a) ESCA spectra of coal sorbent, b) FTIR difference spectra of coal sorbent and untreated coal, and c) titration plot of coal sorbent showing estimated pK$_a$.

Table I

Coal Sorbent Batch Removal Efficiency for
Various Elements

Element	Starting Conc., µg/L	Avg. Conc. after contact, µg/L	Final pH
Cd	7,000	0.0	6.6
	1,100	0.0	6.6
Ni	42	0.0	6.4
	50	10.0	7.1
	75	0.0	6.4
	890	0.0	7.3
	1000	9.0	6.6
	10,600	0.0	6.6
	25,200	0.0	7.2
Pb	97	0.0	6.0
U	478	6.4	6.2
	1,200	2.2	6.2
	1,325	2.0	6.3
	1,450	5.5	6.5
	14,800	5.1	6.2

Isotherms for nickel and uranium were created using a flow-through system consisting of a peristaltic pump, a 0.45 µm filter, and a glass column with a frit. The test amount of coal sorbent (typically 0.1g) was loaded as slurry into the column and is allowed to compact under a deionized water flow. Test solutions were prepared and pH adjusted. The test solutions containing the nickel (1,000 and 10,000 µg/L) and uranium (1,500 and 5,000 µg/L) were pumped through the filter upstream of the coal sorbent bed. In this manner the only removal mechanism would be via ion exchange since any precipitate would have been removed prior to reaching the adsorber. A test sample was taken just after the filter to determine the exact concentration passing through the coal bed. Based on these experiments the maximum capacity of the CS-WA for nickel and uranium was 31.3 mg (3.1%) and 11.1 mg (1.1%), respectively, per gram of coal sorbent in the absence of background divalent ions. In an experiment to determine the capacity of the coal sorbent for uranium in the presence of typical groundwater divalent elements such as calcium and magnesium, a solution of calcium, magnesium and uranium at 100,000, 1,000 and 1,500 µg/L, respectively, was made. When this solution was pumped through the coal sorbent the uranium capacity was determined to be 11.1 mg/g (1.1%). The uranium concentration in the effluent was below the detection limit (0.03 ug/L) of the ICP-MS used to analyze the test solutions demostrating the capacity and removal efficiency of CS-WA in a simulated groundwater environment.

While TRW's coal sorbent is an excellent ion exchange material, the MCL process also modifies the coal sorbent to produce a material that has many of the properties of an activated carbon. An industry standard test using methylene blue was used to compare the adsorption capacities of TRW's coal sorbent and standard activated carbon. Samples of activated carbon and TRW's coal sorbent were placed in separate test tubes with water. After agitation to thoroughly wet the solids, 0.25 mL portions of a solution of methylene blue (0.15%) were incrementally added to each tube. The tubes were agitated and the disappearance of the blue color was noted after each incremental addition. Approximately 15 minutes elapsed between additions of methylene blue. Under these conditions both the activated carbon and TRW's coal sorbent adsorbed 93 mg of methylene blue per gram of solid before the color no longer disappeared. In a separate experiment the raw coal did not adsorb any methylene blue. Other tests where the coal was pre-loaded with methylene blue and then exposed to Ni solutions, showed that equivalent capacity to the coal sorbent without the methylene blue present. These tests indicate that the organic removal and ion exchange capability are independent and would not interfere with each other. A direct comparison of TRW's coal sorbent (derived from both brown and black coals) with commercial ion exchange resin and activated carbon is presented in Table II. TRW's coal sorbent compares favorably with commercial products at a much lower cost.

3. Coal Sorbent System

TRW's coal sorbent system is a process for removal of heavy metals (including uranium) and organic compounds from wastewaters and groundwater which also destroys the organic compounds and encapsulates the metals for disposal while attaining 4 to 5 orders of magnitude volume reduction. The process employs coal sorbent derived from the MCL process to remove contaminants present in groundwater and then provides a method to combust/encapsulate the coal sorbent with slagging agents to encapsulate the metals and destroy the organics. Contaminated water is contacted with MCL-processed coal sorbent in a continuous fixed bed contactor. The contaminated water is pH adjusted and pumped through the fixed bed unit until the bed reaches saturation of contaminants and breakthrough occurs. At this time the unit is taken off-line and the water flow switched to the readied second fixed bed unit. The cleaned water is filtered for removal of coal sorbent fines and is sent to disposal as nonhazardous waste or pumped back into the aquifer down gradient from the contamination site or injected up-gradient to add additional volume to flush the contaminants from the saturated zone.

Table II

Comparison of the Properties of TRW Coal Sorbent to Ion Exchange Resin and Activated Carbon

Property	TRW's Coal Sorbent From:		Ion Exchange Resin	Activated Carbon
	Czech Brown Coal	U.S. Pittsburgh #8		
UO_2^{+2} (as U), mg/g	--	11	~10^a	N/A
Nickel, mg/g	21	--	33.5^b	--
Methylene Blue Sorption, mL	5.1-5.7	7.6-8.1	N/A	min $18-22^c$
Iodine Number, mg/g	428	570	N/A	min 700-850^c
Surface Area, m^2/g (10% N_2/He)	393	438	N/A	$650-900^c$
Sulfur, %	0.32	0.47	--	~1-2
Ash, %	0.84	0.44	--	7-18
Volatiles, %	28.1	26.3	--	N/A
Cost, $/ton	330^e	330^e	$15,700^a$	~$1,600^d$

[a] Dowex 21K ($150/ft^3); [b] Dowex CCR-3; [c] Hainowka (CR); [d] Calgon Carbon Company; [e] 2 TPH plant model

To this point the coal sorbent system is similar to most ion exchange processes. However, because the metals and organics are trapped on a low cost, combustible material, a unique and safe disposal process is possible. The coal sorbent is a high energy, clean fuel (no SO_x controls required) which can be combusted in a number of way. Because of the low cost of the coal sorbent, the process can be designed as a "once-through" system where the loaded coal sorbent goes directly to a vitrification or combustion plant designed to destroy the organics and encapsulate the metals in an environmentally safe and stable glass. In a proof-of-concept test, a portion of the coal sorbent loaded with uranium was mixed with 10% by weight of ground soft glass and fired in a crucible at 1000°C. Within the limits of the analysis, 100% of the uranium was found in the resulting glass after the coal sorbent had burned off indicating the potential for trapping non-volatile metals in a glassy matrix. Examples of available vitrification/combustion technologies include:

- <u>Combustion in a rotary kiln</u> - the reduced gas velocities will reduce the generation of fine particles and reduce particulate control/costs. A commercial system for non-radioactive waste such as the Waste Technologies System is an example of this type of disposal.
- <u>Vitrification</u> - the spent coal sorbent is a low volume feed stream. Conditions and controls in the vitrification process would burn off the carbon matrix and capture the metals in the glass matrix. This option can be easily implemented at many of the DOE sites which will have a vitrification process to stabilize waste. A commercial vendor would be a company such as Duratek.
- <u>Specialized vitrification technologies</u> - for example, Vortec Corporation under DOE sponsorship has developed a fossil fuel fired vitrification process.

The vitrification/combustion step enables the coal sorbent system to exhibit exceptional capability to concentrate metals in a safe disposable matrix. For example, water containing 1,500 µg/L uranium can be cleaned to levels below 5 µg/L with the coal sorbent retaining up to 30 mg/g uranium. Under these conditions one ton of coal sorbent can be expected to treat from 34,000 cubic meters of contaminated water. This represents a waste volume reduction of 42,000:1. The waste volume can be further reduced by the vitrification/combustion of the coal under controlled conditions. Glass mineral additives would be used during combustion as a means of vitrifying and encapsulating the waste materials in the slag. Organic materials would be destroyed by the high combustion temperatures. Assuming that adsorbed contaminants, glass mineral additives, and remaining coal mineral matter comprise 10 percent of the coal sorbent, there is a further waste volume reduction of 10:1 for the vitrification/combustion step. This represents overall waste volume reduction of up to approximately 420,000:1 for the complete process. Waste minimization is particularly important for radioactive waste because of the high cost of disposal based on volume and also because of severe limitations on storage area.

4. Coal Sorbent System Process Economics

Cost studies indicate that TRW coal sorbent can be produced for $330 per ton in a 2 ton per hour MCL plant and at even lower costs for larger sized MCL plants. In comparison, activated carbons for organic removal can cost as high as $2000/ton depending on quality and intended application. Unlike the TRW coal sorbent, activated carbon derived from coal contains sulfur which precludes burning without sulfur oxide emissions controls. Activated carbon derived from coconut shells is low in sulfur but is difficult to burn. Although efficient for organics removal, activated carbon is far less effective than TRW coal sorbent for metals (such as uranium) removal from water.

Ion exchange resins used for the removal of heavy metals and uranium from water, such as Dowex 21K, typically cost about $16,000 per ton and a single resin is not effective for the removal of both metals and organics. The high cost of ion exchange resins prohibits once-through use, therefore, ion exchange resins are periodically regenerated in batch operations to elute the contaminants into another aqueous stream which requires additional processing for recovery and disposal of wastes.

For the specific application of uranium removal from wastewater, a process (Potts, 1993) has been proposed based on the co-precipitation of uranyl hydroxide $[(UO_2)_3(OH)_5]^+$ with potassium ferrate. The process uses a chemical treatment step to remove dissolved and suspended solids, radionuclides and priority pollutant metals from aqueous waste streams. Potassium ferrate and magnesium salts are then used to enhance contaminant removal.

A comparison of the CSS to the potassium ferrate process was made on a the basis of volume generated (Table III) and the treatment costs (Table IV) for a groundwater source containing uranium at a concentration of 490 µg/L. The encapsulated product quantity includes the 10 weight percent ground borosilicate glass added to the spent coal prior to combustion or vitrification. In calculating the disposal costs for the CSS process, the waste monitoring and disposal charges are assumed to be the same as for the potassium ferrate process on a volume basis. The cost of the combustion/ vitrification operation is based on information supplied by Vortec Corporation.

Based on chemical costs alone, treating the wastewater with potassium ferrate costs $0.036/gal. However, when the costs for monitoring, analyzing, treating, transporting,

Table III

Comparison of the Volume of Solid Materials Generated by the Potassium Ferrate Process and the CSS Process

	Potassium Ferrate[1]	Coal Sorbent System
Average uranium concentration (stormwater retention basin)	490	490
Volume treated in one year, L	24,000,000	24,000,000
Mass of water treated, kg	24,000,000	24,000,000
Nominal case coal usage, kg	N/A	2,400
Sludge produced, m³	70	None
Kiln dried material disposed, m³	49	N/A
Vitrified glass disposed, kg	N/A	240
Estimated volume of vitrified material, m³	N/A	0.2*
Volume of waste material disposed, m³	*49.0*	*0.2*

[1] Nominal uranium loading of 5 mg/g of coal sorbent, borosilicate glass density estimated at 2.3, glass added at 10% by weight of the spent coal sorbent.

Table IV

Annualized Cost Comparison of Potassium Ferrate Process with CSS Process

Chemicals/Operation	Potassium Ferrate	CSS Process
Potassium Ferrate [1]	$217,000	-
Sodium Hydroxide [2]	$8,010	$490
Hydrochloric Acid [3]	-	$2,650
TRW coal sorbent [4]	-	$834
Monitoring [5]	$1,745	$75
Rotary Kiln Operation [6]	$17,450	-
Combustion/Vitrification Operation [7]	-	$1,922
Disposal [8]	$258,640	$1,060
Total Annual Cost	$502,846	$7,032
Total Treatment Cost per gal.	*$0.080*	*$0.001*
[1] $35/lb	[5] $0.68/cu. ft. (wet sludge or spent coal)	
[2] $2.45/l	[6] $6.80/cu. ft. (wet sludge)	
[3] $0.06/lb (37.2 wt.% sol'n)	[7] $0.25/lb (Vortec Corp., cost/lb spent coal)	
[4] $0.17/lb	[8] $145/cu. ft. (dry sludge/encapsulated product)	

and disposing of the resulting solids are included, the total cost of the potassium ferrate treatment process is $0.08/gal. The total cost of treating the same quantity of wastewater using the CSS process, however, is $0.001/gal. The main cost savings are a direct consequence of the reduced volume of waste generated in the CSS process: 23:1 for wet sludge:spent coal and 244:1 for dry sludge:encapsulated product when compared to the potassium ferrate process.

The CSS process is also less expensive than ion exchange systems requiring regeneration and processing of the regeneration liquor. An estimated operating cost for an ion exchange system for removing uranium from drinking water in Colorado was given at $0.006/gal (Jelinek, 1989). However, this cost only includes the cost of trucking the regeneration brine containing the uranium to a waste water treatment plant where it is dumped into the process treatment stream. No provision or cost is included for stabilization and disposal.

5. Other Applications For The Coal Sorbent

In a broad sense, the CSS process has the unique ability to simultaneously remove both organic compounds and heavy metals from aqueous streams. This is because the TRW coal sorbent has both the high surface areas found in activated carbon and the ionic surface functionalities of ion exchange resins. Because of this dual capability, capital and operating costs of a water treatment facility would be reduced as only a single adsorber system would be necessary to remove both types of contaminants. Because TRW coal sorbent is inexpensive and thus can be used on a once-through basis, it may have applications in other areas (e.g., wastewater, process water, acid mine drainage, etc.) where activated carbon or ion exchange would be considered. For example:

- The ability to slurry the coal and its affinity for water would enable the TRW coal sorbent to be used as a semipermeable barrier that would retain the metals and organics yet allow the flow of water.
- A hybrid system using specialized ion exchange resins coupled with the CSS as the regeneration loop to concentrate and dispose of the metals without the generation of sludge (soil washing systems).
- Future applications in acid mine waters will be enabled by a new variant of the coal sorbent having strong acid functional groups added to the sorbent surface. The net effect is a 3 fold or more increase in capacity with a laboratory demonstrated use range down to a pH of 2.8.
- For selected metals requiring high purity coke, the coal sorbent could act as a metallurgical coal
- TRW has demonstrated mercury vapor capture from a gas stream at flue gas temperatures (250 and 400°F). This opens the use of the coal sorbent as a low cost air emission control material.

6. Summary

TRW's CSS process is unique as it provides for the simultaneous removal of heavy metals, radionuclides and organics in pumped water and washed-soil leachate, with the additional benefit of providing an effective waste concentration and stabilization method for metals together with thermal destruction of hazardous organic chemicals. Waste minimization is attained by the coal sorbent system in three areas:

- The coal sorbent system adsorbs both metals (radionuclides) and organics on a low cost, easy to handle material.
- There is no generation of primary wastes such as alkali sludges which are hard to handle and dewater or secondary acid or brine streams derived from the regeneration and elution of ion exchange columns and which require neutralization and precipitation.
- Due to the low cost and capacity, the coal sorbent is used as a once-through sorbent.
- The coal sorbent has virtually no ash or sulfur and can be destroyed by current combustion/vitrification technologies.
- Trace metals and radionuclides are encapsulated in the vitrified glass and the coal sorbent while any absorbed organics are destroyed.
- Waste volume reductions as high as 420,000:1 have been demonstrated for uranium from simulated Fernald groundwater at an estimated cost <u>through disposal</u> of $0.001/gallon water treated.

References

Maddalone, R. F.: December 8, 1992, "Metal Ion and Organic Contaminant Disposal," US Patent 5,169,534.

Jelinek, R.T. et.al: March 1989., "Uranium Removal from Drinking Water Using a Small Full-scale System," EPA Report No. EPA/600/2-89/012, NTIS No. PB89-169890.

Potts and Hampshire: January 1993, "New Treatment for Uranium in Wastewater," *Water Environment and Technology.*

THE EPRI STATE-OF-THE-ART COOLING WATER TREATMENT RESEARCH PROJECT
A TAILORED COLLABORATION PROGRAM

Babu Nott
Electric Power Research Institute
3412 Hillview Avenue
P.O. Box 10412
Palo Alto, California 94303-0813

David Morris
Southern Company Services
42 Inverness Parkway
Birmingham, Alabama 35242

K. Anthony Selby
Puckorius & Associates, Inc.
P.O. Box 2440
Evergreen, Colorado 80437-2440

S. R. Pate
Georgia Power
333 Piedmont Avenue
Atlanta, Georgia 30303

Tammy Brice
Entergy Operations- River Bend Station
P.O. Box 220
St. Francisville, Louisiana 70775

Dale Smay
Penelec - Keystone Station
R.D. #1 Box 269
Shelocta, Pennsylvania 15774

Don Goldstrohm
Salt River Project - Coronado Station
P.O. Box 1018
St. Johns, Arizona 85936

Terry Spaulding
PSI - CINergy
1000 East Main Street
Plainfield, Indiana 46168

Abstract The EPRI Tailored Collaboration State-of-the-Art Cooling Water Treatment Research Program has been initiated with several electric utility participants. Started in January 1995, the program will provide O&M cost reduction through improved cooling water system reliability and operation. This effort is discussed along with the objectives and goals, the participants and project timetable. The program will provide three (3) main results to the participating utilities: cost effective optimization of cooling water treatment, production of a new Cooling Water Treatment Manual and updating of two (2) EPRI software products - SEQUIL and COOLADD. A review of the specific objectives, project timetable and results to date will be presented.

Keywords: cooling water, cost reduction, cooling water treatment manual, COOLADD, SEQUIL

1. Background

The Electric Power Research Institute (EPRI) recognized the critical role of cooling systems in power production and began funding research on improved control of cooling water chemistry in 1978. Since that time, EPRI's R&D efforts in this area have resulted in a variety of valuable products. These have included:

- Design and Operating Manual for Cooling-Water treatment (CS-2276, March 1982)[1]. The manual was originally developed to assist zero-discharge plants with the design and operation of recirculated cooling water systems at very high cycles of

Water, Air and Soil Pollution **90**: 173-181, 1996.
© 1996 *Kluwer Academic Publishers.* .

concentration (>10). This manual offered the utility industry the first computer code (DRIVER) for predicting the saturation limits of potential scaling species (calcium carbonate, calcium sulfate, calcium phosphate and silica) based on actual liquid-liquid and solid-liquid ion pairing relationships instead of generic rules of thumb.

- SEQUIL: An Inorganic Aqueous Chemical Equilibrium Micro-computer Code (GS-6234-CCML, September 1990)[2]. This is an IBM PC code that is an updated version of the earlier mainframe computer code, DRIVER. SEQUIL made it easy for plant chemists and engineers to evaluate the impacts of changes in cooling water chemistry and to predict system scaling potential at different operating conditions.

- COOLADD: A Database of Generic Chemical Additives Usage in Cooling Water Systems (CS-5133, April, 1987)[3]. The COOLADD database provided utilities with the first comprehensive information on industry use of cooling water scale and corrosion inhibitors, making it possible for power plant staff to independently review and compare chemical treatment programs recommended by vendors.

By today's standards, however, these EPRI products have cumbersome user interfaces making them somewhat more difficult to use than most people would like. Equally important, from a potential user's perspective, is the lack of integration between all three products and with other EPRI cooling water chemistry related information, as well. Since each product was developed independently by different EPRI contractors over a ten year period, overall integration depended upon EPRI's continuing research and technical staff guidance in this area.

2. Objectives

This project is sponsored by EPRI as a Tailored Collaboration (TC) program. An EPRI TC program is cofunded by EPRI and the participating utilities. There are several EPRI member utilities that are participating in the program.

The project has three basic objectives:

- To update and enhance COOLADD and SEQUIL as EPRI software products.

- To develop a new Cooling Water Treatment Manual as a state-of-the-art guidance document by integrating EPRI research results and products with related information from other reliable technical sources.

- To demonstrate the advantages of using EPRI's revised products from improved cooling water chemistry control and to document the industry benefit based on site-specific case studies.

Meeting these objectives will assure that EPRI's member utilities will have improved access to cooling water research results and integrated products that encompass current technology.

3. Task Descriptions

The tasks are designed to result in an integrated series of EPRI products that will have tangible benefits to the program participants and other EPRI members.

Software Update and Enhancement

The purpose of this task is to revise both the COOLADD database and the SEQUIL code. The primary focus of these revisions will be incorporation of new and more current information on cooling water additives usage, modification of the user input/output (I/O) interface, and preparation of an internal applications interface for simultaneous access of the two software packages. All software revisions will operate in the Microsoft Windows 3.0+ environment, which is consistent with EPRI's goal of maintaining a common "Look and Feel" based on a visual, graphics-oriented interface (EPRI Software Guidelines - Contractor Summary, July 1992).

COOLADD

The initial step in revising the COOLADD database has been the development of a new survey of the utility industry to determine any changes in the information collected during the earlier 1986 survey. This repeat survey is similar to the original survey with two exceptions. First, some information from the existing COOLADD database will already be complete so that respondents need only verify or correct these data. Second, additional information (not included on the original survey) is requested, such as type of cooling tower fill, metallurgy of auxiliary heat exchangers linked to the main steam condenser cooling loop, and the length of time (years) the current chemical treatment program has been in use. The survey has been sent to the plant manager at all power plants having one or more units with a total capacity equal to or greater than 500 MW (based on the Utility Data Institute's Power Statistics Database).

Following a response quality check, the data will be entered into a new set of data files that can be accessed, searched and sorted using commercial database management software (DBMS). Microsoft ACCESS will be used as the DBMS.

EPRI and the participating utilities will test the beta version of COOLADD as part of the proposed case studies. Each party will receive a complete COOLADD package consisting of diskettes, documentation, installation instructions and test cases. When the beta testing has been concluded, each party will be asked to complete an evaluation form and to provide suggestions for software / documentation revisions. The beta testing results will be used to develop the final version of COOLADD and the associated documentation.

SEQUIL

EPRI's Inorganic Aqueous Chemical Equilibrium code has undergone extensive technical evaluation since it was first developed as a mainframe computer code. SEQUIL has proven to be a reliable predictor of scaling potential over a wide variation of cooling water chemistries. Therefore, only two revisions are proposed for this code: an improved user I/O interface and an internal applications interface that links it with COOLADD database.

SEQUIL's entire code is presently written in ANSI FORTRAN '77, and consequently has severe limitations with regard to the user I/O interface. Data entry is accomplished through a cumbersome question-and-answer format. Input corrections and file modifications are not particularly straightforward, and output (which is typically too long to be summarized on a screen) scrolls by quickly and cannot be conveniently viewed on a monitor. All file and execution commands are in DOS.

All of these difficulties can be overcome by operating SEQUIL from a Windows 3.0+ environment. DOS commands can be replaced by pull-down menus. Data entry can be simplified for the user through input screens that are mouse-activated. And output that is more than one screen can be sectioned into multiple windows that can be viewed simultaneously and moved about on the desktop. Therefore, without modifying any of the technical calculations performed by SEQUIL, the code can be given a more user-friendly "Look and Feel" by making it a Windows application.

Since both COOLADD and SEQUIL will be converted to the Windows environment, linking the two software product will be easier. With applications software, a user will be able to access the database and the equilibrium code without storing results and exiting either product. For example, a user that has just used SEQUIL to determine the calcium carbonate scaling potential for a particular cooling water may want to search COOLADD to learn about the water treatment programs commonly used for these conditions. The user can temporarily reduce the size of the SEQUIL window and move it to another area of the desktop, then open the COOLADD database from the applications software and search it for the necessary information. Through the Windows environment, the user will be able to go back and forth between the two software products as often as necessary by simply changing the size of the software window and rearranging the desktop.

Following internal review of an alpha version, EPRI and the TC utilities will test the beta version of the SEQUIL applications code as part of the proposed case studies. Each party will receive a complete SEQUIL applications package consisting of diskettes, documentation, installation instructions, and test cases. When the beta testing has been concluded, each party will be asked to complete an evaluation form and to provide suggestions for software / documentation revisions. The beta testing results will be used to develop the final version of the SEQUIL applications code and the associated documentation.

Cooling Water Treatment Manual Development and Demonstration

The purpose of this task is to develop a state-of-the-art cooling water treatment manual based on a combination of EPRI research results and products in addition to current industry information, standards, and practices. To achieve this goal, this task will focus not only on the preparation of the document, but also on the demonstration of the document's direct application to utility cooling water chemistry control through case studies conducted for each of the TC utilities. These case studies will include demonstration of the EPRI software updated and enhancements completed under the software portion of the project.

The Cooling Water Treatment Manual will be application-oriented. As such, it will be organized to help a reader:

• Understand the importance of cooling water chemistry control.

• Recognize problems related to cooling water chemistry and determine probable causes.

• Evaluate possible corrective actions.

• Implement corrective actions and follow-up as necessary.

The draft outline for the manual has been reviewed by EPRI and the TC utilities. Based on this initial review, the outline was revised as a basis for preparing a preliminary draft manual.

Each TC utility has been visited to discuss and select a case study for a cooling system at one plant. These case studies will have two goals: 1) to demonstrate EPRI software, and 2) to develop an actual site-specific cooling water treatment approach using the draft manual as guidance and document the associated benefits. In most cases, the case studies are being conducted for cooling systems that have experienced or continue to experience operating problems, or for situations in which a successful cooling water treatment program is place but program costs are high.

A typical case study began with a system audit in which historic cooling system data are evaluated to determine the cause(s) of current/potential problem(s). The contractor personnel are spending time on-site working with a utility staff representative(s) to review these data, define the need for additional data, and assist in using EPRI software and the draft manual. As a result of the on-site evaluations (in conjunction with the utility staff) a recommended cooling water treatment program will be developed.

At this point, the participating TC utility has the option of deciding whether to continue with their current cooling system treatment program or to change to the program recommended by the system audit. Should the TC utility decide to change treatment programs, the contractor will assist in implementation by:

- Identifying chemical formulations and supply sources.

- Specifying chemical storage and feed equipment.

- Defining monitoring instrumentation needs (equipment and locations).

- Participating in program startup and performance monitoring over a limited period.

- Preparing a letter report evaluating the success of the new treatment program and documenting the benefits in terms of cooling system performance and / or total program costs.

Participation in the case study activity does not obligate a TC utility to modify an existing cooling water treatment program or to switch to a different treatment program.

When the case studies have been completed, the preliminary draft manual will be revised to incorporate the results. Based on review comments and suggestions from EPRI and the TC participants the final Cooling Water Treatment Manual will be submitted to EPRI for publication.

Technology Transfer
The purpose of this task is to assist EPRI in the introduction of new and updated products developed as a result of this project. Such assistance will consist of preparing selected publications, conducting a training workshop, and providing telephone consultation to product users.

Because the new products will include software, a training workshop will be provided in which attendees are instructed in the use of the Cooling Water Treatment Manual and provided hands-on experience with the software. During the workshop, attendees will review case studies presented in the Manual and evaluate new cases based on site-specific information they provide regarding current problems and/or concerns in their systems.

In addition, telephone hotline consulting support will be provided to software users for a period of six(6) months following product delivery to EPRI. The telephone consulting will focus on user support for installing and applying the software.

4. Program Participants

There are currently five (5) EPRI member participants in the program. These are shown in Table I.

Table I

State-of-the-Art Cooling Water Treatment Program Participants

Utility	Station	Utility Contact
Entergy	River Bend	Tammy Brice
CINergy	Gibson Station	Terry Spaulding
Georgia Power	Plant Bowen	S. R. Pate/David Morris*
Penelec	Keystone Station	Dale Smay
Salt River Project	Coronado Station	Don Goldstrohm

*Mr. Morris is with Southern Company Services

The intent of the program is to use the on-site evaluations and EPRI products to optimize water treatment operations at the individual sites. For the most part, this applies to the condenser cooling systems but service water systems may also be involved. Each site is different, however, and the program is designed to address the needs of the individual participants. River Bend, Plant Bowen, Keystone and Coronado all are cooling tower systems. Of these, Plant Bowen and Keystone both have experienced problems with the fouling of the cooling tower film fill. Program optimization will focus on this subject. A goal of the program at River Bend and Coronado will be on minimization of water treatment program cost while maintaining program effectiveness. Gibson Station utilizes a cooling lake. The dynamics of condenser cooling system are therefore different. Like the other sites, however, a primary goal is water treatment cost minimization and optimization of program cost.

5. Results to Date

At the present time, the State-of-the-Art Cooling Water Treatment Program is only ten months into a three year program. In spite of the short time span, tangible results to date have been achieved:

- The survey forms that will be used to update the COOLADD database have been sent to 422 generating stations representing 128 utilities.

- Work has begun on reviewing the SEQUIL source code in order to develop plans for interface improvements.

- Initial site visits have been made to all participating generating stations. In some cases follow-up visits have also been made.

- Specific initial recommendations were made at Plant Bowen regarding the on-line cleaning and maintenance of fouled cooling tower film fill. These recommendations consisted of an aggressive chlorination program to clean up the fouled fill. A dechlorination program was implemented in conjunction with the chlorination program. The preliminary recommendations also included the discontinuation of one treatment chemical which was not compatible with the dispersant program. The treatment program has been effective in meeting the objective of full load operation during the summer of 1995.

- An evaluation has been conducted at Coronado regarding lime softener operation, cooling system scaling characteristics and overall program economics. Initial efforts have been on optimizing lime softener operation. The cooling system scaling issues relate primarily to magnesium silicate. For these efforts, SEQUIL will be used extensively and it's equilibrium calculations reconciled with other approaches to estimating magnesium silicate deposition.

- Assistance has been provided to River Bend regarding the selection of clarification polymers. Polymer performance data was developed, screening tests were run and bid specifications were developed. Through these efforts, the plant is expected to achieve a cost reduction on the order of 25%.

- The overall goal at Keystone is to reduce chemical and monitoring costs for the treatment of circulating and service cooling waters without sacrificing equipment life of efficiency. One area that has been evaluated is the development of an on-line cleaning procedure for calcium carbonate deposits that uses carbon dioxide to reduce pH.

- The program goals at Gibson Station also are to reduce treatment costs while protecting equipment life and efficiency. Gibson Station is an impounded lake cooled station that has had a history of calcium carbonate deposition problems.

6. Summary

The State-of-the-Art Cooling Water Treatment Program is an EPRI Tailored Collaboration research program designed to update existing EPRI products, develop new products and perform site specific studies in order to assist participants with selecting optimum chemical treatment programs for their cooling water systems. The products will consist of a Cooling Water Manual and enhancement of the software products SEQUIL and COOLADD. These will be used as part of the site evaluations and recommendations for specific problem solving at the participant's sites.

The TC program is less than a year into a three year effort but has already achieved some significant results. The problem of film fill fouling is being addressed at two (2) sites. Initial steps have been taken to reduce water treatment program costs at other sites.

Even though this program is now underway with the initial five (5) TC participants, the authors encourage additional participants to join. The benefits will include participation in the development of the EPRI products, early access to the products and the direct benefits resulting from the on-site activities. A specific target will be reduced O&M costs through improved water treatment operations.

7. Acknowledgments

The State-of-the-Art Cooling Water Treatment Program is a team effort that has benefited from the dedicated efforts of many individuals and their companies. The authors wish to thank Winston Chow of EPRI for his support in initiating the project. We also wish to thank Wayne Micheletti for his efforts in helping to establish the program as well as handling the software aspects of the project.

References

[1] Electric Power Research Institute (EPRI): 1990, *SEQUIL, An Inorganic Aqueous Chemical Equilibrium Code for Personal Computers*, GS-6234-CCML-q.

[2] Electric Power Research institute (EPRI): 1982, *Design and Operating Manual for Cooling-Water Treatment*, CS-2276.

[3] Electric Power Research institute (EPRI): 1987, *A Database of Generic Chemical Additives Usage in Cooling Water Systems*, CS-5133.

THE USE OF RECLAIMED WATER IN ELECTRIC POWER STATIONS AND OTHER INDUSTRIAL FACILITIES

K. Anthony Selby
Paul R. Puckorius
P.O. Box 2440
Evergreen, CO 80437

Kris R. Helm
West Basin Municipal Water District
17140 S. Avalon Blvd.
Carson, CA 90746

Abstract There has been much recent progress in the use of reclaimed water (treated municipal sewage plant effluent) for use in the cooling circuits of electric utility plants and other industrial facilities. In the greater Los Angeles area, reclaimed water has been used industrially for over 25 years but some major new projects have been initiated in 1995. By using reclaimed water, electric utility generating stations and other industrial facilities can reduce their need for water from higher quality water sources which can then be conserved for other purposes, such as municipal drinking water. This paper presents an overview of the factors required to successfully use reclaimed water as makeup to recirculating cooling systems. The primary focus is on the possible effects on equipment relative to corrosion, deposition and biological fouling and on the required changes in water treatment. Implementation of the use of this water in some new projects began in May 1995. The paper provides some of the latest available results on the use of this water. The pretreatment process for ammonia removal and chlorination practices is also discussed.

Keywords: reclaimed water, sewage plant effluent, water reuse, water recycling, conservation, refinery cooling, electric utility cooling system

1. Introduction

With the demand for high quality water increasing and the available resources decreasing, greater attention is being given to the use of reclaimed water (treated municipal plant effluent) for a number of purposes, including makeup to industrial process cooling systems. In Los Angeles county, industrial facilities are being supplied with enough reclaimed water to meet some, if not all, of the makeup requirements to their recirculating cooling water systems, in addition to other plant requirements such as boiler makeup and process use. By using reclaimed water, these plants can reduce their need for water from other higher-quality sources which can then be reserved for other purposes, such as supplying municipal drinking water.

Reclaimed water can successfully be utilized in electric utility power plant, refinery and commercial heating, ventilation and air conditioning (HVAC) cooling systems. Technically, the success of the reclaimed makeup cooling water program for these industrial facilities depends on a number of factors such as cooling system design and operating characteristics, cooling water quality and treatment requirements, reclaimed water quality and environmental constraints. In Southern California, the crude oil refining industry is taking a leading position in the use of reclaimed water[1]. Some electric utility plants are using reclaimed water and more will be in the future. For this reasons, the "lessons learned" by the refineries should be closely examined

Water, Air and Soil Pollution **90**: 183-193, 1996.
© 1996 *Kluwer Academic Publishers.*

The crude oil refining industry is a large user of fresh water, primarily for process cooling and boiler makeup. By switching to reclaimed water, particularly for the plant process cooling systems, these refineries can significantly reduce their need for fresh water. Since fresh water is a valuable resource in Southern California, the West Basin Municipal Water District has undertaken a major program by having industry utilize treated municipal sewage effluent (Reclaimed Water) that is currently released into the ocean. Known as the West Basin Water Recycling Project, this program is designed to treat and recycle 70,000 acre-feet/year (nearly 23 billion gallons/year) of treated waste water for industrial cooling (a major user), plus other uses such as landscape irrigation to replace fresh water for drinking (potable) water.

However, differences in fresh water and reclaimed water qualities could make direct substitution for process cooling problematic. Reclaimed water qualities plus refinery design, materials of construction and operating characteristics, along with the proper water treatment, will strongly influence the potential for successful use of reclaimed water for process cooling. Therefore, it is important for each reclaimed water user to thoroughly examine and evaluate the possible effects of using reclaimed water on cooling system equipment protection and its life expectancy relative to cooling water related corrosion, scaling, deposition, and biofouling

The West Basin Water Recycling Project in Southern Los Angeles County receives treated effluent from the Hyperion municipal sewage treatment plant for further processing (flocculation, filtration and disinfection) and produces a reclaimed water quality consistent with criteria established by the California Department of Health Services in Administrative Code Title 22[2]. For this reason, the reclaimed water from the West Basin Recycling Project is frequently referred to as "Title 22 Water".

The proximity of the Chevron USA El Segundo Refinery to the West Basin Water Recycling Project resulted in it being the first refinery to receive reclaimed water through a new supply pipeline. The Mobil refinery in Torrance was the second refinery to receive reclaimed water. Each of the refinery cooling water systems were reviewed as to reclaimed water applicability and potential problems with Reclaimed Water use.

There are several electric utility plants in this region that are candidates to receive reclaimed water. These include the Scattergood Generating Station of City of Los Angeles Department of Water and Power (DWP), El Segundo Generating Station of Southern California Edison (SCE) and Redondo Beach Generating Station of SCE.

2. Characteristics of Reclaimed Water

Since the original source of the reclaimed water is potable water, the reclaimed water quality is very similar to potable water in many respects. There are some significant differences, however. Total dissolved solids (TDS) is typically about 20 to 30% higher than the original potable water. Phosphate concentration in reclaimed water can range from 4 to 20 mg/L depending on the source of the raw wastewater. Ammonia can also be present in the reclaimed water at significant concentrations of 10 to 30 mg/L.

The increased TDS may impact cooling tower cycles of concentration, boiler cycles of concentration and makeup water treatment systems. In cooling systems, the phosphate increases the tendency toward the formation of calcium phosphate deposits. The treatment program must be adjusted to prevent these deposits. Ammonia can attack copper alloys. The most serious type of attack is the stress corrosion cracking of brasses, such as Admiralty.

Because of the concern for the effect of ammonia on Admiralty brass, nitrification plants were installed at two of the refinery locations (Chevron and Mobil) to remove ammonia. The nitrification process removes ammonia and increased nitrate. The process also results in somewhat higher chloride and sulfate levels in the water.

The typical water quality of the Chevron nitrification plant influent and effluent are shown in Table I.

Table I - Reclaimed Water Quality at Chevron Nitrification Plant*

	Nitrification Influent		Nitrification Effluent	
Constituent	Mean	Maximum	Mean	Maxi
Conductivity; μS/cm	1252	1660	1240	1840
pH	7.6	7.8	7.6	7.8
Total Hardness; mg/L as $CaCO_3$	161	220	143	214
Calcium; mg/L as Ca	40	65	41	67
Total Alkalinity; mg/L as $CaCO_3$	155	244	78	133
Total Phosphate; mg/L as PO_4	5.2	7.4	5	7.7
Ammonia; mg/L as NH_3	14	24	<0.5	1.3
Chloride; mg/L as Cl	201	276	215	302
Sulfate; mg/L as SO_4	108	184	124	294
Nitrate; mg/L as NO_3	25	68	84	116
Nitrite; mg/L as NO_2	2.6	5.9	<0.5	0.7
Silica; mg/L as SiO_2	21	84	19	27.6
Total Organic Carbon; mg/L	9.1	10.0	8.3	8.9
Chemical Oxygen Demand; mg/L	29	52	22	41

*Operating Data from May 17 to July 14, 1995

3. Specific Reclaimed Water Projects

3.1 Cooling Systems at the Chevron El Segundo Refinery

The Chevron El Segundo Refinery has eleven (11) recirculated cooling water systems (cooling towers) located throughout the plant. The design and operating characteristics for individual towers very widely. The El Segundo Refinery has a significant number of Admiralty-tubed heat exchangers as well as mild steel exchangers and shell side cooling, all of which are more prone to tube failure due to high heat loads combined with low-flow or stagnant areas that can accelerate deposition and corrosion.

During the initial startup phase, two (2) cooling towers are operating on reclaimed water. More towers will go on line soon. The current total cooling tower makeup usage is about 5,000 gpm or 7.2 million gallons per day. This is approximately 2.6 billion gallons per year or about 7800 acre-feet/year.

3.2 Cooling Systems at the Mobil Torrance Refinery

A total of five (5) cooling towers have been converted to nitrified reclaimed water. Though the predominant metallurgy is mild steel tubed heat exchangers, there are sufficient Admiralty tubed exchangers to require ammonia removal. All five cooling towers currently are operating on reclaimed water. Usage is in the range of 17 million gallons per day.

3.3 City of Burbank Generating Station

There are six (6) generating units at Burbank. Cooling tower system are used to cool the steam surface condensers. The condensers all contain copper alloy tubes. The City of Burbank Generating Station has used reclaimed water from the City of Burbank Waste Treatment Plant since approximately 1966[3]. Five (5) of the six (6) units use the reclaimed water. Failures of the copper alloy tubes have been minimal.

3.4 Service Water Cooling System at DWP Scattergood Generating Station

There are three (3) generating units at Scattergood. All use seawater for condenser cooling but utilize an evaporative cooling tower system for service water. The makeup to the cooling tower system is potable water. The service water systems contain numerous heat exchangers dedicated to tasks such as turbine lubricating oil cooling, hydrogen cooling, bearing oil cooling, etc. Many of these heat exchangers contain Admiralty brass tubes.

The use of reclaimed water at Scattergood currently is being carefully investigated. Estimated usage is approximately 250,000 gallons per day. The successful use of reclaimed water containing ammonia at Burbank is being evaluated to determine if conditions are sufficiently similar to allow the use of reclaimed water without ammonia removal. In addition, several ammonia removal options are being explored. These include air stripping and nitrification.

3.5 SCE El Segundo Generating Station

This location will use reclaimed water for bearing cooling water. Developments in reverse osmosis (RO) pretreatment and RO membrane technology currently are being studied which will allow reclaimed water to be used for boiler makeup. El Segundo will also use reclaimed water for landscape irrigation.

4. Chemical Treatment Programs for Predicted Cooling Water Compositions

4.1 General Requirements

An effective cooling water chemical treatment program must simultaneously control a number of potential problems: corrosion, inorganic scaling/ deposition, fouling and microorganisms (capable of causing both corrosion and deposition). Depending upon makeup water quality, cooling system design, metallurgy and operation, any one of these problems may be the primary focus of a treatment program. Only a treatment program that simultaneously addresses all four problems is acceptable.

Chevron Requirements. At the El Segundo Refinery, the major cooling system factors of concern are: (1) the large number of heat exchangers with Admiralty tubing, (2) the existence of shell side cooling in certain heat exchangers, (3) high exchanger water temperatures, (4) the potential for process leaks (which is true for all refineries) and (5) mild steel tubed heat exchangers.

Admiralty metallurgy in refinery heat exchangers is not a problem with <u>ammonia-free</u> water, since good general corrosion control of Admiralty is achieved by adding copper corrosion inhibitors (such as tolyltriazole or specialized azoles). However, in the presence of ammonia, these inhibitors cannot absolutely prevent stress corrosion cracking of copper alloys. Ammonia and chloramines cause cracking of Admiralty brass.

Mobil Requirements. At the Torrance Refinery, the major cooling system factors of concern are: (1) a number of heat exchangers with Admiralty tubing, (2) the existence of high exchanger water temperatures and (3) good corrosion control of mild steel tubed heat exchangers.

Other Considerations. Refinery heat exchangers with shell side cooling have low cooling water velocities of less than one (1) foot per second (fps) and often high water temperatures (above 120^0 F, 50^0 C) that make these units more likely to suffer from scaling, corrosion and fouling. These problems can be reduced by adding specific scale control agents (particularly some of the more recently developed synthetic copolymers and terpolymers).

4.2 Review of Reclaimed Water Quality for Cooling System Makeup

When evaluating a water source for makeup to a refinery cooling system, typically the parameters of primary concern are hardness (total and calcium), alkalinity, silica and total suspended solids. In special situations (such as in the use of reclaimed water), a number of other parameters may also be important, including ammonia, phosphate, iron, compounds, total dissolved solids, chlorides, microorganisms and organics. Reclaimed water qualities (mean and maximum concentrations) from the Chevron nitrification plant were shown in Table 1.

4.3 Water Quality Parameters

Ammonia. Ammonia particularly is important when considering reclaimed water as makeup to the refinery cooling systems. Ammonia can be extremely corrosive to copper alloys causing both metal loss and stress corrosion cracking. In addition, ammonia is a nutrient for many microorganisms. The ammonia concentration in Title 22 Reclaimed Water is substantial with a mean concentration of 14 mg/l NH_3 and a maximum concentration of 24 mg/l NH_3. When used in the cooling tower system, at 5 cycles of concentration (5 times the ammonia level) the average cooling water concentration could be over 70 mg/l NH_3 and possibly as high as 120 mg/l at pH levels of 7.0-7.5.

A great deal of technical literature has been published regarding ammonia as a cause of stress corrosion cracking in certain copper alloys, specifically Admiralty brass[4,5,6,7]. Even at relatively low ammonia concentrations (2.0 mg/l), the documented cases of catastrophic failure (cracks) are numerous. While most of this information is based on laboratory data rather than field results, it still raises concern about using ammonia containing water in systems with Admiralty brass. At the anticipated cooling water ammonia levels that would result from the use of Reclaimed Water, current chemical treatment technology does not exist that can assure operation of copper alloy heat exchangers without the danger of rapid, sudden failure due to stress corrosion cracking. Control of copper corrosion (other than stress corrosion cracking due to ammonia) is possible with copper corrosion inhibitors.

Phosphate. The potential for calcium phosphate scaling increases with increasing phosphate, calcium, pH and temperature. When the cooling water calcium and phosphate levels are chemically controlled, phosphate can be an effective carbon steel corrosion inhibitor without creating scale[8]. However, at high levels (calcium>1000 mg/l as $CaCO_3$ and phosphate>20 mg/l as PO_4), calcium phosphate deposits will occur, particularly at high temperature points in the system (such as heat exchanger surfaces) and in low-flow areas (such as shell side heat exchangers) without the proper scale inhibitors. Phosphate will also combine with soluble iron to form a deposit of iron phosphate requiring specialized scale inhibitors. In addition, phosphate provides a nutrient for algae growth.

Title 22 Water has an average of 5.2 and a maximum of 7.4 mg/l phosphate (as PO_4). This is considered a high level and will require specialized chemical treatment at the cooling towers operating at five (5) cycles of concentration, which results in 26 to 37 mg/l phosphate (as PO_4) using 100% reclaimed water.

Total Dissolved Solids (TDS) and Chlorides. Dissolved solids affect the corrosion reactions by increasing the electrical conductivity of the water. At higher TDS levels, the conductivity is higher and corrosion is greater. Dissolved chloride has a particularly strong influence on the corrosion of most metals, especially mild steel, but also for copper and stainless steel alloys. Chloride will increase the general corrosion, any galvanic effects, and possibly the propensity for localized (pitting) corrosion. High chlorides (over 200 mg/l as Cl) can create stress corrosion cracking and/or pitting corrosion of some stainless steels. Title 22 Water has both high levels of total dissolved solids and chlorides. This does increase the water corrosivity to all metals; however, the magnitude is not sufficient to require excessive corrosion inhibitors. However, some increase is often required.

Oxygen Demand. Oxygen demand, usually measured as biological oxygen demand (BOD) and chemical oxygen demand (COD), is important because it reflects an increased organic content and the associated increase in biological organism growth and the demand for biocides. Cooling systems require effective biofouling control to prevent reductions in heat transfer performance and potential, serious underdeposit corrosion.

4.4 Cooling Water Treatment When Using Reclaimed Water

The anticipated reclaimed water that would be supplied as cooling system makeup will contain ammonia. The mean ammonia concentration in Title 22 Water is 14 mg/l as NH_3; the maximum ammonia concentration can be as high as 24 mg/l as NH_3. Because of the presence of a significant amount of Admiralty brass, coupled with low flows and high temperatures, nitrification for ammonia removal was considered the most prudent approach to successful operation with reclaimed water.

4.5 Review of Nitrified Reclaimed Water Quality

Eliminating the ammonia in the reclaimed water can be successfully accomplished via three (3) methods: (1) nitrification followed by breakpoint chlorination, (2) breakpoint chlorination, and (3) chloramine production. The treatment approach selected was nitrification that consisted of dechlorination with sodium bisulfite (to protect the downstream biological treatment process), nitrification in biological filters followed by (3) breakpoint chlorination (to assure the destruction of any residual ammonia). In addition, this treatment approach would completely convert any nitrite to nitrate and significantly reduce the level of dissolved iron in the Reclaimed Water. The Nitrified Reclaimed Water qualities (mean and maximum concentrations) are also shown in Table 1.

While nitrification followed by breakpoint chlorination will reduce the concentrations of certain constituents, the levels of some other components will be increased. For example, the nitrate level will be raised substantially due to the conversion of ammonia and nitrite. Similarly, the chloride concentration will increase as a result of the final breakpoint chlorination step. The addition of the required chemicals for this process will produce an overall increase in the TDS level by about 200 mg/l for average and about 300 mg/l for maximum water quality. Chlorides will increase by about 20 mg/l.

4.6 Cooling Water Treatment When Using Nitrified Reclaimed Water

Corrosion Control - Carbon Steel. The Nitrified Reclaimed Water treatment program is based on utilizing both the existing phosphate and nitrate (from the ammonia) in the makeup water as the primary carbon steel corrosion inhibitors. Should the available phosphate level drop below 15 mg/l in the concentrated cooling water, then it would be necessary to add phosphate (usually as a 50/50 blend of orthophosphate and polyphosphate) to maintain the cooling tower water recommended levels of 20-25 mg/l total filtered phosphate for adequate carbon steel protection. Given that the mean phosphate concentration (as PO_4) in the Nitrified Reclaimed Water is 7 mg/l, operating at 5 cycles yields 35 mg/l in cooling water. This results in a considerable potential for calcium phosphate scaling and would require any of the following in order to operate scale free: (1) additional scale inhibitor, (2) operating at a lower pH and/or (3) operating at lower cycles.

Corrosion Control - Copper Alloys. Tolytriazole (TT) used at very low concentrations (1-2 mg/l as active) will form an extremely tenacious protective film on copper alloy surfaces. It is required when phosphate is present, and higher dosages may be necessary if the phosphate levels are above 30 mg/l (as PO_4) and total dissolved solids are higher.

Scale Control - Calcium/Iron Phosphate. To avoid phosphate scaling, it will be necessary to maintain a cooling water pH at or below 7.5 (range 7.0 - 7.5) and to add a combination of specialized polymers. If desired or necessary, higher concentrations of filtered total phosphate (up to 80 mg/l as PO_4) and calcium hardness (up to 2000 mg/l as $CaCO_3$) can be accommodated by lowering the pH (to 6.0-7.0) and by increasing the specialized polymer when greater than 25 mg/l phosphate (as PO_4) is present in the cooling tower water.

Scale Control - Calcium Sulfate. The solubility of calcium sulfate will be exceeded when the cooling systems are operated at 5 cycles of concentration and a pH of 7.5 or lower. To avoid calcium sulfate scaling, a specific polymer for calcium sulfate scaling, should be added to the cooling tower water.

Scale Control - Calcium Carbonate. No calcium carbonate will form at the pH (7.0 - 7.5) and in the presence of phosphate.

Microbiological Fouling Control. An oxidizing halogen such as chlorine, sodium hypochlorite, or the use of sodium bromide or other bromine-release agent can be used for microbiological fouling control. Biocide usage will increase relative to current treatment because of the organics (TOC, BOD, and COD) and higher levels of nitrates (produced by the nitrification treatment). A biodispersant should be used to improve biocide effectiveness.

5. Cooling Water Treatment Program Costs

5.1 Costs Using Fresh Water

The costs for a cooling water treatment program as supplied by a water treatment service company depend upon a number of factors, including site-specific water equipment qualities, cooling system operation and special client-vendor arrangements. Thus, predicting the exact cost of a given chemical treatment program is dependent upon these factors. However, a general cost estimate can be developed based on a knowledge of the chemical costs obtained from the chemical suppliers and chemical usage relative to cooling system water and equipment requirements.

5.2 Estimated Costs Using Reclaimed Water

The use of reclaimed water at the City of Burbank Generating Station resulted in a chemical cost increase of approximately 20%. This is in line with some of the estimates for the 1995 nitrified reclaimed water applications in refineries. Other estimates have shown that there is a potential for an overall decrease in chemical costs.

The are several components to the cost factors. Since the reclaimed water contains phosphate and nitrate, these can be used as the basis of a carbon steel corrosion control program. It is not necessary to add much in the way of additional carbon steel corrosion protection although some water treatment service companies have elected to supplement with zinc. Another factor is the cost of water which may be lower than potable water.

The main potential problem is scale control due to excessive phosphate levels (above 25 mg/l). This could increase polymer costs by 20 to 50%.

At the outset of the reclaimed water program, it is prudent to increase the levels of copper corrosion inhibitors. The monitoring program will provide the information needed to optimize the azole program.

The bio-control portion of the program may require additional biocide. It is recognized that the reclaimed water will contain organic constituents which will add to biocide demand and provide nutrients for biological growth.

6. Monitoring To Assure Treatment Performance

It is critical to evaluate the performance of the cooling water treatment rapidly and thoroughly. This must include corrosion, deposition and biological action. Ammonia impact on Admiralty stress corrosion, copper alloy corrosion and bio-control are critical. The use of rapid or instantaneous monitoring techniques are necessary to detect corrosion and deposition, but also to optimize treatment programs.

The two refineries, in preparation for Nitrified Reclaimed Water use, established excellent monitoring techniques and developed good baseline performance with fresh water use. This continued with the conversion to Nitrified Reclaimed Water to assure continued effective equipment protection and treatment modification, if needed.

The cooling water monitoring equipment for each system should include a corrosion coupon rack with coupons for each metallurgy present in the system and instantaneous corrosion monitor probes for Admiralty and mild steel, a heat transfer/deposit monitor, an oxidation-reduction potential (ORP) probe to monitor and control oxidizing biocide concentration in the cooling water, a bio-fouling sessile bacteria monitor. This should be supplemented by the monitoring of actual heat exchanger performance.

In refinery applications, additional monitoring of any critical stainless steel heat exchangers is advisable since the cooling water chloride level may be as high as 1800 mg/l as Cl. At these chloride levels, severe stainless steel pitting and stress corrosion cracking can occur unless fouling and deposition are well controlled.

7. Conclusions

Reclaimed water is being produced by tertiary treatment of municipal sewage plant effluent. This reclaimed water can be a significant resource for use in industrial cooling water systems. In California, reclaimed water is being used in electric utility plants, refineries and other industrial facilities.

Reclaimed water has some characteristics that are different from the water previously used for cooling in these facilities. It is somewhat higher in total dissolved solids. In addition, it contains phosphate and ammonia. The phosphate can be used as a corrosion inhibitor for steel components but can be a foulant, if not controlled. Ammonia can attack certain types of copper alloys. In some applications such as refineries, the ammonia has been removed in a separate nitrification process. In some electric utility applications, the plant has operated for several years and no ammonia-induced damage has occurred.

The use of reclaimed water may require an increase in the quantities of chemicals used for control of deposition and copper alloy corrosion. To some extent, this potential cost increase is offset by the fact that the phosphate in the water provides the corrosion inhibitor for steel alloys. At the City of Burbank Generating Station, which has used reclaimed water for several years, a chemical cost increase of approximately 20% has been observed.

The use of reclaimed water for industrial cooling uses technically is feasible and cost effective. This frees up existing fresh water supplies for use as potable water.

References

[1] Puckorius, P.R., Helm, K,R., Spurrell, C.: 1995, *Reclaimed Water as Cooling Tower Makeup for Refinery/Petrochemical Plants - Southern California's Activities and Time Table*, Cooling Tower Institute.

[2] California Code of Regulations, CCR Title 22, Chapter 3.

[3] Moran, D.K., Johnston, B.A., Kolessar, K.M.: 1995, *Corrosion Reduction in a Cooling System Utilizing Title 22 Reclaimed Water*, International Water Conference.

[4] Thompson, D.H.: 1959, *Corrosion* 15, 433-436.

[5] Thompson, D.H.: 1961, *Materials Research & Standards*, 108-111.

[6] Dillon, C.P.: 1974, *Materials Performance*, 49.

[7] Lynes, W.: 1965, *Corrosion* 21, 125-131.

[8] Puckorius, P.R., Strauss, S.D.: 1995, *Power* 140 (July), S.1-S.30.

REACTIVE SOLUTE TRANSPORT IN ACIDIC STREAMS

ROBERT E. BROSHEARS

U.S. Geological Survey, MS 415, Denver Federal Center, Denver, Colorado 80225

Abstract. Spatial and temporal profiles of pH and concentrations of toxic metals in streams affected by acid mine drainage are the result of the interplay of physical and biogeochemical processes. This paper describes a reactive solute transport model that provides a physically and thermodynamically quantitative interpretation of these profiles. The model combines a transport module that includes advection-dispersion and transient storage with a geochemical speciation module based on MINTEQA2. Input to the model includes stream hydrologic properties derived from tracer-dilution experiments, headwater and lateral inflow concentrations analyzed in field samples, and a thermodynamic database. Simulations reproduced the general features of steady-state patterns of observed pH and concentrations of aluminum and sulfate in St. Kevin Gulch, an acid mine drainage stream near Leadville, Colorado. These patterns were altered temporarily by injection of sodium carbonate into the stream. A transient simulation reproduced the observed effects of the base injection.

Key words: Reactive solute transport modeling, acid mine drainage, geochemical modeling

1. Introduction

Acid mine drainage degrades water quality in many of our nation's streams. Effective remediation at these sites requires an understanding of the diverse physical and biogeochemical processes that control spatial and temporal profiles of concentrations of metals and other acid constituents. This paper presents an approach to site characterization that includes tracer-dilution experiments, synoptic sampling, and reactive transport modeling. Tracer experiments quantify physical transport, and synoptic sampling defines the spatial distribution of constituent concentration in inflow and instream waters. The primary interpretive tool is a reactive solute transport model that characterizes the interplay of transport and chemical reactions producing the observed spatial profiles of concentration. These spatial patterns were altered by an injection of sodium carbonate into the stream. Transient simulations were conducted to interpret the dynamic response of pH and other stream constituents during the base injection.

2. Site and Methods

We applied our site characterization methods to a 1804-meter reach of St. Kevin Gulch, a mountain stream receiving acid mine drainage near Leadville, Colorado (Figure 1). The stream receives metal-rich, acidic water in a series of springs that discharge at the base of a mine dump. Pools, riffles, and cascades in this small (Q < 20 L/s) shallow stream cause a rapid mixing of inflows. The 1804-meter reach was divided into 13 subreaches. Water samples were collected at many stream sites, both upstream and downstream from eleven sampled inflows. Upstream from the acid inflows (0-363 m), streamwater was affected by mine drainage, but loads of metals were relatively small. Between 363 and 484 m, the stream became much more acidic, and pH decreased. Downstream from the acidic inflows, Shingle Mill Gulch entered St. Kevin Gulch (at 501 m). This confluence doubled the discharge and increased the pH. Additional loading of some acidic constituents from ground-water seepage occurred in the next downstream reach (526-781 m). In the remaining subreaches, loading was minimal.

Water, Air and Soil Pollution **90**: 195-204, 1996.
© 1996 *Kluwer Academic Publishers.*

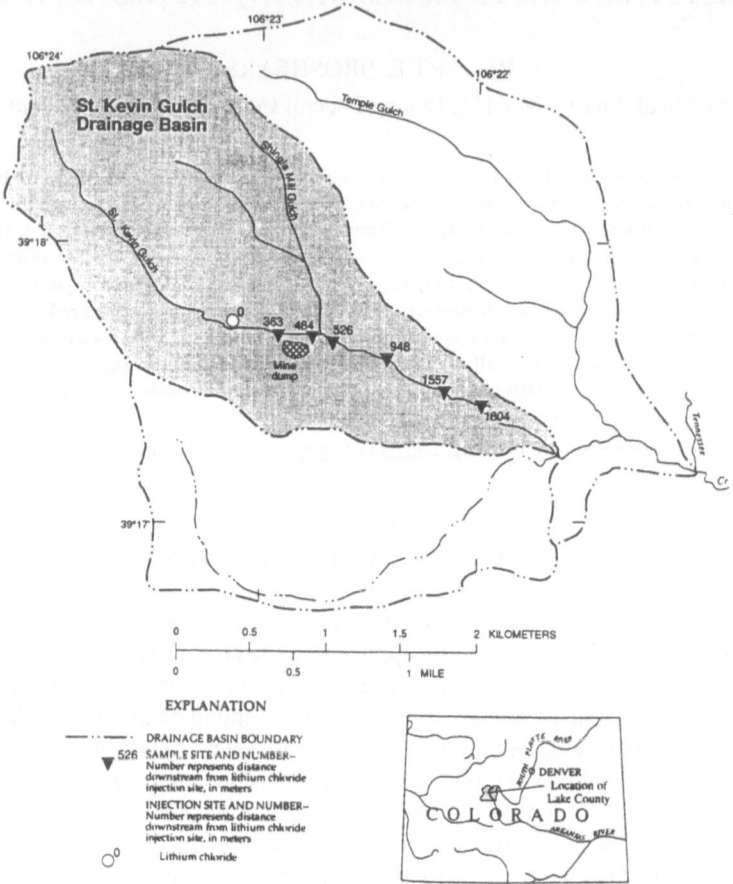

Fig. 1. Location of St. Kevin Gulch drainage basin, indication injection point and sampling sites

2.1 1986 Tracer Injection and Synoptic Sampling.

In August 1986, a tracer solution of 4.7 M LiCl was injected into St. Kevin Gulch at a rate of 27 mL/min for 52 hours. Sampling sites and inflow locations are referenced by their distance downstream from the injection. The arrival of the LiCl pulse and the development of a plateau concentration were observed at six sites, ranging from 26 to 1804 m downstream from the injection (Figure 1). During the plateau period, additional samples were collected from these six sites, from ten surface inflows, from St. Kevin Gulch upstream and downstream from each inflow, and from a pit dug in an area of ground-water seepage. Pressure filtration, using 0.1 μm nitrocellulose membrane filters, operationally defined a dissolved sample. Analysis of acidified samples for metals was by ICP-AES. Ferrous iron was measured in filtered, unacidified samples using the 2,2'-bipyridine colorimetric method. Filtered, unacidified samples were analyzed for chloride and sulfate by ion chromatography. Specific methods were reported by Kimball *et al.* (1994b).

2.2 1988 Transient Modification of Instream pH

On August 25, 1988, instream pH in St. Kevin Gulch was increased in step-wise fashion from 3.5 to 5.8 by injecting a concentrated solution of sodium carbonate. The sodium carbonate injection site was 1306 meters downstream from the LiCl injection site for the 1986 experiment. Sodium chloride was injected simultaneously to enhance the sodium pulse, which was used as a conservative tracer for definition of subreach travel times and transient storage parameters. The injection began at 9.3 hours and continued at a constant rate until 11.9 hours. An increasing rate of injection was sustained until 14.9 hours, when the injection was stopped. As the pulse of increased pH moved downstream, the response of major ions and trace metals was documented by analyzing water samples collected at sites located 24, 70, 251, and 498 meters from the sodium carbonate injection site. This paper will focus on the response at site 24 only.

2.3 Reactive Solute Transport

Physical transport and mixing were simulated by a module that includes advection-dispersion and transient storage (Bencala and Walters, 1983; Runkel and Broshears, 1991). These processes are described by the following equations.

$$\frac{\partial C}{\partial t} = -\frac{Q}{A}\frac{\partial C}{\partial x} + \frac{\partial}{\partial x}\left(DA\frac{\partial C}{\partial x}\right) + \frac{q_L}{A}(C_L - C) + \alpha(C_S - C)$$

$$\frac{\partial C_S}{\partial t} = -\alpha\frac{A}{A_S}(C_S - C)$$

where C = solute concentration in the stream $\{ML^{-3}\}$
t = time $\{T\}$
Q = volumetric flow rate $\{L^3T^{-1}\}$
A = cross-sectional area of the stream $\{L^2\}$
x = distance $\{L\}$
D = dispersion coefficient $\{L^2T^{-1}\}$
q_L = inflow rate per unit stream length $\{L^3T^{-1}L^{-1}\}$
C_L = solute concentration in lateral inflow $\{ML^{-3}\}$
α = stream-storage exchange coefficient $\{T^{-1}\}$
C_S = solute concentration in the storage zone $\{ML^{-3}\}$
A_S = cross-sectional area of the storage zone $\{L^2\}$

Kimball et al. (1994b) described the removal of metals and other acid constituents from the dissolved phase in terms of first-order rate constants. Figure 2 shows observed and simulated steady-state profiles of aluminum in St. Kevin Gulch both with and without simulation of a first-order removal process. This approach to site characterization is relatively easy to apply and provides engineering parameters that are valuable in site remediation. Instream travel times and stream lengths necessary for various levels of metal removal can be estimated and compared among different stream systems. However, this approach does not describe in rigorous fashion the actual reactions that are causing metal removal.

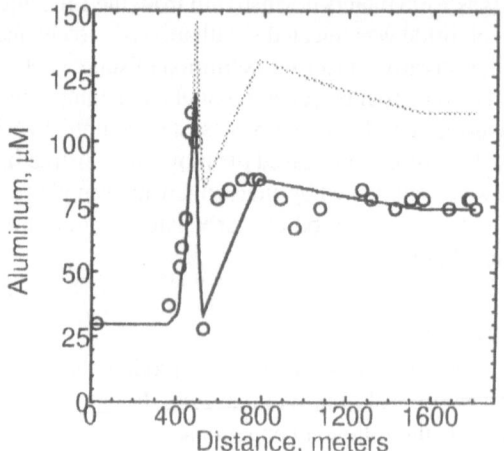

Figure 2. Observed (open circles) and simulated steady-state profiles of aluminum at St. Kevin Gulch in August 1986, assuming conservative (dotted line) and reactive (solid line) behavior. The removal rate coefficient was varied from 1.0×10^{-3} to 1.5×10^{-3} sec^{-1}.

In this paper we explore the use of equilibrium thermodynamic concepts in site characterization. Chemical reactions were simulated by a module based on the geochemical speciation model MINTEQA2 (Allison *et al.*, 1991). This module was coupled with the physical transport module by a sequential iteration method (Runkel, 1993). The geochemical module applies equilibrium thermodynamic concepts in calculations of chemical speciation, including sorption reactions and the formation of aqueous complexes and solid species. This more rigorous geochemical approach has utility in site remediation efforts as it can quantify the behavior of metals under the chemically altered conditions of various remediation scenarios.

3. Results

3.1 1986 Tracer Experiment and Synoptic Sampling

The spatial profile of lithium concentration during the 1986 study is shown in Figure 3. Discharge at each site was calculated by the dilution required to match the measured plateau concentration of the lithium tracer. The downstream increase in discharge was from 6.9 to 19.7 L/s. Discharge declined to 14.7 L/s after a losing subreach near the downstream part of the study area. Travel time for the conservative tracer through the 1804-meter reach was about 5.6 hours. Stream cross-sectional areas were adjusted to reproduce tracer behavior consistent with observed arrival and departure times. The cross-sectional areas of the storage zones and stream-storage exchange coefficients were adjusted to accommodate deviations from traditional advective-dispersive behavior in the observed concentration profiles. A summary of fitted parameters is presented in Table 1. The chemical character of inflows is shown in Table 2.

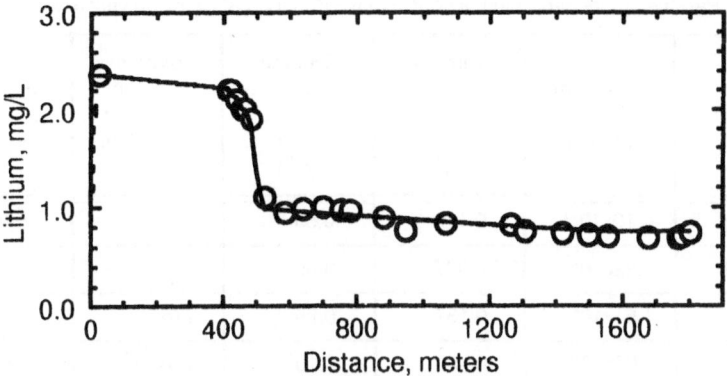

Figure 3. Steady-state profile of lithium from tracer dilution at St. Kevin Gulch, August 1986.

3.2 Steady-State Reactive Transport Simulations

Steady-state reactive transport simulations were conducted to interpret the observed profiles of pH, aluminum, and sulfate (Figure 4). In the first simulation, no solids were specified. In the latter two simulations, aluminum hydroxide or hydroxysulfate solids were permitted to precipitate. When no solid-phase controls were defined, simulated concentration profiles reflected loading near the mine dump and dilution by Shingle Mill Gulch, but remained higher than observed profiles at downstream sites. This discrepancy is a clear indication of chemical reactions that remove these constituents from the dissolved phase. Common controls on metal solubility include the formation of insoluble metal hydroxides. Aluminum, for example, undergoes the following hydrolysis reaction:

$$Al^{3+} + 3H_2O \rightarrow Al(OH)_3 + 3H^+$$

The formation constant for amorphous $Al(OH)_3$, $K_f = \{H^+\}^3\{Al^{3+}\}^{-1}$, has been reported at $10^{-10.8}$; formation constants of more crystalline forms of $Al(OH)_3$ (gibbsite) range from $10^{-9.35}$ to $10^{-8.11}$ (Nordstrom *et al.*, 1990). In the downstream subreaches simulated aluminum concentration was higher than observed concentration even when the least soluble solid phase of gibbsite ($K_f = 10^{-8.11}$) was specified (Figure 4b). This result may reflect use of an unrepresentatively large value for aluminum concentration in inflow seepage in the subreach between 526 and 781 m, or a different solid phase may control aluminum concentration. Sulfate concentration decreased with downstream distance (Figure 4c). This decrease may be explained by the formation of aluminum hydroxysulfate solids. Sulfur is a component of stream-bed sediment in St. Kevin Gulch, where it has been measured as high as 3% by weight (Smith, 1991). Nordstrom (1982) described solid phase controls on aluminum concentration in waters of high sulfate concentration and low pH. A variety of aluminum hydroxysulfate minerals may control aluminum concentration, including jurbanite

Table 1: Parameters for simulation of conservative transport at St. Kevin Gulch, August 1986

Reach boundaries (m)	Maximum discharge (m^3s^{-1})	Stream cross-sectional area (m^2)	Storage zone cross-sectional area (m^2)	Stream-storage exchange coefficient (s^{-1})	Dispersion coefficient (m^2s^{-1})
0-26	6.19×10^{-3}	0.120	0.05	3.0×10^{-5}	0.02
26-484	7.92×10^{-3}	0.097	0.05	2.0×10^{-5}	0.02
484-526	1.51×10^{-2}	0.137	0.25	2.0×10^{-5}	0.02
526-948	1.68×10^{-2}	0.199	0.10	1.5×10^{-5}	0.02
948-1557	1.97×10^{-2}	0.152	0.20	5.0×10^{-5}	0.02
1557-1804	1.47×10^{-2}	0.153	0.10	1.5×10^{-5}	0.02

Table 2: Concentration of sulfate and aluminum and pH at the upstream boundary and at lateral inflows, St. Kevin Gulch, August 1986

Location (m)	Sulfate concentration (mg/L)	Aluminum concentration (mg/L)	pH
0	55	0.81	4.94
372	860	17	2.69
417	980	14	2.66
424	1075	14	2.69
449	1110	16	2.60
459	926	16	2.66
469	794	25	2.72
501	31	0.08	6.51
570	330	22.7	--
851	40	0.37	5.98
1281	72	1.2	3.87
1391	22	0.19	6.19

$[Al(SO_4)OH]$, alunite $[KAl_3(SO_4)_2(OH)_6]$, and basaluminite $[Al_4(SO_4)(OH)_{10}]$. Based on solubility diagrams presented by Nordstrom (1982), we selected jurbanite as a possible solid phase control within the ambient range of pH and aluminum and sulfate concentration in St. Kevin Gulch. Nordstrom (1982) cited a solubility product, K_{sp} = $\{Al^{3+}\}\{SO_4^{2+}\}\{OH^-\} = 10^{-17.8}$ for jurbanite, which at the 15° C average temperature of St. Kevin Gulch is equivalent to a formation constant of $10^{3.45}$. Figure 4 also presents the results of simulations conducted with jurbanite specified as the solid phase control for aluminum. The aluminum (Figure 4b) profile was not reproduced unless the formation constant for jurbanite was increased to $10^{4.25}$, which is somewhat higher than the reported value. Even with this less soluble value for jurbanite, the simulated sulfate concentration remained higher than the observed profile (Figure 4c).

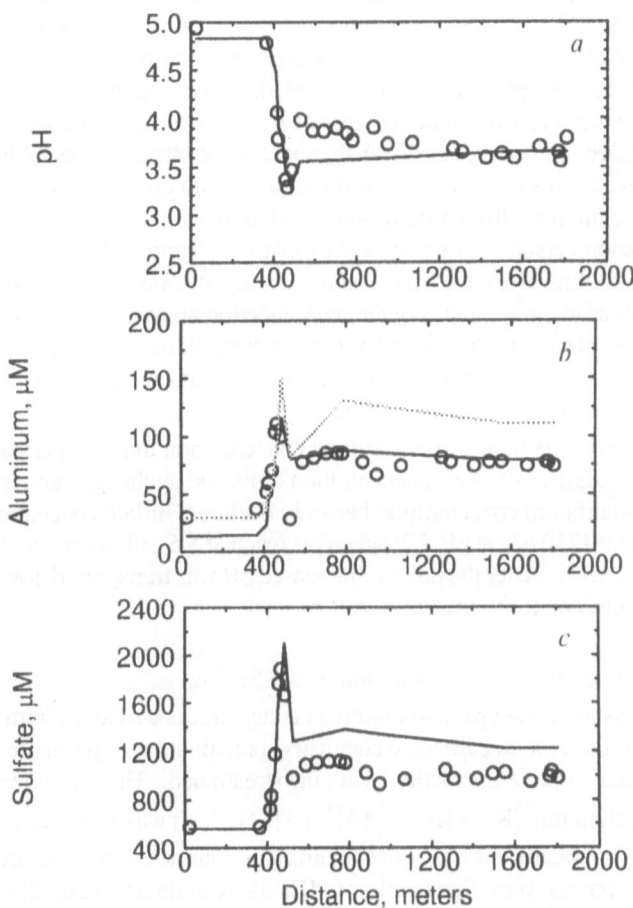

Figure 4. Observed (open circles) and simulated (lines) values of (a) pH and concentrations of (b) aluminum with the controlling solid specified as $Al(OH)_3$, $K_F = 10^{-8.11}$ (dotted line) or jurbanite [$Al(SO_4)OH$, $K_F = 10^{-4.25}$ (solid line)] and (c) sulfate with the controlling solid phase specified as jurbanite, St. Kevin Gulch, August 1986.

3.3 Profiles of pH, aluminum, and sulfate during the 1988 base injection

pH. Prior to the sodium carbonate injection, background pH values were about 3.5. The profile of pH at site 24 during the experiment is shown in Figure 5a. The injection quickly resulted in a pH of 4.2 at site 24. An increased injection rate begun at 11.9 hours eventually resulted in a pH of 5.8 at this site.

Aluminum. Background concentration of aluminum was spatially uniform at about 115 μM before the injection began. When pH increased from 3.5 to 4.2 at site 24, the aluminum concentration at this site did not change (Figure 5b). As pH increased above 5.0, the concentration of dissolved aluminum declined substantially, to less than 1.0 μM at pH 5.8. This decrease in dissolved aluminum was accompanied by an increase in particulate aluminum to more than 100 μM. Thus, the concentration of aluminum in the whole water sample declined only slightly. These observations were consistent with the formation of a slowly settling precipitate. After the injection stopped, dissolved aluminum concentration eventually returned to pre-injection values (Figure 5b). However, for approximately thirty minutes after the injection ended, dissolved aluminum concentration was higher than the background level. This post-injection spike in aluminum concentration indicated the recruitment of aluminum from a finite source. A flow weighted integration of total aluminum concentration versus time at site 24 showed that 27 grams of aluminum were removed from the water column along the first subreach during the injection interval (Kimball *et al.*, 1994a). A similar integration during the post-injection aluminum spike showed that 28 grams of aluminum were recruited back into the water column. This mass balance indicated that the finite source of aluminum during the post-injection spike was dissolution of freshly settled aluminum from the injection period.

Sulfate. Before the pH modification, sulfate concentration in the experimental reach was about 1330 μM (Figure 5c). This concentration remained unchanged until pH exceeded 4.7 and dissolved aluminum concentration began to decline. Sulfate concentration then also declined, to about 1230 μM at pH 5.2; when pH reached 5.6, sulfate concentration increased slightly (to 1270 μM). After the pulse of increased pH was transported downstream, sulfate concentration returned to its ambient level.

3.4 Transient Simulations of pH, Aluminum, and Sulfate

Figure 5 displays simulated pH and dissolved concentrations of aluminum and sulfate. These simulations included carbonate chemistry, precipitation of jurbanite and amorphous basaluminite, and sorptive interactions with the stream bed. The formation constant for amorphous basaluminite ($K_F = \{H^+\}^{10} \{Al^{3+}\}^{-4} \{SO_4^{2-}\}^{-1}$) was specified as $10^{-27.46}$ (Nordstrom, 1982). The stream bed was conceptualized as a surface of hydrous iron oxides undergoing double-layer complexation reactions. Kinetic restraints were applied to all interactions between the water column and the stream bed, including sorption-desorption and the dissolution of mass settled to the stream bed during the interval of higher pH.

The simulations reproduced the general features of the pulse of modified pH as it was transported downstream and the accompanying changes in aluminum and sulfate concentrations. Both in the simulations and in the field observations, aluminum was removed from the dissolved phase during the interval of higher pH. Particulate aluminum was removed from the water column by settling and was available for dissolution after the pulse of higher pH had passed. Two stages of sulfate removal were simulated, including an initial stage representing jurbanite precipitation (at about pH 4.3) and a second stage when amorphous basaluminite acted as the solid phase control (at about pH 4.6).

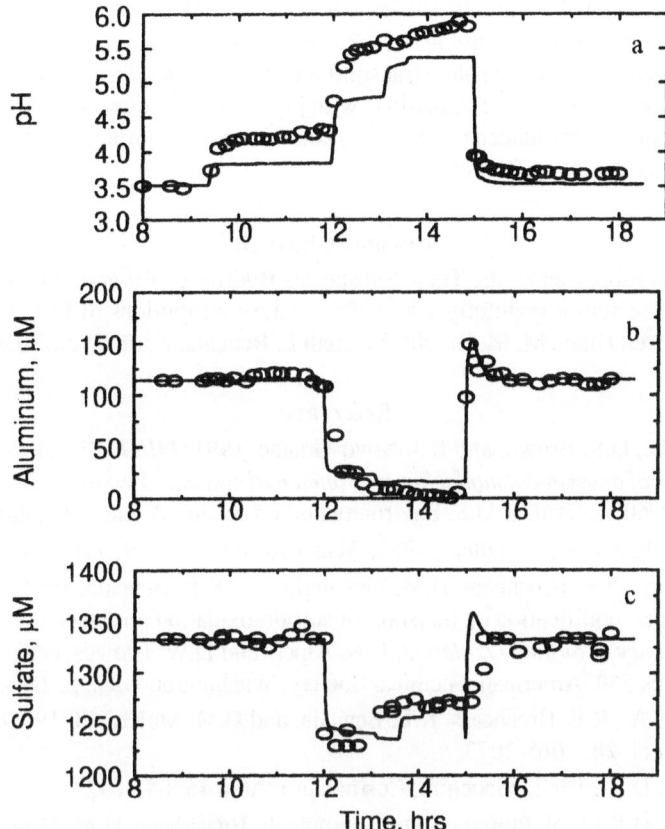

Figure 5. Observed and simulated transient profiles of pH, aluminum, and sulfate 24 meters downstream from a base-injection at St. Kevin Gulch, August 1988, with jurbanite and amorphous basaluminite specified as solid phase controls.

4.0 Conclusions

We have presented an approach to site characterization in streams affected by acid mine drainage. Because interactive processes control the behavior of acid constituents at these sites, a rigorous definition of site hydrology and spatial patterns of contaminant loading is necessary. A reactive solute transport model is offered as an internally consistent means of integrating our knowledge about a site and for testing hypotheses about physical and chemical processes that determine contaminant profiles.

An application at St. Kevin Gulch, a mountain stream receiving acid mine drainage near Leadville, Colorado, has demonstrated the reactive nature of aluminum and sulfate in the stream. Observed profiles of these constituents were reproduced by specifying hydroxysulfate controls on metal solubility. The aluminum profile could not be produced by simulations using the reported range in solubility for $Al(OH)_3$. The aluminum profile could be reproduced only when the formation constant for jurbanite $[Al(SO_4)OH]$ was somewhat higher than a reported value. Lack of fit between observed and simulated concentrations of aluminum and sulfate also may be attributable to an unrepresentative value for inflow concentrations in a subreach of major ground-water seepage.

The robustness of the reactive solute transport model was tested further by simulations of a pH modification experiment. Simulations with jurbanite and amorphous basaluminite as solid-phase controls reproduced the observed behavior of pH, aluminum, and sulfate during a step-wise injection of sodium carbonate.

Acknowledgments

This study was supported by the Toxic Substances Hydrology Program of the U.S. Geological Survey. The author gratefully acknowledges the contributions of Briant A. Kimball, Robert L. Runkel, Diane M. McKnight, Kenneth E. Bencala, Katie Walton-Day, and Tracy B. Yager.

References

Allison, J.D., D.S. Brown, and K.J. Novo-Gradac, 1991, *MINTEQA2/PRODEFA2, A geochemical assessment model for environmental systems: Version 3.0 User's Manual*, Rep. EPA/600/3-91/021, U.S. Environmental Protection Agency, Washington, D.C.

Bencala, K.E. and R.A. Walters, 1983, Water Resour. Res. **19**, 718-724.

Kimball,B.A., R.E. Broshears, D.M. McKnight, and K.E. Bencala, 1994a, Effects of instream pH modification on transport of sulfide-oxidation products, in *Environmental Geochemistry of Sulfide Oxidation*, C.N. Alpers and D.W. Blowes, eds., ACS Symposium Series 550, American Chemical Society, Washington D.C., p. 224-243.

Kimball, B.A., R.E. Broshears, K.E. Bencala, and D.M. McKnight, 1994b, Environ. Sci. Technol. **28**, 2065-2073.

Nordstrom, D.K., 1982, Geochim. Cosmochim. Acta **46**, 681-692.

Nordstrom, D.K., L.N. Plummer, D. Langmuir, E. Busenberg, H.M. May, B.F. Jones, and D.L. Parkhurst, 1990, Revised chemical equilibrium data for major water-mineral reactions and their limitations, in *Chemical Modeling of Aqueous Systems II*, D.C. Melchior and R.L. Bassett, eds., American Chemical Society, Washington D.C., p. 398-413.

Runkel, R.L., 1993, *Development and application of an equilibrium-based simulation model for reactive transport in small streams*, Ph.D. dissertation, Dept. of Civil, Environmental, and Architectural Engineering, University of Colorado, Boulder, 202 p.

Runkel, R.L. and R.E. Broshears, 1991, *One dimensional transport with inflow and storage (OTIS): A solute transport model for small streams*, Technical Report 91-01, Center for Advanced Decision Support for Water and Environmental Systems, University of Colorado, Boulder, 85 p.

Smith, K.S., 1991, *Factors influencing metal sorption onto iron-rich sediment in acid-mine drainage, Ph.D. dissertation*, Department of Chemistry and Geochemistry, Colorado School of Mines, Golden, 239 p.

CONSTRUCTED WETLAND TREATMENT SYSTEMS APPLIED RESEARCH PROGRAM AT THE ELECTRIC POWER RESEARCH INSTITUTE

J.W. Goodrich-Mahoney

Electric Power Research Institute, 3412 Hillview Avenue, Palo Alto, California 94303

Abstract. The Electric Power Research Institute's (EPRI) Environmental and Health Sciences Business Unit, within the Environment Group, has initiated a multi-disciplinary applied research program to develop constructed wetland treatment systems as a cost-effective technology for the treatment of metal-bearing electric utility aqueous discharges. EPRI's program involves the building of constructed wetland treatment systems, collection of field data from these systems, and conduct of controlled laboratory experiments to more fully understand their functions and the factors affecting these functions. Both data collected through this program and existing data will be used to develop and deliver design criteria for the effective use of this technology to reduce or eliminate the risk that electric utilities will not meet regulatory-imposed effluent discharge limits. Currently, EPRI along with one of its members, is funding the construction of a state-of-the-art constructed wetland treatment system to treat a discharge from a closed dry ash management facility in Pennsylvania. This constructed wetland treatment system, along with existing ones located in California and Tennessee, will be used to collect data on the cycling of trace metals. Controlled laboratory experiments are underway to develop trace element uptake curves for wetland plant species, to determine which plant species one should plant in a wetland in a particular geographic area to maximize trace element removal. Plants also will be identified that are high volatilizers for selenium, arsenic and lead. As part of this research, the best plant/microbe associations (i.e., best plant species with the best microbe species) will be identified for achieving the highest rates of trace metal removal. Once this work is completed, these plants will be introduced into the three constructed wetland treatment systems mentioned above and the wetlands will be monitored to determine if any improvement in trace metal uptake is occurring. Additionally, EPRI is coordinating its research program with the Tennessee Valley Authority's (TVA) program, where design criteria for manganese rock drains for the removal of manganese and successive alkalinity-producing constructed wetlands for the treatment of acidic aerobic discharges will be developed. TVA is also supporting the development of EPRI's Wetland Environmental and Management (WEM) Model and is conducting research on anoxic limestone drains.

Keywords. Wastewater treatment, biogeochemistry, water chemistry and constructed wetland treatment systems.

1. Factors Necessitating Research

The objectives of the Clean Water Act (CWA) are the restoration and maintenance of the physical, chemical and biological integrity of the waters of the United States through the elimination of pollutant discharges. In 1996, the CWA may be re-authorized and the objectives of the Act strengthened through the addition of various requirements, including new ones for point source discharges. The Act may also begin to address the more difficult problem of controlling non-point source discharges (i.e., stormwater discharges), by adding enforcement provisions to guidance currently under development by the United States Environmental Protection Agency (EPA) under the authority of the Coastal Zone Act Reauthorization Amendments of 1990. Additionally, under current authority, states are beginning to impose, at the time of permit renewal, point source discharge effluent limitations that are based on water quality criteria that can be very stringent.

Critically, the Federal government's wetland policy appears to be moving in the direction of a "no net loss of wetlands" standard, as articulated by President Bush. While there is some uncertainty as to what such a standard means, this new policy has resulted in a heightened awareness of wetland issues both on the part of the Federal government and the general public. The EPA already has initiated a research program that will develop information to provide a better understanding of the values and functions of natural wetlands and constructed wetland treatment systems.

The electric utility industry has an opportunity to benefit from this research by building upon and augmenting it in areas that are of particular importance to the industry. In as much as the generation of electric power by steam requires cooling water from lakes, rivers, bays or the ocean, it is probably not an understatement to say that wetland use and electric utilities go hand-in-hand. Many electric utilities have successfully used wetlands for their electric generating needs (e.g., through pond construction and water level manipulation) and, at the same time, have gained the additional benefits of community good will through their positive stewardship of these areas (Benjamin, 1991).

For all of the foregoing reasons, EPRI believes that the timing is critical for constructed wetland treatment system research and that the policy, scientific and technical issues relating to these systems will become increasingly important for the electric utility industry in the foreseeable future.

2. Constructed Wetland Treatment Systems

2.1. OVERVIEW

Constructed wetland treatment systems are conceptually simple systems that allow for the establishment of a wide variety of control mechanisms to retain or degrade contaminants in wastewater. Typically, in treatment cells oxygen is translocated into flooded soils by marsh plants to sustain root respiration, creating localized areas of aerobic metabolism and associated chemical reactions near the rhizosphere. Simultaneously, organic matter accumulations derived either from wastewater input or from in situ organic production create zones of anaerobic metabolism and associated chemical reactions (including sulfate reduction and formation of insoluble metal sulfides). These biogeochemical processes allow different types of control mechanisms to operate simultaneously down a long treatment flow path, and are unique to both natural and constructed wetland treatment systems. The capacity of many wetlands to reduce the levels of BOD, nutrients and metal concentrations and other pollutants in water passing through them has been well documented (Kadlec, 1979, Nixon, 1986 and Moshiri, 1993).

Essentially, constructed wetland treatment systems are high energy throughput ecosystems where the energy derived from gravity flow of varying intensities, in situ primary production and degradation of allochthonous organic carbon, provides for diverse, high-intensity biogeochemical reactions that act to retain or degrade toxic substances present in wastewaters. As a result, contaminants of differing chemical and physical characteristics are physically removed, adsorbed to surfaces or chemically transformed and stored within the wetland matrix. In general, the integration of many different control processes in constructed wetland treatment systems is desirable because it is imprudent to rely on any single biogeochemical mechanism for the control of toxic substances in wastewaters.

To date, constructed wetland treatment systems have been built to treat municipal and home wastewater, acid mine drainage, landfill and industrial wastewater and non-point source pollution. At this time, there are more than 200 municipal and stormwater constructed wetland treatment systems (Reed, 1990) and more than 140 acid mine drainage systems in the eastern U.S. (Wieder, 1989). Within the electric utility industry there have been some successes in the use of these systems to treat acidic drainage with its associated dissolved metals (most notably 10 of the 13 systems built by the Tennessee Valley Authority), there also have been a number of failures (Brodie, 1991). Most of these systems have been constructed since 1984, an indication of the rapid rise of interest in this technology.

3. Outline of EPRI's Applied Research Program

3.1 SCOPE AND OBJECTIVES OF RESEARCH

EPRI has designed a multi-disciplinary research approach to develop information on wetland functions and factors affecting these functions, for application in developing and delivering design criteria for the effective use of this technology to reduce or eliminate the risk that electric utilities will not meet regulatory-imposed effluent discharge limits.

The scope of the research entails information development in the following subject areas.

- Geochemistry: Information will be gathered to determine and quantify the most important retention and transformation processes, including their rates, seasonality, governing factors, etc. for the trace elements listed in Table 1.
- Microbiology: Since wetlands are living systems, information will be gathered to determine and quantify the most important retention and transformation processes driven by bacterial and plant activity, including their rates, seasonality, governing factors, etc.
- Hydrology: Hydraulic residence time is very important in terms of efficiency of contaminant removal. Information will be developed to determine and quantify those factors that affect hydraulic residence time, such as clogging, channelizing and others that develop as the system ages.
- Engineering: The environmental process knowledge gained in the aforementioned areas will be used to develop design criteria for the construction and operation of constructed wetland treatment systems (improved designs may need to be field tested).

The objectives of this research are as follows.

- To gather existing information and collect new information (cited above) on constructed wetland treatment systems through laboratory and field studies, in order to better understand the functioning of these systems.
- To evaluate this information to determine the opportunities for and the effectiveness of using constructed wetland treatment systems for the treatment of discharges from electric utility facilities.
- To develop and deliver data and design criteria for the effective use of constructed wetland treatment systems for the treatment of discharges from electric utility facilities.

3.2 EPRI'S APPLIED RESEARCH APPROACH

The applied research is being carried out in three phases.

- Phase One - Evaluation Phase
  Phase One is a fact finding and evaluative phase, which will encompass a review of the published literature. It also involves an effort to collect data from electric utilities that have operational constructed wetland treatment systems (e.g., Tennessee Valley Authority (TVA) and Allegheny Power System (APS)). This work will emphasize the potential for the effectiveness of constructed wetland treatment systems to treat a variety of electric utility discharges, if properly designed and constructed. Additionally, this work will identify data gaps that will be addressed in Phase Two.

- #### Phase Two - New Information Development Phase
  Phase Two is a monitoring phase, which will be based on the work and recommendations from Phase One. This will include further characterization of electric utility facility discharges and the development and conduct of laboratory and field studies. This phase will close data gaps by providing discharge characterization data, data on wetland systems functions and data on the effect of electric utility operations on these systems.

- #### Phase Three - Integration/Synthesis Phase
  Phase Three is a data analysis and evaluation phase, which will synthesize the data on the potential for the use of constructed wetland treatment systems for the treatment of a variety of electric utility facility discharges and integrate this data into the development of design criteria. An important conclusionary step in Phase Three will be the testing of the design criteria through field studies.

4. Status of EPRI's Applied Research Program

Phase One - Evaluation Phase
EPRI has completed this phase and has published the results in "Constructed Wetland Treatment Systems for the Remediation of Metal-Bearing Aqueous Discharges" (EPRI, 1995).

Phase Two - New Information Development Phase
This phase started in 1994 and will continue through 1998. Following is a brief description of each of the applied research projects.

4.1 EPRI FUNDED PROJECTS

"Use of Constructed Wetland Treatment Systems for the Removal of Toxic Trace Elements from Electric Utility Wastewaters: Role of Vegetation," RP 4163. The project is under the direction of Dr. Norman Terry, Department of Plant Biology, University of California, Berkeley, California. EPRI's project manager is Mr. John W. Goodrich-Mahoney.

Overview
Wetland plants play an important role in the trace elements removal process. It is not known, however, which wetland plant species absorb trace elements at the fastest rates. Such knowledge is essential to the proper design of constructed wetlands. Wetland plants remove pollutants through biological processes, especially uptake and immobilization in plant tissues. In the case of the volatile elements (e.g., selenium, arsenic), an additional and potentially very important pathway of removal is through biological volatilization. Recent research has shown that removal of selenium by biological volatilization from wetlands is substantial. The use of biological volatilization as a method of metalloid trace

element removal may possibly be applied to other trace elements in addition to selenium. Other elements that are known to be volatilized include arsenic, lead, mercury, and tin.

Microbes also are known to play an important role in trace element removal processes in wetlands. Microbes facilitate the removal of toxic trace elements from wastewaters by biosorption, sulfide-precipitation and biotransformation. By combining plants with microbes, the efficiency of trace element removal can be increased greatly. Plants sustain large microbial populations in their rhizosphere by secreting substances such as mucigel, carbohydrates and amino acids through root cells and by sloughing root epidermal cells. In wetlands, plants also provide oxygen to roots, and the immediate root environment, so that certain aerobic microbes can survive around the roots. These microbes mediate specific chemical reactions, which are important to trace element removal. Unfortunately, very little is known about the nature of the interaction between rhizosphere microbes and their host plant. Thus, this research will identify which species of bacteria are found in association with wetland plants that are important in the removal or detoxification of toxic metals and metalloids, and will examine the mechanisms of interaction between plants and microbes by which trace element removal can be facilitated.

Research Goals
The goal of this project is to answer the following questions. Which wetland plant species remove specific toxic trace elements (see Table below) most efficiently under wetland conditions? What is the fate of toxic trace elements entering wetlands in terms of their movement to sediments, waters, plant tissues and the atmosphere? To what extent can biological volatilization be used as a remediation process for the toxic elements selenium, arsenic and lead? To what extent can the removal of selenium and other trace elements be enhanced by manipulation of plant/microbe associations?

Table 1
Trace elements to be studied

Trace element	Symbol	Trace element	Symbol
Aluminum	Al	Manganese	Mn
Antimony	Sb	Mercury	Hg
Arsenic	As	Molybdenum	Mo
Barium	Ba	Nickel	Ni
Beryllium	Be	Selenium	Se
Boron	B	Silver	Ag
Cadmium	Cd	Strontium	Sr
Chromium^{+6}	Cr	Sulfur & Sulfate	S & SO$_4$
Cobalt	Co	Thallium	Tl
Copper	Cu	Tin	Sn
Fluorine	F	Titanium	Ti
Iron: Total & Dis.	Fe	Vanadium	V
Lead	Pb	Zinc	Zn

Experimental Approach

The experimental approach of this project is a multidisciplinary effort, which will include ecological studies conducted in three different wetlands, screening studies to determine trace element uptake characteristics of different wetland plant species grown in growth chambers, as well as field and laboratory microbiological studies. The specific tasks are as follows:

- Task 1: To establish trace element uptake curves for different wetland plant species under controlled environmental conditions. The collected data can be used to determine which plant species one should plant in wetlands in specific locations to maximize trace element removal into vegetation;

- Task 2: To monitor how each trace element cycles through specific constructed wetlands (i.e., those owned by Chevron Oil Company, California, Allegheny Power Service, Pennsylvania and the Tennessee Valley Authority, Tennessee) to determine the extent to which each trace element is retained in the sediments or waters versus the proportion removed by the plants into roots and shoots;

- Task 3: To determine the extent to which biological volatilization of Se, As and Pb can be used as a technique for constructed wetland bioremediation;

- Task 4: To establish new methods that can be used to enhance the rate of removal of trace elements from wastewaters through manipulation of plant/microbe interactions; and

- Task 5: To carry out pilot studies in specific constructed wetlands to determine how much of an enhancement in trace element removal can be obtained by 1) implementing improvements in plant species composition and 2) utilizing more efficient plant/microbe associations, using knowledge developed from this research.

The knowledge gained from this project will be used to develop and test new methods for improving wastewater cleanup in the three constructed wetlands identified in Task 2.

"Allegheny Power System - Springdale Constructed Wetland Treatment System," RP 9065. EPRI's project manager is Mr. John W. Goodrich-Mahoney. Project design is under the direction of Mr. Terry Rightnour, EES Consultants, Inc., Hyde, Pennsylvania. Construction management is under the direction of Mr. John Cenkner, Allegheny Power System, Greensburg, Pennsylvania and project construction is under the direction of Mr. Mike Koestler, R and L Development Company, New Alexandria, Pennsylvania.

Overview

This project involves the design and construction of a state-of-the-art wetland treatment system to treat a metal-bearing discharge from a closed dry fly ash management facility in Pennsylvania. The project is a joint APS/EPRI Tailored Collaboration applied research project. Construction of a pond for oxidizing ferrous iron was completed in 1994 and has brought the drainage into compliance at < 2.6 mg/L. Construction of eight wetland treatment cells for the removal of trace metals was completed during the summer of 1995.

The wetland cells consist of four aerobic cells; two rock drains (Mn removal); one anaerobic sulfate-reducing cell; and one algal subsurface sand filter. An additional four research cells, two of which are covered, have been constructed for use as part of the University of California, Berkeley work described above and other research projects to be defined.

Site: Springdale Power Station (Springdale, Pennsylvania)
Size: 1.2 ha (3 acre)
Flow: 120 L/min (30 gal/min)
Type of Influent: Seepage from a closed ash disposal site
Design Elements: One pond for iron removal; four aerobic cells; two aerobic rock
                            drains (Mn removal); one anaerobic sulfate-reducing cell; and one
                            algal subsurface sand filter; soil and spent mushroom compost;
                            limestone and sand; and cattails and wetland grasses.

Untreated
Discharge Data: pH 6.61 std. units, TSS 19.98 mg/L, Al 0.479 mg/L, Ag
                            n.a.,Be 0.002., B 1.4.23 mg/L, Cd 0.0023 mg/L, Fe dissolved
                            n.a., Fe total 14.409 mg/L, Mn 2.841 mg/L, Hg n.a., Mo
                            0.220 mg/L, Sn n.a., Ti n.a., Tl 0.002 mg/L and Zn 0.049
                            mg/L. Values are an average of four sampling events during
                            March, April, June and August 1995.

Treated
Discharge Data: pH 7.09 std. units, TSS 2.0 mg/L, Al <0.005 mg/L, Ag
                            n.a.,Be <0.001., B 17.01 mg/L, Cd <0.0006 mg/L, Fe
                            dissolved n.a., Fe total 0.337 mg/L, Mn 2.598 mg/L, Hg n.a.,
                            Mo 0.077 mg/L, Sn n.a., Ti n.a., Tl 0.003 mg/L and Zn 0.03
                            mg/L. Value are from a single sampling event in August 1995,
                            seven days after start up of the constructed wetland. Note that in
                            the fall of 1995 a more detailed sampling program will
                            commence.

NPDES Limits: pH 6-9 std. units[*], TSS 100 mg/L[*], Al 0.5 mg/L, Ag 0.0002
                            mg/L,Be n.d., B 1.3 mg/L, Cd 0.002 mg/L, Fe dissolved 7.0
                            mg/L[*], Fe total 2.6 mg/L, Mn 1.7 mg/L, Hg n.d., Mo 0.014
                            mg/L, Sn n.d., Ti 0.11 mg/L, Tl 0.024 mg/L and Zn 0.09
                            mg/L

---

n.a. Not available.
* Current NPDES limits; others are proposed.
n.d. Not detectable using requisite test method.
All values listed are monthly average limits.

Research Goals
EPRI plans on building several constructed wetland treatment systems across the United States to aid in developing design criteria for the treatment of a variety of electric utility wastewaters (discharges). By building these systems at a variety of locations, there will be coverage of a full spectrum of site conditions (i.e., power plant and coal/ash characteristics, climate, soil and plant species) that can influence treatment efficiencies. The focus of the research will be on the development of design criteria for the treatment of trace metals; the reduction of land requirements for the treatment systems and the development of design criteria based on wetland volume instead of area, which is the current standard.

Experimental Approach
Build lined systems to assure good hydrologic control and monitoring these systems to provide data for the development of design criteria. Monitoring devices include weirs, mullet-level sampling ports, and rock baskets in the rock drain cells. Trace metals to be analyzed in water are the same as given in Table 1 above, with additional measurements for pH, temperature, specific conductivity, alkalinity, acidity, total dissolved and suspended solids, suspended solids and dissolved oxygen. Additionally, organic and metal oxides will be analyzed in the substrates.

"Development of the Wetlands Environmental and Management (WEM) Model," PR 3561. EPRI's project manager is Dr. Donald Porcella. Model development is under the direction of Mr. Monsoon, Tetra Tech, Pittsburgh, Pennsylvania and data collection and analysis is under the direction of Dr. Frank Sikora at TVA's Constructed Wetland Research Facility, Muscle Shoals, Alabama.

Overview
This EPRI/TVA Tailored Collaboration funded research project is underway at both natural and constructed wetland treatment systems in Tennessee, and is providing data on influent/effluent constituent concentrations, including accumulation of constituents in wetland plants and soils. This, and other information, is being used for the development of the Wetlands Environment and Management (WEM) Model.

Research Goals
To develop the Wetlands Environmental and Management (WEM) Model to aid electric utilities in evaluating options for wetland trade-offs, designing new (natural and constructed wetland treatment systems) or restored natural wetlands and obtaining permits for activities affecting natural wetlands. The completed mathematical model will simulate hydraulic and chemical behavior of wetlands and vegetation effects.

Experimental Approach
Perform topographic surveys, hydrologic and water quality measurements, sediment characterizations, vegetation surveys and bioaccumulation studies at natural and constructed wetland treatment system to support model development.

"Manganese and Trace Metal Removal in Successive Anaerobic and Aerobic Wetland Environments," RP 3561. EPRI's project manager is Dr. Donald Porcella. System design, data collection and analysis is under the direction of Dr. Frank Sikora at TVA's Constructed Wetland Research Facility, Muscle Shoals, Alabama.

Overview
This EPRI/TVA Tailored Collaboration funded research project is underway at TVA's Constructed Wetland Research Facility. This research is providing data on manganese and trace metal removal rates in successive anaerobic and aerobic wetland treatment systems for the treatment of acidic aerobic discharges from such sources as coal mines, coal pile rejects and coal piles.

Research Goals
To develop design criteria for successive alkalinity producing systems for the treatment of acidic aerobic discharges.

Experimental Approach
The research is being conducted in cells (i.e., lined insulated cattle feeding troughs), which are designed to simulate anaerobic and aerobic wetlands. There are three cells in series, with the first cell constructed as an anaerobic cell (composted chicken litter underlain by crushed limestone gravel) and the remaining two constructed as aerobic cells. The troughs were plumbed with PVC pipe so water entered the surface of the first cell, exited at the bottom of the first cell, entered the second cell at the bottom, exited the surface of the second cell, entered the surface of the third cell and exited on the bottom of the third cell. The two aerobic treatments consisted of reciprocating the top 20 cm of water from one cell to another or not reciprocating. Influent and effluent data are collected for each cell. The anaerobic cell received a simulated acid mine drainage (20 ml/min), with the constituents Mn, Fe(II), Ni, Cu, Pb, Zn, Ca, Mg and S at predetermined concentrations. This flow rate results in loading rates of 0.82 and 0.12 $g/m^2/d$ for Mn and Fe, respectively, and 0.08 $g/m^2/d$ for the trace metals Cu, Ni, Pb and Zn. Initial data after four months of operation indicate that the anaerobic troughs are effective in reducing SO_4^{-2} to S^{-2} and producing alkalinity in the range from 80 to 300 mg/L. Production of alkalinity decreased with time and coincided with seasonal decline in water temperature. Manganese removal in the anaerobic systems decreased with time, which may have been due to solubilization of $MnCO_3$, as a result of decreasing alkalinity. Manganese removal in the reciprocating aerobic cells was quicker than in the nonreciprocating aerobic cells. Removal in the aerated systems appeared to be due to precipitation of Mn oxides. Removal of Cu, Ni, Zn and Pb was very effective in the anaerobic cells with organic matter and was presumed to be due to precipitation of metal sulfides.

4.2 TVA FUNDED PROJECTS

The following two research projects are funded solely by TVA, but are coordinated with EPRI's applied research program.

"Mn Removal in Saturated Gravel Beds: Obtaining Operating Criteria." This project is under the direction of Dr. Frank Sikora at TVA's Constructed Wetland Research Facility, Muscle Shoals, Alabama.

Overview

Manganese is difficult to remove in constructed wetland treatment systems due to the high pH requirement for Mn oxidation and the solubilization of MnO_2 in the presence of Fe^{+2}. To achieve adequate Mn removal to concentrations below the federal requirement of 2 mg/L, gravel bed systems placed after the wetland treatment system have been proposed. Efficient Mn removal occurs in the gravel beds because most of the iron has been removed in the wastewater in the wetland treatment system preceding the gravel bed. Although some data are available on gravel beds, not enough is available to recommend operating parameters for these systems.

Research Goals

To develop design criteria for adequate Mn removal at varying Mn and Fe^{+2} loadings.

Experimental Approach

The outdoor wetland cells at the TVA research facility will be used to study Mn removal in rock bed filters. The eight cells have a surface dimension of 6.1 x 9.1 m^2 with a depth of 0.6 m. The cells are filled with river gravel. Each cell has 9 wells consisting of slotted 4" PVC pipes placed in an equally spaced grid pattern. Mn loading rates of 0.3, 0.8 and 2.0 $g/m^2/d$ will be used. An additional treatment will consist of cells planted with a common wetland species, with a Mn loading rate of 0.8 $g/m^2/d$. At six months into the experiment, Fe^{+2} will be added at a concentration of 3.0 mg/L for six months, and at 18 months Fe^{+2} will be added at a concentration of 10 mg/L for six months. Batch studies also will be conducted in the second and third 6 month periods by exposing Mn coated gravel to various concentrations of $FeSO_4$ to determine the kinetics and extent of MnO_2 solubilization in the presence of Fe^{+2}. Hydraulic loading rate to each cell will be 10 cm/d. This is equivalent to a flow rate of 3.8 L/min and a whole cell retention time of 2.3 d. For comparison, hydraulic loading rates of 25 constructed wetland systems for treating acid mine drainage in Pennsylvania and Tennessee range from 0.9 to 27 cm/d, with an average of 9 cm/d and a median of 6.7 cm/d (11 and 12).

Influent and effluent concentrations will be monitored biweekly, including composites of the wells (3 each) to determine in cell efficiencies. Samples will be analyzed for Mn^{+2}, Fe^{+2}, Fe^{+3}, pH and alkalinity.

"Abatement of Plugging in TVA Anoxic Limestone Drains." This project is under the direction of Dr. Frank Sikora at TVA's Constructed Wetland Research Facility, Muscle Shoals, Alabama.

Overview
Anoxic limestone drains (ALDs) are passive treatment systems used to raise alkalinity in discharges prior to treatment with constructed wetland treatment systems. The effectiveness of ALDs can be reduced if discharges have high concentrations of aluminum, which precipitates as aluminum hydroxide.

Research Goals
To determine the feasibility of using fluorite (CaF_2) to prevent the plugging of ALDs from the precipitation of aluminum hydroxide.

Experimental Approach
Small laboratory experiments will be conducted to determine the kinetics of gravel fluorite dissolution. These experiments will be followed by bench-scale investigations to evaluate the effectiveness of fluorite in keeping aluminum in solution when the solution is exposed to gravel-sized crushed limestone typical of an ALD. If the results are supporting, a pilot-scale test will be conducted at one of TVA's acidic discharge sites.

Phase Three - Integration/Synthesis Phase
If all goes according to schedule, this phase of EPRI's research will commence in 1998 and end in 1999, with the development of design criteria for the construction of wetland treatment systems to treat a variety of electric utility discharges. As previously stated, an important conclusionary step in Phase Three will be the testing of the design criteria through field studies.

5. Conclusion

Through a combination of base funding and additonal support from EPRI members, through Tailored Collabration, EPRI is undertaken and coordinating an extensive constructed wetland treatment systems applied research program to met current and future treatment needs of the electric utility industry. As the industry moves into a more competitive market environment, the utilization of passive cost effective technologies, such as constructed wetlands, will become more attractive

References

Benjamin, S.: 1991, *Longtime Friends*, Electric Perspectives,Edison Electric Institute. 12.

Brodie, G.A.: 1991, *Achieving Compliance with Staged, Aerobic Constructed Wetlands*, In: Proceedings, Annual Meeting of the ASSMR, Durango, CO. 151-174.

Electric Power Research Institute: 1995,*Constructed Wetland Treatment Systems for Remediation of Metal-Bearing Aqueous Discharges*. TR-105487.

Kadlec, R.H. and Kadlec, J.A.: 1979,*Wetlands and Water Quality*, In: "Wetland Functions and Values: The State of Our Understanding", Greeson, P.E., Clark, J.R., and Clark, J.E., eds., American Water Resources Association. 436-456.

Moshiri, G.A.: 1993, *Constructed Wetland for Water Quality Improvement*, Lewis Publishers.

Nixon, S.W., and Lee, V.: 1986, *Wetland and Water Quality: A Regional View of Recent Research in the United States on the Role of Freshwater and Saltwater Wetlands as Sources, Sinks, and Transformers of Nitrogen, Phosphorus, and Various Heavy Metals*, Technical Report Y-86-2, U.S. Army Corps of Engineers, Vicksburg, Mississippi. 229.

Reed, S.C., Kadlec, R.H., and Knight, R.L.: 1990, *Wetlands for Wastewater Treatment: Database Summary*, Proposal submitted to US EPA, Office of Research and Development, Wetlands Research Program.

Wieder, R.K.: 1989, *A Survey of Constructed Wetlands for Acid Coal MineDrainage Treatment in the Eastern United States*, Wetlands. **9(2)**, 299-315.

Toxicity Reduction of Ontario Hydro Radioactive Liquid Waste

D. W. Rodgers, D.W. Evans and L. Vereecken Sheehan

Ontario Hydro Technologies, 800 Kipling Ave., Toronto, Ontario, CANADA M8Z 5S4

Abstract. The radioactive liquid waste (RLW) system in Ontario Hydro's pressurised heavy water reactors collects drainage from a variety of sources ranging from floor drains to laundry waste. RLW effluent was intermittently toxic to rainbow trout and *Daphnia magna* during the first phase of Ontario's Municipal Industrial Strategy for Abatement (MISA) Program, apparently as a result of the interaction of a variety of known and unknown organic and inorganic compounds. Accordingly, we employed a treatment-based approach to reducing its toxicity, supplemented by chemical analysis. Two series of toxicity reduction tests were conducted. The first series explored the potential for sorption of the possible toxicants, while the second series incorporated a wider variety of treatments. Of the 24 samples in the first test series, 17 were toxic (*D. magna* mortality ≥ 50%). Of the toxic samples, only 7 of 17 were still toxic after passage through an activated carbon column, but 5 of 6 samples tested remained toxic after passage through a metal chelating resin column. In the second series, at least one of the treatments was effective in reducing toxicity of all samples which were initially toxic (16 of 24 samples), but no one treatment was effective for all toxic samples. Three treatments (UV/H_2O_2 photo-oxidation with prior pH adjustment, or passage through a column of either a non-functionalized (N-F) resin or a mixture of N-F resin and a weak base (W-B) anion exchange resin), were effective in reducing the toxicity of more than 50% of the toxic samples; yet roughly 25% of these samples remained toxic after treatment. O_2 sparging, UV/H_2O_2 photo-oxidation without prior pH adjustment, and passage through a column of the W-B Resin were less effective, as more than 50% of the samples remained toxic after treatment. Filtering was not effective, as all of the treated samples (9/9) retained their toxicity. There was no obvious correspondence between toxicity and the concentrations of metals (Cu, Zn, Fe, Al and Cd) nor were any simple relationships apparent between toxicity and Total Organic Carbon or NH_3 concentrations. At stations where radioactive liquid wastes are segregated, toxicity was also segregated, suggesting that we may be able to address the problem at source through a combination of Best Management Practices and smaller scale treatment facilities.

Key Words: toxicity reduction, sorption, UV photo-oxidation, *Daphnia magna*, radioactive liquid waste

1. Introduction

Ontario Hydro's pressurised heavy water power reactors (PHWR) differ from the pressurised light water reactors (PWR) prevalent elsewhere in a number of features that are relevant to discharge water quality. Fission product activity in Hydro's PHWR systems is generally lower than in PWR designs, because of the capability for on-power fuelling, facilitating prompt removal of failed fuel elements. Tritium (^{3}H), which is formed by neutron activation of the deuterium oxide (^{2}H$_2$O) moderator and coolant, is a particular problem for PHWRs and is frequently the limiting radionuclide with respect to station emissions.

The radioactive liquid waste (RLW) systems at Ontario Hydro's nuclear generating stations are designed to collect, segregate, monitor, process (as necessary) and discharge the water from numerous drains within the radioactive zones of the plant. Contributing streams include drainage from fuelling machine and mechanical maintenance shops,

Water, Air and Soil Pollution 90: 219–229, 1996.
© 1996 *Kluwer Academic Publishers.*

laboratory and laundry wastes, process equipment drainage and light water wastes from the on-site heavy water upgrading units. Details of the RLW systems at Hydro's five nuclear stations vary, but typically consist of several (5 to 12) large volume tanks, segregated for collection of low (<37 kBq/L) and high (>37 kBq/L) activity water. Individual tank volumes are about 120 m^3 and the total RLW collection capacity ranges from 600 to 1700 m^3 per station. The systems currently incorporate filtration and ion exchange for activity control but these have been rarely required since equalization of the tank contents achieves discharge activity compliance. As ^3H is not removed by the membrane and evaporation processes used for RLW management in other nuclear reactor designs, these systems have not been incorporated into Ontario Hydro's nuclear generating stations. Effluent from the RLW system is discharged to the combined station condenser cooling water outfall duct, with on-line activity monitoring. Grab samples of the individual tanks are also analyzed for radioactivity prior to any pump-outs. The combined condenser cooling water flows at Hydro's multi-unit stations are about 120-170 m^3s^{-1}, yielding dilution factors of at least 10^4 during RLW tank discharges.

Historically, the RLW discharge restrictions were solely based on radionuclide limits established by the Canadian Atomic Energy Control Board. These derived emission limits (DEL's) are determined by pathway analysis (air, water, direct irradiation) to limit exposure of identified critical groups of the public, such that these members of the public do not receive radiation doses in excess of those recommended by the International Commission on Radiological Protection. Ontario Hydro uses a much more restrictive administrative limit, requiring that nuclear station radionuclide emissions not exceed 1% of the DEL. The DEL's are site-specific and range from 7 - 44 x 10^7 GBq per month of ^3H (as tritiated water) and from 8 x 10^2 to 1.1 x 10^5 GBq per month of gross β,γ activity for aqueous discharge pathways. Actual radioactivity emissions via the RLW system are typically <0.1% of the DEL for ^3H and gross β,γ.

Prior to 1989 the RLW effluent was not regulated by, or monitored for, conventional water quality parameters, except for pH. Under the Municipal and Industrial Strategy for Abatement (MISA) program of the Ontario Ministry of Environment and Energy (OMEE), Ontario Hydro undertook a one year (June 1990-91) preliminary monitoring program of aqueous effluents from its nuclear, fossil and hydroelectric generating facilities (Ontario Hydro, 1992). Hydro is now engaged in the second phase of the MISA program, which includes a three year program to monitor and control those effluents identified as toxic. Toxicity testing is integral to the MISA program, with current regulations (OMEE, 1995) requiring monthly or quarterly sampling of specific effluent streams using acute lethality toxicity tests with water fleas, *Daphnia magna*, and rainbow trout, *Oncorhynchus mykiss*, followed by chronic toxicity testing using fathead minnow, *Pimephales promelas*, and a second cladoceran zooplankton, *Ceriodaphnia dubia*. As slightly more than 25% of the samples of RLW effluent were toxic to *D. magna* and/or rainbow trout (Ontario Hydro, 1992), during the initial year of MISA monitoring, continued monitoring is required (OMEE, 1995), with the objective of rendering it non-toxic by 1998.

Despite detailed chemical measurements conducted during the MISA monitoring program, no single or combination of inorganic or organic compounds was consistently detected at concentrations which would be expected to cause toxicity problems. Rather the RLW stream was intermittently toxic, apparently as a result of the interaction of a variety of known and unknown organic and possibly inorganic toxicants. Although the

US EPA and others have developed detailed procedures for Toxicity Identification Evaluations (TIE) (Mount and Anderson-Carnahan, 1988), we felt that a full TIE was not appropriate to our RLW effluents, where the nature of the contamination was expected to vary with time and from site to site, and where any treatment must be technically feasible. Rather, we employed a treatment-based approach to reducing its toxicity, supplemented by chemical analysis. Two series of toxicity reduction tests were conducted. The first (1994) series explored the potential for sorption of the possible toxicants, while the second (1995) series incorporated a wider variety of treatments.

2. Test Procedures

As *D. magna* was more sensitive than rainbow trout to the RLW stream during the initial MISA tests (Rodgers *et al.*, 1996), and as a much smaller volume of effluent is required for the *D. magna* test, all tests were conducted using *D. magna* only. Over two eight-week periods in 1994 (Series 1) and 1995 (Series 2), samples of RLW were collected weekly from Bruce Nuclear Generating Station (NGS) A, Bruce NGS B, Darlington NGS and Pickering NGS. Samples were collected using grab or pump samplers into two plastic pails ($\approx$ 10 L/each) lined with food grade polyethylene bags and transported to the test laboratory within 24 h of collection. The toxicity of the sample was determined by the *D. magna* acute toxicity test (Environment Canada, 1990) conducted with duplicate samples of 100% effluent. Samples were considered toxic to *D. magna* if more than 50% of the animals in each duplicate test died within the 48 h test. Throughout the tests, the response of *D. magna* to a standard toxicant (2-chlorophenol) was consistent. If the sample was toxic, it was then treated and the toxicity of the treated sample was then determined by a second *D. magna* acute toxicity test conducted with duplicate samples of treated effluent. The treatments and the rationale for their use are summarized below.

2.1 SERIES 1

Sorption to the appropriate substrates offered a potentially useful, generic method to remove a wide spectrum of compounds from RLW and thus reduce its toxicity. In this series of tests, we employed either granulated activated carbon (GAC) to selectively remove organic compounds or a metal selective chelating resin to selectively remove potentially toxic heavy and transition metal species, while limiting the depletion of desirable cations, normally present in natural waters. Depletion of desirable cations (ie. Ca^{++}) may lead to problems with deionized water toxicity similar to those observed with boiler blowdown effluent (Rodgers, 1994). Accordingly, toxic RLW samples were subject to the following treatments:

2.1.a. Sorption - activated charcoal. Samples ($\approx$ 2 L) were passed through a 600 mL packed bed (45 mm i.d. plastic column) of the activated carbon (Calgon Filtrasorb 400), under gravity. The initial 800 - 1000 mL of the treated effluent was discarded and the subsequent 1 L retained for testing. Because of the small number of samples treated and the large total sorption capacity available, the activated carbon was used throughout the treatment program.

2.1.b. Sorption - metal chelating resin. Samples ($\approx$ 1.2 L) were treated by passage through a 150 mL of a sodium form chelating resin (Rohm and Haas - Amberlite IRC-718) packed in a 25 mm i.d. glass column. The initial 150-200 mL of the treated effluent was discarded, and the subsequent 1 L sample retained for testing. Fresh resin was used for each sample.

2.2. SERIES 2

In these tests, toxic RLW samples were subject to the following treatments.

2.2.a. Filtration. Although the current RLW Management System design incorporates filters, these are generally coarse filters and have been rarely used. The extent of particulate loading and its potential contribution to RLW toxicity was virtually unknown. If fine filtration (circa 0.45 μm) were effective in de-toxifying the RLW effluents, it would be simple and cost-effective. Samples ($\approx$ 1 L) were initially filtered through a glass fibre pre-filter (Whatman 934-AH, 1.5 μm retention). This was followed by a 0.47 μm cellulose nitrate/acetate filter (Millipore HA type) to remove residual particulate. The filtrations were conducted under suction using all-glass filtration apparatus.

2.2.b. Oxygen sparging. Sparging with oxygen was evaluated to determine if purgeable or readily oxidizable compounds contributed to effluent toxicity. Oxygen for the sparging treatments was supplied by a pressure-swing adsorption oxygen concentrating device, developed for home supply of medical oxygen (AirSep Model AS-10, AirSep Corporation, Buffalo, N.Y.), which delivers 95% O_2 at flow rates of 1 to 3 L/minute. One litre samples of the toxic RLW effluents were sparged for one hour at an O_2 flow rate of 1.5 L/minute. Fritted glass gas dispersion tubes were used to ensure effective oxygenation of the water samples.

2.2.c. Sorption. In previous work on contaminants in heavy water liquid and vapour recovery streams at Ontario Hydro stations, numerous water soluble organics, such as methanol, ethanol, isopropyl alcohol, acetone and ethylene glycol were identified. GAC is known to be a relatively poor sorbent for many water soluble organics, and roughly 40% of the samples treated with GAC were still toxic in the first series of tests. Therefore, several novel organic resin based sorbents were evaluated for the second series of tests. On the basis of the manufacturer's test data, two of these new sorbents were selected. These are reported to have a wider range of pore sizes, while maintaining a high specific surface area, comparable to GAC ($\approx$ 1000 m^2/gram of dry sorbent) and also to have higher operating capacities than a number of current commercial materials. The first was a non-functionalized organic sorbent, while the second had a similar polymeric substrate, but was functionalized as a weak base anion exchanger (tertiary amine). For purposes of this paper the sorbents are designated as N-F Resin and W-B Resin, respectively.

The N-F and W-B Resins were pre-cleaned following manufacturer's recommendations. This entailed three washings with methanol to remove any leachable organics, followed by repeated rinses with deionized water to remove residual methanol. The sorbents were then refluxed in boiling water for 30 minutes. After cooling the supernatant water was decanted off and discarded and the sorbents were given two final rinses with water. The bulk water was decanted off and the sorbents were stored in tightly sealed plastic bottles until used.

Toxic RLW samples were treated by a single passage through a resin column. Disposable polypropylene low pressure liquid chromatography columns (Bio-rad Econo-Pac columns) were loaded with 20 mL of the appropriate sorbent. A flow rate of one bed volume per minute was established using a peristaltic feed pump and about one litre of sorbent-treated solution was collected for subsequent toxicity testing and chemical analysis. Fresh sorbent was used for each toxic effluent sample. All samples were treated individually with both the N-F Resin and W-B Resin. In light of the relative success of the sorbent treatment, for Weeks 6 through 8 the sorbent test matrix was extended to include a mixture of the two sorbents. Equal volumes of the N-F and W-B Resins were hand-mixed in the moist state and 20 mL columns of the inter-mixed resins were evaluated as a third sorbent treatment.

2.2.d UV/H₂O₂ Photo-oxidation.

2.2.d UV/H$_2$O$_2$ Photo-oxidation. UV/hydrogen peroxide (H$_2$O$_2$) photo-oxidation was adopted as a generic method for destruction of total organic carbon (TOC) and other readily oxidizable compounds such as hydrazine (N$_2$H$_4$), while minimizing the impact on other water quality parameters. Ontario Hydro's nuclear generating stations use hydrazine (20-200 µg/kg) extensively for oxygen scavenging in the steam cycle and at higher concentrations (up to 200 mg/kg), for corrosion protection in steam generator lay-up and in near-stagnant systems, such as the emergency coolant injection system. Although photo-oxidation may also affect other oxidizable species, such as lower oxidation state metallic ions (eg. $Fe^{2+} \rightarrow Fe^{3+}$; $Cr^{3+} \rightarrow Cr^{6+}$), the concentrations of dissolved iron and chromium in the RLW effluent is low (Ontario Hydro, 1992) and photo-oxidation of these metals should thus have little impact on toxicity.

The UV/hydrogen peroxide treatment is based on the photolytic generation of hydroxyl radicals:

$$H_2O_2 + h\nu \rightarrow 2 \; OH\cdot$$

The resulting hydroxyl radical, OH·, is a much more powerful oxidant than the parent hydrogen peroxide. Many, but not all, organic compounds may be fully degraded to CO_2, water and corresponding inorganic anions (Cl⁻, SO_4^{2-}, etc) by UV/H$_2$O$_2$ photo-oxidation. This process has been extensively used for the remediation of organically contaminated groundwaters and wastewaters and for purification of boiler make-up water (Tyldesley and Knowles, 1989; Bellamy *et al.*, 1991; Ollis *et al.*, 1991; Clayton, 1992). Commercial units are available from a number of vendors (eg. Solarchem Environmental Systems; Vulcan Peroxidation Systems). Several PHWR facilities currently use UV/hydrogen peroxide technology for the removal of TOC from recovered, downgraded heavy water, prior to upgrading by fractional distillation.

The UV/peroxide photo-oxidation treatments were performed in a standard glass and quartz photochemical system (Ace Glass Inc., Vineland, NJ). This consists of a water-cooled quartz immersion well that houses the UV source, surrounded by a concentric, cylindrical reaction vessel containing the sample. Batches (250 mL) of the RLW effluent waters were spiked with hydrogen peroxide to yield an initial concentration of 20 mg/L H_2O_2. The samples were then irradiated for 30 minutes, with magnetic stirring. Hydrogen peroxide concentrations were monitored after 30 minutes, to confirm that the peroxide concentration had decreased to less than 2 mg/L, as determined using colorimetric test strips (Merkoquant 10011; MDL = 0.5 mg H_2O_2/L). To ensure that residual hydrogen peroxide did not enhance *D. magna* mortality, irradiations were, if necessary, continued until residual peroxide levels were ≤2 mg/L. Acute toxicity tests with *D. magna* showed no mortality, in reference water (dechlorinated municipal water) dosed with 2 mg/L H_2O_2, suggesting that this 2 mg/L limit was reasonable.

The UV source was a 450 W (electric) medium pressure mercury lamp (Hanovia), with a nominal radiant power of 27 watts in the far UV (220-280 nm) and 29 watts in the mid UV (280-320 nm). The absorbance of hydrogen peroxide increases steadily towards shorter wavelength and only the far UV portion of the emission is expected to give significant photolytic production of hydroxyl radicals. The TOC destruction processes are not expected to be UV light-limited, because of the relatively high intensity and short optical path length (≈ 1 cm) in the irradiation apparatus.

From Weeks 6 through 8, an additional set of UV/H_2O_2 treatments was performed on samples acidified with H_2SO_4 to a pH of 2.5 to 4.0, prior to UV/H_2O_2 treatment. Acidification prior to UV/peroxide treatment provides a means of improving the photo-oxidative destruction of organics by removing the bicarbonate ion (HCO_3^-), which is known to effectively scavenge hydroxyl radicals (Glaze and Kang, 1988) After irradiation the pH of the samples was re-adjusted with sodium hydroxide to the initial value, prior to the *D. magna* acute lethality test.

2.3 CHEMICAL ANALYSIS

The pH, total hardness, conductivity and dissolved oxygen of the test waters were analyzed routinely during the tests following standard methods (APHA, 1989; Environment Canada, 1990). In addition, samples were taken and appropriately preserved for determination of TOC, NH_3 and total Kjeldahl nitrogen using standard methods (APHA, 1989). In Series 1, metal concentrations of toxic and treated samples were determined by ICP spectrophotometry (Cu, Zn, Fe and Al) and GFAAS (Cd). In addition, in weeks 7 and 8 of Series 1, we utilized a solid phase microextraction preparation technique, (using 7 μm polydimethylsiloxane and the 85 μm polyacrylate fibres for semi-volatile and volatile compounds, respectively) coupled with GC-MS in an attempt at qualitative and semi-quantitative characterization of the organic compounds in toxic RLW samples.

3. Results and Discussion

3.1 Series 1

Of the 24 samples in the first test series, 17 were toxic ($\geq$ 50% of the *D. magna* died) and 7 were not toxic (Table I). Of the toxic samples, only 7 of 17 were still toxic after passage through an activated carbon column but 5 of 6 samples were still toxic after passage through the metal chelating resin column. There was no obvious correspondence between toxicity and concentrations of the metals (Cu, Zn, Fe, Al and Cd) which occasionally exceeded Ontario Water Quality objectives in the first round of MISA testing (Ontario Hydro, 1992). In particular, there was no consistent pattern of change in metal concentrations, for pairs of samples where treatment reduced the toxicity of the effluent (range in untreated toxic samples Cu <0.005-0.4 mg/L; Zn <0.003-0.5 mg/L); nor was any consistent pattern apparent for those samples where treatment did not change the toxicity of the effluent. Similarly, although passage through an activated charcoal column reduced TOC, no simple relationship between TOC and toxicity was apparent. No semi-volatile compounds were detected using the 7 μm polydimethylsiloxane fibre micro-extraction coupled with GC-MS and only one volatile compound was detected using the 85 μm polyacrylate fibre micro-extraction coupled with GC-MS. Butanoic acid, ethenyl ester was tentatively identified (estimated concentration $\approx$ 20 μg/L), in the Week 7 sample from Darlington NGS and not detected in the sample after passage through activated carbon. It is unlikely this compound was responsible for the sample's toxicity, however, as the sample remained toxic after treatment, .

3.2 Series 2

In the second series of tests, 16 of 26 samples were toxic before treatment (Table II). Although at least one of the treatments was effective in reducing toxicity of all 16 toxic samples, no one treatment was effective for all samples. Three treatments (UV/H_2O_2 photo-oxidation following pH adjustment and passage through either the non-functionalized N-F Resin or the combined resins), were effective treatments in reducing the toxicity of more than 50% of the toxic samples; yet roughly 25% of these samples remained toxic after any of these treatments. O_2 sparging, UV/H_2O_2 photo-oxidation and passage through the functionalized W-B Resin were less effective as more than 50% of the samples remained toxic after treatments. Filtering was not effective, as all of the treated samples (9/9) retained their toxicity. There was no obvious correspondence between toxicity and the concentrations of NH_3, with no consistent pattern of change either in pairs of samples where treatment reduced the toxicity of the effluent (range in untreated toxic samples NH_3 <0.01-2.0 mg/L) or for those samples where treatment did not change the toxicity of the effluent. Similarly, although TOC was generally reduced by passage through a sorbent column (W-B resin, N-F resin, and both) and by UV/H_2O_2 photo-oxidation, no simple relationship between toxicity and either TOC or total N was apparent. In many cases, toxicity reduction was accompanied by a substantial decrease in TOC. In other cases, however, toxicity was largely unchanged following a substantial decrease in TOC, or conversely toxicity was reduced without an appreciable change in TOC. Although low ionic strength (conductivity <8

Table I: Toxicity[a] of Untreated and Treated RLW Effluent: Series 1.

	Untreated	Treated	
		Activated Charcoal	Metal Chelating Resin
Station A	7/7[b]	1/7[b]	3/3
Station B	3/4	2/3	1/1
Station C	3/6	2/3	0/1
Station D	4/7	2/4	1/1
All Stations	17/24	7/17	5/6

[a] Samples considered toxic if *Daphnia magna* mortality $\geq$ 50% in duplicate tests
[b] # of samples remaining toxic after treatment / total # of samples tested

milliSiemens/m) may have contributed to the toxicity of up to 30% of the samples, it was not the sole factor, as some of the treatments reduced toxicity of these samples without affecting ionic strength.

In further, discussions with the stations, we have also determined that hydrazine was present in some of these RLW samples at concentrations of 1 mg/L or greater and, at these concentrations, will persist for several days in the samples. As hydrazine is reported as acutely toxic to *D. pulex* at concentrations of 0.19 to 0.16 mg/L (Velte, 1984), it may well have contributed to the observed toxicity of the samples. Consequently, the effectiveness of the UV/H_2O_2 photo-oxidation in reducing toxicity may in part derive from destruction of hydrazine rather than TOC. Unfortunately, the concentrations of hydrazine were not determined in either the initial MISA tests or our present work, and further studies to determine the contribution of hydrazine to RLW toxicity are planned for the near future.

These tests thus confirm our earlier suspicions that RLW effluent is a complex mixture in which many, possibly all, of the toxic components are organic, but these components vary significantly among stations, among tanks within stations, and over time. In particular, there was no "smoking gun", and we were unable to identify specific compounds or classes of compounds which were responsible for the observed toxicity in the large majority of cases. Consequently, none of the treatments we have tried to date represents a "silver bullet" which will completely remove toxicity from RLW effluent.

In the second series of tests, however, at least one treatment was effective in reducing the toxicity of all the toxic RLW samples. Thus, we feel that a combination of

Table II: Toxicity[a] of Untreated and Treated RLW Effluent: Series 2.

Treatment	Station				
	A	B	C	D	All
Untreated	7/7[b]	3/6	1/6	5/7	16/26
Filtered	4/4	2/2	1/1	2/2	9/9
O_2 Sparged	4/6	2/3	1/1	2/4	9/14
W-B Resin	5/7	3/3	1/1	1/5	10/16
N-F Resin	1/7	1/3	0/1	3/5	5/16
Mixed Resins	1/3	1/1	0/1	0/2	2/7
UV/H_2O_2 photo-oxidation	1/7	3/3	1/1	3/5	8/16
pH adjustment + UV/H_2O_2 photo-oxidation	0/3	0/2	1/1	1/3	2/9

[a] Samples considered toxic if *Daphnia magna* mortality $\geq$ 50% in duplicate tests
[b] # samples remaining toxic after treatment / # samples tested

treatments, such as the combined resins and UV/H_2O_2 photo-oxidation following pH adjustment, would be effective in reducing the toxicity of RLW effluent in the large majority of cases. It is also possible that one or both of these treatments could be further refined so that it alone would be effective in reducing the toxicity of the large majority of RLW effluent. If only one treatment is selected, there will be a need to balance the relatively high initial capital costs of UV/H_2O_2 photo-oxidation units versus the continuing costs of waste disposal or resin regeneration if the sorption option is selected. In all cases, however, considerable effort will be required to expand our bench-scale treatments to process-level treatments. Further studies will also be required to determine if one or more chemical, physical or biological parameters can serve as markers of toxicity to monitor process efficiency and potential breakthrough on a routine basis.

At some stations, the segregation of radioactive liquid waste is such that discharges from specific tanks will be consistently toxic or consistently non-toxic. For example, at Station D, Tank 9 was consistently non-toxic, while the contents of Tank 12 were invariably toxic to *D. magna* (Table III). In the near future, we will be conducting a detailed study of the chemical composition and toxicity of the various streams which supply the RLW system for at least one station. We hope that this approach may allow us to identify the principal streams contributing to RLW toxicity and to address the problem at source through a combination of Best Management Practices and smaller scale treatment facilities.

Table III. Toxicity[a] of Station D RLW Samples (as-received) by Tank Number

Tank	# toxic samples / total # samples
TK 5	1/2
TK 6	1/2
TK 9	0/4
TK 11	2/2
TK 12	5/5
ALL	9/5

[a] Samples considered toxic if *Daphnia magna* mortality $\geq$ 50% in duplicate tests

Acknowledgments

We thank Reid Mowatt for technical assistance, and P. Wiancko, M. Greenall, and J.Sharma for their help and support throughout this study. We also thank the many people from Ontario Hydro Stations who provided samples in a timely and efficient manner.

References

American Public Health Association, American Water Works Association and Water Pollution Control Federation (APHA): 1989, *Standard Methods for the Examination of Water and Wastewater. 17th ed.* Washington, D.C..

Bellamy, W.D., Hickman, G.T., Mueller, P.A., Ziemba, N.: 1991. *Research J. Water Pollution Control Fed.*, **63**, 120-128.

Clayton, R.: 1992, The Chemical Engineer, May 14, 1992, 23-26 (1992).

Environment Canada: 1990, *Biological test method: Reference method for determining acute lethality of effluents to Daphnia magna*, Environmental Protection Publications, Conservation and Protection, Reference Method EPS 1/RM/14.

Glaze, W.H., Kang, J.W.: 1988, *J. Amer. Water Works Assn.*, May 1988, 57-63.

Mount, D.I., Anderson-Carnahan, L.: 1988, *Methods for Aquatic Toxicity Identification Evaluations, Phase I. toxicity Characterization Procedures.* U.S. Environmental Protection Agency, Duluth, MN, Report EPA-600/3-88/034.

Ollis, D.F., Pelizzetti, E., Serpone, N.: 1991, **Environ. Sci. Tech.**, **25**, 1523-1529.

Ontario Ministry of Environment and Energy (OMEE): 1995, *Effluent Monitoring and Effluent Limits - Electric Power Generation Sector*, Ontario Regulation 215/95 made under the Environmental Protection Act.

Ontario Hydro: 1992, *A review of Ontario Hydro's MISA monitoring results.* Ontario Hydro, NOCS-IR-07291-OUO/TISD-07291-0001.

Rodgers, D.W.: 1994, *Chemical Speciation and Bioavailability 6*, 23-26.

Rodgers, D.W., Schröder, J., Vereecken Sheehan, L.: 1996, *Water, Air and Soil Pollution*

Tyldesley, J.D., Knowles, G.: 1989, Paper #62, Water Chemistry of Nuclear Reactor Systems, British
    Nuclear Energy Society, London.
Velte, J.S.: 1984, *Bull. Environ. Contam. Toxicol.*, 33, 598-604.

ELECTRICITY AND WATER DESALINATION:
SEPARATE SITES OFFER VALUE

DR. I. MOCH, JR.[1], FRED DEPENBROCK[2], and YUSUF MUSSALLI[3]

[1] DuPont Co., Wilmington, Delaware, [2] Stone & Webster Management Consultants, Inc., 7677 E. Berry Ave., Englewood, Colorado, [3] Stone & Webster Saudi Arabia, Inc., Khobar, Saudi Arabia

Abstract. For the fuel type and water situation in the Middle East, the case is strong for the use of combined cycle technology for power generation and reverse osmosis for potable water production, where each are sited for their maximum economic benefit and interconnected by electric power transmission. Because of the fuel efficiency of Combined Cycle generation technology, its use of liquid/gas fuels and its low need for cooling water, it can be optimized for cost away from cities. Conversely, water desalination by reverse osmosis can be sited in optimal locations to take advantage of its modularity and to minimize water pipeline needs. Electric power transmission provides an inexpensive and flexible means to connect these two technologies. Together these technologies may offer an overall minimum cost approach, better than the combining of electric power and water desalination at one location, where power to water ratios must be fixed, independent of need, for optimum efficiency. The use of reverse osmosis with power generation has other, important ancillary benefits over using distillation and power combinations. These advantages include abatement of environmental pollution, delivery of potable water at reasonable drinking temperatures, lower total energy consumption, more efficient land use and less demanding operator skills.

Keywords: Desalination, reverse osmosis, multi-stage flash, combined cycle power plant, electric transmission, plant siting, steam power plant, multiple effect distillation, cost comparison

1. Introduction

Two basic staples of modern society are water and electricity. Upon these two commodities humankind has established an industrial society that continues to thrive, even as the number of people climbs.

But because these two essential staples are used together to feed our modern, industrial world does not mean they must be produced together. Indeed, it is the thesis of this paper that there are significant opportunities for both electricity generation and water production by desalinization in which the processes should be separated for best economy. The separation of production facilities should be tested for the use of the most economically efficient combination of both electric and water production facilities and their respective delivery systems. The economics should consider not only respective cycle efficiencies, but also relative environmental effects, land values, joint facilities use and fuel/water/ power delivery requirements.

Water, Air and Soil Pollution **90**: 231-241, 1996.
© 1996 *Kluwer Academic Publishers.*

2. Electric Power Efficiency Improvements

The continued increase in efficiency of combustion turbine-based electric power production facilities, especially in combined cycle operation, suggests that a move away from boiler-based electric power production with steam extraction for water production may be in order. This is true whether the combustion turbine waste heat is captured by a heat recovery steam generator (HRSG) for use in power generation in conjunction with reverse osmosis (RO), or in supplying a desalination facility. The move to separate power production and desalination by using combined cycle power generation and RO would also have the important benefit of freeing up the problem of siting from the double constraint of both fuel sourcing and seawater access. The low electric power consumption of reverse osmosis water desalination further suggests that the capital requirements for the combined electric and water transmission and their inherent transmission power costs could be reduced by shifting the burden of transmission to the more flexible electric power network. The locating of reverse osmosis facilities, which have very little air and water environmental impact, near seawater and population centers would provide a potential improvement in city quality of life over combination facilities while keeping highly efficient but pollutant-emitting fuel burning at more remote sites.

A subtle but valuable advantage of separating water from electricity production is the ability to size the individual production facilities to suit their specific needs, not having to compromise the technical efficiency of a single plant for both purposes on the one hand, or the right balance of water needs and electricity demands by the community on the other.

3. Technology Overview

Steam boiler-based thermal desalination technologies include multi-stage flash (MSF), multiple effect distillation (MED), with mechanical vapor compression (MVC) and thermal vapor compression (TVC). The steam was formerly derived from conventional boiler and steam turbine power generation cycles with extraction. Presently, the most efficient way to provide the steam needs for MSF or MED is through steam developed by a heat recovery steam generator at the exhaust of a combustion turbine power generation cycle. The steam from the HRSG can be used with an extraction turbine or as supply to a desalination process exclusively. MSF has traditionally been the most widely used technology, though MED is now emerging with an increased market share. All of these systems essentially work the same way, by using steam from a boiler to vaporize and then condense seawater. A typical single-purpose distillation plant might be comprised of an oil-fired steam plant or a combined cycle power generation island, and a distillation/condensation unit within the same facility. A single-purpose distillation plant directs all of the steam generated by the boiler to the distillation unit, while a dual-purpose plant diverts only part of the steam and leaves the rest for power generation.

Most of the largest MSF and MED plants are dual-use, as the design offers better efficiency than single-use plants, of course, assuming that the power generation can be effectively used.

A reverse osmosis configuration, by comparison, separates the power generation and water desalination processes. This allows the two functions to be performed in separate locations, with each function using the technology optimal for the task. Power can be generated using a modern, efficient technology such as combined cycle (CC) gas-fired plants, while the desalination is performed using RO, a clean safe process, close to a large population center.

The use of a combined cycle plant would provide economic advantages over a traditional steam boiler and turbine arrangement, both in capital construction costs and in operation and maintenance costs. Using recent estimates, a modern 250 MW gas-fired CC plant would cost approximately 40% less to operate in the Middle East than a traditional steam plant of comparable size. The total economic efficiency arises from lower construction capital costs and lower operation and maintenance costs, as well as a lower heat rate. In addition, the CC plant also enjoys the advantages of lower emissions and shorter plant installation time. [See Reference 1.]

These capital cost advantages found in a CC facility used only for power generation would also apply to the combustion turbine and HRSG portion of a facility supplying steam to a single or dual-purpose desalination plant. However, the operating efficiencies would be comparable to a conventional steam power plant. Therefore, this use would provide economics that fall between the traditional plant and the power only CC case modeled later in this paper, but not provide the advantages of optimizing separate siting and sizing of power and desalination facilities.

TABLE IV

**Comparison of Desalination Plant Capital Costs
For 6 MGD (22,700 CMD) RO and MSF plants on the Arabian Gulf***

	RO		MSF		MED	
	US $(000s)	Percent	US $(000s)	Percent	US $(000s)	Percent
Direct Capital Costs						
Permeators, Installed	6,000	15%				
Process Equipment, Installed	16,500	40%	33,500	57%	41,500	63%
Site Development	1,000	2%	1,000	2%	1,000	2%
Intake and Outfall Systems	7,000	17%	9,065	15%	6,593	10%
Total Direct Costs	30,500	74%	43,565	74%	49,093	74%
Indirect Capital Costs						
Interest During Construction	1,525	4%	2,178	4%	2,455	4%
Contingency	2,440	6%	3,485	6%	3,927	6%
A&E Fees, Project Mgmt.	5,185	13%	7,406	13%	8,346	13%
Working Capital	1,525	4%	2,178	4%	2,455	4%
Total Indirect Costs	10,675	26%	15,247	26%	17,183	26%
Total Capital Cost	41,175	100%	58,812	100%	66,276	100%
US $/GPD	$6.86		$9.80		$11.05	
US $/CMD	$1,813		$2,590		$2,918	

* Note - Values for MSF and MED plants are scaled to 6 MGD from 25 MGD plants.

4. Capital and Operational Costs

The capital cost of a 6 million gallons/day (MGD) [22,700 cubic meters/day (CMD)] RO plant is estimated to be approximately 2/3rds the cost of an MSF or MED plant of the same size. (See Table IV above, Comparison of Desalination Plant Capital Cost.)

Non-fuel operation and maintenance expenses are also estimated to be lower for the RO plant. Conversely, the station service power consumption for an RO plant in a Middle East application would be approximately 20-25 KWh per 1000 gallons (5.3-6.7 kWh/M^3), while a single-use MSF or MED plant would require approximately 55% or 32% of that, respectively. (See Table V, Comparison of Desalination Energy Requirements.)

TABLE V

Comparison of Desalination Energy Requirements
For 6 MGD (22,700 CMD) RO and MSF plants on the Arabian Gulf

	RO		MSF		MED	
	KWH/KGals	KWH/CM	KWH/KGals	KWH/CM	KWH/KGals	KWH/CM
Seawater Supply	1.5200	0.4016	3.2625	0.8620	1.5833	0.4183
Booster Pump	1.5200	0.4016				
Chemical Dosing Pumps	0.0700	0.0185				
High Pressure Pumps	34.0500	8.9960				
Energy Recovery	-13.3600	-3.5297				
Degasser Blower	0.0700	0.0185				
Product	0.6100	0.1612				
Recirculating Pump			10.8750	2.8732	6.5250	1.7239
Distillate Pump			0.0400	0.0106	0.0400	0.0106
Plant Services	1.0000	0.2642				
Total	25.4800	6.7318	14.1775	3.7457	8.1483	2.1528

It should be noted that most dual-use MSF plants use low-pressure exhaust steam from the electrical generator and hence have a lower energy requirement. This is particularly true of MED technology, which utilizes lower temperature steam than that required by MSF. The use of waste heat, however, takes away some power that can be generated from the residual steam. This electric power loss must be included in any process energy economic comparisons, including the capital cost to achieve the same electrical output while supplying adequate steam for the desalination process.

Operational costs of the various desalination technology plants are comprised of labor, maintenance and parts, chemicals and consumables, and in the case of RO, membrane replacement. General maintenance and parts cost is higher for RO than MSF/MED, but labor cost is lower, as constant monitoring of the equipment is not required. However, RO technology requires periodic replacement of the membrane. The membranes are guaranteed for 5 years, but experience has shown they may last 10 years or more. Under a reasonable 8-year life estimate, membrane replacement makes up approximately 21% of total O&M expenses, excluding capital carrying costs, and 10% of total costs, including capital. (See Table VI, Comparison of Total Water Costs, which is developed from information in Tables 1, 2, and 3, which are located at the end of the article.)

TABLE VI

Comparison of Total Water Costs
For 6 MGD (22,700 CMD) RO and MSF plants on the Arabian Gulf

Assumptions

Financial Assumptions:		Conversion Factors:	
Cost of Capital	8.0%	Gallons → Cubic Meters	0.00379
Plant Amortization Period	30	Cubic Meters → Gallons	264.2
Fixed Charge Rate	8.8827%		

Plant Assumptions:	Units	RO	MSF	MED
Plant Capacity	MGals/Day	6	6	6
Desalination Capacity Factor		90%	90%	90%
Annual Production	MGals	1,971	1,971	1,971
Process Power Used	KWH/KGals	25.48	14.18	8.15
Cost of Electric Power	$/MWH	$19.87	$23.37	$23.37
Cost of Power after 2.5% Transm. Losses	$/MWH	$20.37	$23.37	$23.37
Average Membrane Life	Years	8.0		
Steam Equiv. Power Consumption	KWH/KGals		0.00	0.00
Total Equiv. Power Consumption	KWH/KGals	25.48	14.18	8.15
Plant Capital Cost	$000	$41,175	$58,812	66,276

Annual Operating Costs

	RO		MSF		MED	
	US $(000s)	Percent	US $(000s)	Percent	US $(000s)	Percent
Process Electric Power	$998	14%	$0	0%	$0	0%
Steam	---		$0	0%	$0	0%
Consumables, Chemicals	341	5%	410	6%	200	3%
Maintenance & Parts	915	13%	500	7%	500	7%
Supervision & Labor	283	4%	800	12%	800	11%
Membrane Replacement	700	10%	---		---	
Subtotal	3,237	47%	1,710	25%	1,500	20%
Fixed Charges	3,657	53%	5,224	75%	5,887	80%
Total Annual Costs	$6,894	100%	6,934	100%	7,387	100%

Net Water Cost:			
US $/1000 Gals	$3.50	$3.52	$3.75
US $/Cubic Meter	$0.92	$0.93	$0.99

In total, the production cost of comparable RO, MSF, and MED facilities would typically be about $3.50-$4.00 per 1000 gallons of distilled product, respectively ($0.90 to $1.00 per M^3).

4.1. Transmission Costs

A hypothetical RO/CC configuration would use a combined cycle plant located away from populated areas, for instance in the desert, transmitting the electric power through the electric transmission network to a RO plant located near a major city. The incremental transmission to deliver the incremental power for in-city RO operation is not likely to be more than that required to interconnect reliably the electric power plant itself. However, this type of use will cause incremental power transmission losses, which normally average about 2 - 3% of increased power delivery on the transmission network (the Comparison of Total Water Costs , Table VI, uses a loss factor of 2.5%). On the other hand, locating a MSF plant away from populated areas will also incur capital and operating costs for transmission of desalinated water.

4.2. Fuel & Water Supply Issues

One concern with MSF technology for the Middle East is that it requires large quantities of water for cooling as well as the desalination process. As such, the power production facility would need to be located near the coast, and either shipping or pipeline facilities would be required to fuel the plant. A 250 MW CC plant with dry cooling for its steam cycle, by comparison, requires much less water, on the order of 5,000 to 8,000 gallons per day. Hence, by trucking or piping in the water needed, the plant could be located in a desert at the most useful fuel supply and electric transmission location, perhaps close to a source of natural gas, for example.

4.3. Pollution Issues

A major argument in favor of separating the power generation and desalination functions is the concern over pollution from electric plant emissions. Many Middle Eastern countries have populations that are concentrated in small areas, and the effects of pollution from additional nearby power plants are certain to be detrimental. Typically for gas or oil fired electric power plants, the pollution most encountered is ground level ozone that results from the chemical interaction of nitrogen oxides (power plant emissions) and volatile organic compounds (automobile exhausts) in the presence of sunlight. This is the classic "smog," which is the mark of many modern cities. Although it is difficult to place an objective estimate on the value of placing a power plant away from populated areas in the Middle East, it is certainly a factor that merits consideration.

4.4. Combined Plant Versus Separate Plant Size Efficiency

Another concern with the distillation technologies is the inefficiency of mismatched scale economies. Steam plants tend to be large, with 500-600 MW being a typical figure for a scale-efficient plant. At a typical 8:1 or 10:1 power-to-water ratio for distillation technology, this results in a 50-75 MGD (190,000 - 280,000 CMD) desalination facility. A 250-300 MW combined cycle plant with MSF or MED desalination would have a water supply range of 20 to 30 MGD (76,000-114,000 CMD). This is in comparison to a typical desalination requirement of 15-50 MGD (57,000 - 190,000 CMD) for most population centers under consideration. As such, a steam boiler-based MSF plant located near an urban area might have considerable excess desalination capacity. Of course, water could be transported to other markets, but this would incur additional transmission costs. A CC plant with MSF or MED would be a much better fit to typical needs.

In comparison, the use of a local RO plant would best match production to local consumption requirements. A scale-efficient RO plant can be as small as 6 MGD (22,700 CMD), which would better serve the needs of many small to medium sized cities. At the same time, the generation facilities could also be sized to achieve scale economies, whatever the technology involved for power production.

There will be some cases in which other considerations may favor the use of distillation technology over RO. For instance, an area might already have surplus steam available from an existing power plant. In such cases, it would be difficult to justify installation of an RO facility when a distillation unit could be retrofitted to the existing power installation. Conversely, the more likely situation will be that in which sufficient electric power generation is in place, but there is inadequate water supply. Addition of an RO facility will be very cost effective in this situation.

5. Cost Evaluation

Ultimately, the cost and power issues need to be examined together within the framework of a fully integrated analysis. In its past experience in the Middle East, Stone & Webster has utilized tools such as EPRI's EGEAS(tm) electric power planning software (for which Stone & Webster is the developer and commercializer for EPRI) to model and optimize the economies of various technologies. EGEAS is a state-of-the-art modular production costing, generation expansion software package developed under EPRI sponsorship for user by utility planners to evaluate integrated resource plans, independent power producers, avoided costs and plant life management programs. Only through such a formalized evaluation process can the merits of various scenarios be accurately determined. DuPont's expertise as the world's foremost and largest membrane supplier offers significant assistance in design, operations and maintenance of RO plants.

In order to illustrate some of the issues just presented, we have performed a basic spreadsheet cost calculation comparing a hypothetical RO plant versus comparable MSF and MED plants, all sited in the Middle East. The RO configuration uses the DuPont Permasep(tm) 6880T membrane paired with a 250 MW CC power plant. The MSF and MED packages are assumed to be powered by dual-purpose 300 MW steam plants.

Some assumptions had to be made in order to streamline the calculations for the purposes of this paper. For comparability, the capital and operating cost figures for all three technologies were based on a capacity of 6 MGD (22,700 CMD). In reality, the MSF and MED plants would have larger capacity matched to the accompanying power generation facilities. Under the assumptions made in this analysis, such capacity would be around 25-35 MGD (94,600 -132,500 CMD).

Total Water Cost	RO	MSF	MED
Cost, US \$/1000 Gals Produced	\$3.50	\$3.52	\$3.75
Cost, US \$/M^3 Produced	\$0.92	\$0.93	\$0.99

The analysis indicates that under the hypothetical scenario, RO technology enjoys a small cost advantage over MSF, and a slightly larger advantage over MED. (See Table VI, Comparison of Total Water Costs.) Closer inspection of the calculations indicates that the water cost for both MSF and MED are burdened by the higher capital cost of those technologies. In addition, MSF and MED also incur substantial costs for the steam that is bled from the generator. This shows up in the analysis in the form of higher capital costs for the power plant and a penalty on heat rate, but no other steam costs. Charges were taken for additional auxiliary power needs for the desalination plant that were not included in the CC plant net plant calculation.

In addition to the cost calculations, there are also other assumptions that bear on the analysis results. They include the plant's load factor, RO membrane replacement rate, and the cost of power, all of which are subject to change based on local conditions. Still, the example illustrates the kind of analysis that can be conducted to evaluate the suitability of a particular technology to local needs.

6. Conclusion

In comparing the advantages of RO desalination with combined cycle electric production versus the same power technology with MSF or MED desalination, we have examined several factors both tangible and intangible, which need to be taken into account:

- The efficiency of more modern generation technologies such as combined cycle plants regardless of how they are coordinated with desalination.
- The choice of single-use vs. dual-use thermal distillation facilities. Dual-use facilities can lower the power cost of the distillation process markedly over single-use plants.
- The capital and operational costs of both thermal distillation and reverse osmosis technologies.
- Transmission costs and losses associated with transmitting both power and water.
- Fuel and water supply issues related to siting power plants.
- The societal cost of emissions, and the desirability of locating a generation plant away from populated areas.

- Efficiencies involved in sizing power plant and desalination capacity according to local requirements and scale efficiencies.

Only when all of these factors are evaluated simultaneously can a technology be selected as the best choice for any particular application. Therefore, we believe reverse osmosis in conjunction with efficient remote power generation offers enough benefits for the Middle East region that it merits further site-specific investigation.

TABLE I
BASIS OF DESIGN
LOCATION ARABIAN GULF

Plant Capacity - 6 MGD, 22,700 CMD
DuPont Permeator Model - B-10 TWIN™, 6880T
Product Quality - Less than 500 mg/l delivered
Open Sea Feedwater Intake
Operating Hours per Year - 7884
Annual Capacity - 1971M Gallons, 7460 KM^3
Number of Trains - 4
Number of Blocks per Train - 2
Capacity per Train - 1500 KGD, 5675 CMD
Feed Water Composition -Standard seawater at 45,000 mg/l.
Temperature Range - 17°-35°C.

Conversion - 35%
Single Pass RO System.
Term - 5 years
RO Feed Pressure - 1200 psig maximum
Efficiency of Motors - 92%
Efficiency of High Pressure Pump - 80%
Efficiency of Energy Recovery Device - 82%
Energy Cost - US $0.05 per kWh
Product Pressure - 10 psig
RO Feed pH - 8.3
Piping:
    Seawater - SMO 254 or equivalent
    Product - 316L or plastic

TABLE II

Seawater RO Total Water Costs
For A 6 MGD (22,700 CMD) Plant on the Arabian Gulf

Annual Operating Costs	US $(000s)	Percent
Electric Power at $0.019875/KWH	998	14%
Consumables, Chemicals	341	5%
Maintenance and Parts [1]	915	13%
Supervision and Labor [2]	283	4%
Membrane Replacement [2]	700	10%
Total Direct Costs	3,237	47%
US $/1000 Gallons	$1.64	
US $/CM	$0.43	
Fixed Charge (Amortization) at 8%, 30 Years	3,657	53%
Total Annual Cost	6,894	100%
US $/1000 Gallons	$3.50	
US $/CM	$0.92	

Notes:
1. 3% of Direct Capital Costs
2. Assumed 12% per year replacement rate

TABLE III

Comparison of Generation Technologies
For Combined Cycle Power Plants with Extraction for MSF/MED or Standalone

Assumptions

Technology Assumptions		CC, 250 MW Net Output	
		w/ Extraction	w/o Extraction
Capital Installation Cost	$/KW	800	663
Fixed O&M	$/KW-Year	13.09	13.09
Variable O&M	$/MWH	0.71	0.71

Economic Assumptions	
Cost of Fuel - $/MMBtu	$1.00
Cost of Capital	8.0%

Results at 75% Capacity Factor

	CC, 250 MW Net Output	
	w/ Extraction	w/o Extraction
Levelized Generation Cost - $/MWH	$23.372	$19.875

Plant Type: **COMBINED CYCLE w/Extraction for MSF or MED**

ISO Capacity (MW)	253	Book Life	30
Derated Capacity (MW)	197	Capital Recovery Factor	0.08883
Capital Cost ($/KW)	$800.12	Annual Fixed Cost	$14,016
Variable Cost ($/MWH)	$0.71		
Fuel Cost ($/MMBTU)	$1.00	Average Heat Rate (BTU/KWH)	9,852
Fixed O&M ($/KW-Yr)	$13.09	Fuel Cost ($/MWH)	$9.85

Heat Rate Blocks	100%	75%	50%	25%	10%
[At 105° Ambient Temp]	9,602	10,110	11,457	14,988	15,889

Capacity Factor	Annual Fixed O&M ($000'S)	Annual Energy (MWH)	Annual Variable Cost ($000'S)	Annual Fuel Cost ($000'S)	Annual Total Cost ($000s)	Levelized Power Cost $/MWh
70%	2,581	1,209,296	859	11,914	29,371	24.287
75%	2,581	1,295,674	920	12,765	30,283	23.372
80%	2,581	1,382,052	981	13,616	31,195	22.572
85%	2,581	1,468,431	1,043	14,467	32,108	21.865

Plant Type: **COMBINED CYCLE 250MW**

ISO Capacity (MW)	253	Book Life	30
Derated Capacity (MW)	197	Capital Recovery Factor	0.08883
Capital Cost ($/KW)	$662.96	Annual Fixed Cost	$11,614
Variable Cost ($/MWH)	$0.71		
Fuel Cost ($/MMBTU)	$1.00	Average Heat Rate (BTU/KWH)	8,209
Fixed O&M ($/KW-Yr)	$13.09	Fuel Cost ($/MWH)	$8.21

Heat Rate Blocks	100%	75%	50%	25%	10%
[At 105° Ambient Temp]	7,959	8,380	9,496	12,423	13,170

Capacity Factor	Annual Fixed O&M ($000'S)	Annual Energy (MWH)	Annual Variable Cost ($000'S)	Annual Fuel Cost ($000'S)	Annual Total Cost ($000s)	Levelized Power Cost $/MWh
70%	2,581	1,209,296	859	9,927	24,981	20.657
75%	2,581	1,295,674	920	10,636	25,751	19.875
80%	2,581	1,382,052	981	11,345	26,522	19.190
85%	2,581	1,468,431	1,043	12,054	27,292	18.586

References

Stone & Webster Engineering Corporation: 1993, *Development of Engineering, Cost, and Performance Data for Generation Supply Options for New England*, 541 pages.

Glossary

Combined Cycle (CC) - A generation technology based on combustion turbines that uses heat recovery steam generation to recover waste heat from the exhaust and uses that waste heat to make steam to drive a steam turbine generator. Cycle efficiencies above 50% and relatively low capital costs characterize this technology.

EGEAS - An integrated resource planning (IRP) model developed and maintained for EPRI and its members by Stone & Webster. The model is a world-wide leading innovation for performing production cost and capital cost optimization analysis for electric power supply. As a major part of the EPRI IRP Workstation, EGEAS integrates the cost analysis for supply-side (generator) resources with demand-side management processes and works with financial planning and risk assessment tools to support power supply business decisions.

Multi-Effect Distillation (MED) - A thermal desalination process in which evaporation takes place as a thin film of feed water passes over a heat transfer surface (outside of horizontal tubes - HTE or inside of vertical tubes - VTE). Vapor formed in the last effect is recompressed thermally (TVC) or mechanically (MVC).

Multi-stage Flash Distillation (MSF) - Also a thermal desalination process in which a stream of brine flows through the bottom of up to 25 chambers (stages). The pressure in each chamber is maintained at a lower level than the saturation vapor pressure of the water and a proportion of it "flashes" into steam and is then condensed.

Reverse Osmosis (RO) - A pressure greater than the osmotic pressure of the feed water is applied to one side of a semi-permeable membrane, producing a flow of desalinated water through the membrane.

PART V

MEASURING QUALITY

IMPACT ON DISCHARGE MONITORING OF RECENT EPA INITIATIVES IN WATER QUALITY MEASUREMENT

R. F. Maddalone[1], J. W. Scott[1], J. K. Rice[2], and B. R. Nott[3]

[1] TRW, One Space Park 01/2040, Redondo Beach, CA 90278 USA. [2] 17415 Batchellors Forest Road, Olney, MD 20832, USA. [3] EPRI, P.O. Box 10412, 3412 Hillview, Palo Alto, CA 94303, USA

Abstract. Human health and aquatic life requirements can result in the establishment of effluent limitations on power plant aqueous discharges that are below the quantitation and even the detection capability of many of the current sampling and analytical procedures. To meet this situation for compliance monitoring, EPA has developed analytical methods that reportedly lower detection and quantitation levels by one to two orders of magnitude. Procedures to develop appropriate statistically-based detection and quantitation estimates for these methods are being developed by the Electric Power Research Institute (EPRI) and other industry and regulatory groups. EPA is also in the process of developing guidelines for performance based methods, which they believe will encourage innovation and speed the process of bringing new methods into use. This paper will discuss the impact of these proposed changes on the end user.

Keywords. Water Quality, Discharge Monitoring, Water Analysis

1. Introduction

Many of the water quality criteria established by the United States Environmental Protection Agency (EPA) and the states are at levels which currently available analytical methods are unable to measure. EPA has recently published (USEPA, 1995b) a series of analytical methods (part of the 1600 series) with a claimed sensitivity adequate for making measurements at water quality criteria concentrations for twelve metals. For some elements a chelation/concentration technique has been incorporated to achieve better detection limits, but most of the new analytical methods simply impose additional quality control and sample handling requirements on already existing methods. Extra attention to clean techniques for both sampling and analysis is stated to be essential to achieving these low levels of detection and quantitation. EPA has said that they plan to formally propose these methods for adoption at 40 CFR Part 136 in the near future. Until they do so, it is not clear under what circumstances, if at all, it is appropriate to use those methods in the National Pollution Discharge Elimination System (NPDES) process.

The EPA's measures of detection and quantitation levels are the method detection limit (MDL) and minimum level (ML), respectively. As defined in 40 CFR Part 136, Appendix B, the MDL is simply 3.14 times the standard deviation of seven replicate analyses performed in the same laboratory (even by the same analyst). The ML is then defined as 3.18 times the MDL to yield a value that the EPA assumes to be ten standard deviations above zero. EPRI and the Utility Water Act Group have been on record opposing the use of MDLs (and any value derived from them) for compliance monitoring (Koorse, 1995). The primary reason for this opposition to the MDL and ML is the lack of a technical basis for their computation and the fact that both the MDL and ML are single laboratory estimates of detection and quantitation. Compliance monitoring is fundamentally a comparison of results between the permittee and the permitter so that any estimate of the detection and quantitation level must include the effects of interlaboratory (between laboratory) bias. Between laboratory bias is caused in a large measure by the

Water, Air and Soil Pollution **90**: 245-255, 1996.

lack of any common, absolute (low level) standards that can be used to calibrate analytical instruments.

In this paper the impact of the new 1600 series sampling and analysis methods will be discussed along with the problems associated with EPA's MDL and ML and a summary of the status and issues associated with the Performance Based Methods System.

2. New ICP-MS Method for Measurements at Water Quality Criteria Levels

The new ICP-MS Method 1638 (USEPA, 1995a) was developed for the stated purpose of providing reliable measurement of nine metals at EPA water quality criteria concentrations. It is based largely on EPA Method 200.8 (USEPA, 1994b). As the EPA data in Table I illustrate, the Method 1638 MDLs are significantly lower in most cases than those for Method 200.8. The MDLs cited in Method 200.8 appear to be based on pooled single operator estimates of the precision from thirteen laboratories, but the results of Method 1638's validation study (USEPA, 1995c) involved only two research quality laboratories.

Table I

Comparison of MDLs for Method 200.8 and Method 1638 with Water Quality Criteria and Reference Material Concentrations

| Metal | Method 200.8 MDL, µg /L | Method 1638 | | Lowest Ambient Water Quality Criterion, µg /L | National Research Council of Canada | |
		MDL, µg/L	ML, µg /L		SLEW-2, µg/L (1)	SLRS-3, µg /L (2)
Antimony	0.4	0.0097	0.02	14		
Cadmium	0.5	0.025	0.1	0.32	0.019	0.013
Copper	0.5	0.087	0.2	2.5	1.62	1.35
Lead	0.6	0.015	0.05	0.14	0.027	0.086
Nickel	0.5	0.33	1	7.1	0.709	0.83
Selenium	7.9	0.45	5	5		
Silver	0.1	0.029	0.1	0.31		
Thallium	0.3	0.0079	0.02	1.7		
Zinc	1.8	0.14	0.5	28	1.1	1.04

(1) Estuarine Reference Material

(2) River Water Reference Material

The newer Method 1638 does not call out any specific instrumental improvements to achieve its lower MDLs. The source of the improvements made in the MDLs appears to lie mainly in control of contamination during all phases of sampling and analysis. The changes incorporated into the newer method include sections on selection of materials to avoid contamination, specific callouts for analyses of field blanks to detect contamination, glassware cleaning procedures, and a provision for operating in a class 100 clean room environment if necessary.

Table II
Comparison of EPA Methods for ICP-MS Analysis

Key Features	Method 200.8	Method 1638
Applicability	Dissolved elements in groundwater, seawater and drinking water; re-coverable in waters, soils, sludges, wastewaters	Dissolved elements in ambient waters (total recoverable digestion procedure included)
Elements listed	21	9
Contamination Precautions	In discussion of reagents, labware	Defined in detail
Working concentrations	Mid ppb to high ppt	Low ppb to low ppt
"Performance Based"	No	Yes
Discussion of Interferences	Yes	Identical to 200.8
ICP-MS Specification	Yes	Identical to 200.8
Instrument Standardization	Procedure given	Identical to 200.8
Quality Control Sample (independent)	Yes	Identical to 200.8
Ongoing Precision/Recovery Checks Specified	Yes	Yes
Sample Analysis Procedure	Yes	Almost identical to 200.8
Blanks: Calibration, Laboratory Reagent, Rinse	Yes	Identical to 200.8
Field Blanks	No	Yes
Sampling Equipment and labware Cleaning Procedures	No	Yes
Alternative total recoverable digestion procedure	No	Yes (Section 12.2.8)
Published Instrument DLs	Yes	No
Published MDLs	Yes	Yes; new MDLs average 1/20 of Method 200.8 MDLs
Method Validation	Interlaboratory study data (13 labs) regression equations for precision and bias included in method; however, the MDL was supported by only pooled single operator precision data.	Two laboratories

A comparison of other features of the two methods is presented in Table II. EPA has published another method in the new 1600 series: Method 1669, "Sampling Ambient Water for Trace Metals at EPA Water Quality Criteria Levels" (USEPA, 1995d) which contains procedures to control contamination during sampling and transport. This document addresses the level of protection necessary to preclude contamination during collection, transport, and analysis. In addition to specifying cleaning procedures, use of non-metal equipment, and additional quality control samples required to verify cleanliness, the use of two samplers, one designated "clean hands" and the other "dirty hands" is described.

Although both methods specify the use of a Quality Control Sample (QCS) from an independent source outside the laboratory, dilution of the sample to less than 100 mg/L (or less than 500 mg/L for selenium) is the only requirement specified in the use of the QCS to assess performance. This process of diluting high concentration standards to low concentrations is fraught with potential problems. Laboratory contamination from the water, glassware and laboratory environment can alter what is supposed to be an absolute standard used to check the performance of the methods and the laboratory performing the work. The importance of having a low level standard is illustrated by some data produced during the EPRI RP1851 (Maddalone, 1988) round robin validation of GFAAS by 20-30 qualified utility laboratories (Table III). During this validation program, standard EPA QC vials were sent to the participating laboratories. The laboratories were instructed to dilute the vials to one liter and then to analyze the solution by GFAAS. The data in Table 4 show that for a number of elements (As, Pb, and Se) significant biases were found. Note also that these results were obtained at concentrations a factor of a hundred or more above the WQC levels sought by the EPA and in samples requiring no digestion. It is highly likely that the variability and the amount of relative bias would escalate as the concentration of the sample decreases and approaches the background contamination in the laboratories, particularly when a digestion is involved. These data illustrate a fundamental problem that the EPA faces in driving to lower detection and quantitation levels: laboratories must calibrate their instruments using dilutions of higher level standards. In the course of diluting concentrated standards errors and biases are introduced. EPRI has recognized this problem and has consistently requested that the EPA use interlaboratory based estimates for detection and quantitation levels so as to include these biases. So far the EPA has resisted taking this approach and relies on either single laboratory or pooled single operator (intralaboratory) data from as few as 2 laboratories (Method 1638) to support their work.

Table III

Summary Of GFAAS Analysis Of Standard EPA QC Vial During The EPRI RP1851 Round Robin Validation Program (Results Of 23-29 Laboratories)

Element	Number of Labs	True Value, mg/L	Amount Recovered, mg/L	Bias, mg/L	% Recovered	%, Bias
Arsenic	29	39.7	37.76	-1.94	95.1	-4.9
Chromium	23	60	59.14	-0.84	98.6	-1.44
Copper	26	40	40.04	0.04	100.1	0.1
Lead	24	80	88.92	8.92	111.2	11.16
Nickel	25	70	70	0.5	100.7	0.71
Selenium	28	19.8	17.74	-2.06	89.6	-10.43

These problems could be addressed if reference standards at these low concentrations were available for use as calibration standards and QCS. Standard reference materials that could be used directly (without dilution) are not readily available at WQC concentrations and the EPA is not developing any. One source for at least some of the elements is the National Research Council of Canada Institute for Environmental Research and Technology (NRC), two of whose reference materials are listed in Table I. These limited samples only cover four matrices (the two in Table I plus open ocean and near shore seawater) when a range of matrices and concentrations are required for calibration and QCS. Another approach would be to validate the concentration of a large sample of the desired water by a round robin of peer laboratories. Aliquots of the verified material are then available for checking calibrations in participating laboratories.

3. Costs of Implementing 1600 Series Methods

In 1994 an estimate of the costs associated with implementation of trace metals analyses at lower concentrations was published by EPA's Engineering and Analysis Division (USEPA, 1994c). Cost estimates were based on the experience of the United States Geological Survey (USGS) in implementing changes to enable measurements to be made reliably at the 1 ppb level with a detection level of 0.1 ppb. The USGS has plans to move to a detection limit of 0.01 ppb. The EPA estimated, based on information from the USGS and other laboratories, that in order to achieve that detection level, a laboratory would have to invest in a class 100 clean room, ICP-MS with hydride attachment, mercury analyzer, and stabilized temperature platform graphite furnace atomic absorption spectrophotometer. EPA's estimate of those capital costs was $325,000, which does not reflect the real cost of maintaining the instrumentation and the laboratory. In this same document, it was stated that even the 0.01 ppb detection limit targeted by the USGS would be inadequate for EPA purposes because EPA and State ambient water quality criteria for some metals are more than ten times lower. No cost estimate was provided for making measurements at even lower concentrations.

Table IV contains the cost estimates received from a major manufacturer of ICP-MS and stabilized GFAAS instrumentation. This analysis assumes that a laboratory would not have any of the hardware and would have to purchase the listed hardware to totally up-

Table IV
Cost estimate for installing EPA recommended analysis hardware

Cost Element	Estimated Cost, $K	Comments
Class 100 Clean Room hardware	50	USGS estimate, but without labor included.
Labor estimate for Clean Room	25	Estimated labor cost (50% of hardware cost)
ICP-MS[a]	215	With computer, printer, and water recirculator, installation and three-and half day training for one operator. Also includes shipping cost.
Hydride Generator for ICP-MS[a]	17	With shipping, installation and onsite training for one operator. Also includes shipping cost.
Mercury Analyzer[a]	30	With autosampler, computer, printer, installation and onsite training for one operator. Also includes shipping cost.
Stabilized temperature platform GFAAS[a]	84	With computer, printer, installation and onsite training for one operator. Also includes shipping cost.
Ion Chromatograph	30	Estimate for IC capable of Cr(VI) analysis
Improved water system[a]	7	Milli-Q Plus system. Includes shipping and installation.
Training costs[b] ICP-MS	4	3 operators at San Jose
Stabilized Temperature Platform GFAAS	3	3 operators at San Jose
Mercury Analyzer and Hydride Generator.	2	3 operators at San Jose
Total:	$467K	

[a] The price includes 8.25% Los Angeles county sales tax.
[b] The prices includes only tuition fee.

grade their laboratory including facility modifications. Costs have been added for improved water supplies and for training. The resulting value is ~$467,000 and is substan-
tially higher than the EPA's estimate. These costs are only for the initial investment and
not for the ongoing cost of employing upgraded laboratory personnel (i.e., MS and Ph.D.
operators or supervisors) to run the equipment, additional costs for maintenance of sophisticated instrumentation (maintenance contracts cost about 10% to 15% of the sales
cost per year or for this case about $35,000 to $45,000 per year for this set of instruments), and the general costs for maintaining this level of laboratory. The costs of making measurements at these water quality criteria levels will also increase due to a)the clean
hands/dirty hands additional sampling manpower required and b) the costs of collecting,
preparing, and analyzing an increased number of quality control samples. The cumulative
increase in non-instrument cost per analytical result is estimated to be 30 to 50 percent
(Horowitz, 1995). It is likely that these costs are underestimated too by the EPA, since
only a few commercial vendors will have this capability and will be able to charge what
the market will bear for these required services.

4. Impact of the Use of the EPA MDL and ML

Quantitation has been defined by various authors over the years as a concentration at
which a specific level of precision can be obtained. The measure of precision is usually
stated in terms of a relative standard deviation ("RSD") which is simply the standard deviation at the concentration divided by the concentration and expressed as a percentage.
The RSD most often quoted as indicative of quantitation is 10%. Experience also has
shown that RSD diminishes only slightly as concentrations increase above the foregoing
quantitation level.

The quantitation level also has been expressed in terms of a factor times a standard deviation, and is essentially the way EPA proposed to derive its MLs. Generally the factor
is 10 and so the common way of stating the quantitation level (L_Q) is as a "10 sigma" limit
where sigma actual refers to the standard deviation of a "blank" and not the true population standard deviation. If one assumes the "sigma" to be constant versus concentration,
the 10 sigma quantitation level can be shown to give an RSD of 10% because:

$$RSD = (s/L_Q) \times 100\%$$
$$RSD = (s/10s) \times 100\%$$
$$RSD = (0.10) \times 100\%$$

or

$$RSD = 10\%$$

The EPA minimum level (ML) was derived from the "10 sigma" definition by assuming
that the standard deviation at the MDL and the ML are the same and therefore:

$$MDL = 3.14s$$

and

$$ML \sim 10s$$

so

ML/MDL = 10s/3.14s = 3.18

thus:

ML ~ 3.18 x MDL

In the compliance monitoring situation, there are a number of problems with the 10% RSD definition and with EPA's estimate of sigma in the "10 sigma" approach used to compute an ML. The ML itself is nothing more than a multiple of the MDL, which itself is a flawed statistic for a number of reasons:

- MDLs are based on intralaboratory data - Compliance monitoring is an inter-laboratory comparison between the permittee and the permitter, yet the MDL can be based on only one analyst in one lab performing the MDL analysis in reagent grade water. Data developed from EPRI RP1851 validation programs and EPA Method Study data show that the ratio of interlaboratory (s_t) to intralaboratory (s_o) standard deviation is a factor of 2 or more depending on the method, matrix and analyte.

- The MDL is not reproducible - The EPA specifically states in 40 CFR Part 136, Appendix B that MDLs "are not necessarily reproducible on a routine basis in a given laboratory, even when the same analytical procedures, instrumentation and sample matrix are used." What this means in practice is that an MDL computed from day to day or analyst to analyst will vary.

- MDLs assume constant variance - The implicit EPA model of the MDL assumes that the standard deviation at zero is the same as that at the MDL. However, since variance is not constant with concentration, the MDL will be highly de-pendent on the spiking concentration. The lower the spike concentration the lower the MDL will become. In effect the MDL is not anchored statistically.

- MDLs are not statistical predictors of laboratory performance - MDLs are based on the t-statistic which establishes a distribution of parameter estimates. It is not meant to predict multiple future events.

To address these problems with the MDL and the standard deviation data used to compute it, EPRI initially developed the concept of the compliance monitoring detection limit (CMDL) (Scott, 1994). The CMDL addressed the problem of needing an interlaboratory based estimate for a detection limit directed at the compliance monitoring case where the discharge permit was set at no permissible discharge (zero). In that situation the EPRI CMDL was equivalent to the concept behind Currie's (Currie, 1968) critical level (L_C) (which is simply the lowest concentration that is distinct from zero to a specific level of confidence, thereby avoiding false positive errors). In EPRI's case, the CMDL was com-puted from the interlaboratory precision data from round robin validation studies with 20-30 laboratories and expressed in the form of a curvilinear regression equation. A toler-ance statistic was used rather than the t statistic that EPA used. This change in the com-putational process (Scott, 1994) addressed the problems described above with the MDL.

In its draft guidance (USEPA, 1994a) for dealing with detection and quantitation levels in the National Pollutant Discharge Elimination System (NPDES) the EPA proposed the ML as the method for calculating quantitation levels. The ML, as we have shown above, is a 10 sigma approach to computing a quantitation level. The EPA, however, has not developed a rigorous procedure for calculation, but instead has simply used a factor times the MDL. All the deficiencies associated with the MDL are thus carried over to the ML. EPRI, as part of an inter-industry coalition, has recently developed the alternative minimum level (AML). The AML is also a "10 sigma" estimate of the quantitation level, but with significant computational improvements and statistical rigor to avoid the problems identified with the ML. The AML is 10 times the <u>interlaboratory</u> standard deviation at the lowest concentration that is differentiable from zero. The lowest concentration that is differentiable from zero can be statistically defined and, in EPRI computational approach, it is anchorable, unlike the MDL. Once the lowest concentration differentiable from zero is calculated, the standard deviation is computed at that concentration (using the regression expression for standard deviation versus true concentration) and multiplied by 10. The final steps in the computation process correct the raw "10s" value for errors in estimating the standard deviation and true concentration. The resulting approach has been presented to the EPA at a recent (August 2 and 3, 1995) public meeting on behalf of a coalition of interested industries (Koorse, 1995).

The opinion has been expressed in some quarters that treated wastewater is equivalent to reagent grade water, so that additional method validation studies in treated wastewater are not necessary. Figure 1 is based on the EPRI validation data (Scott, 1994) for various metals methods from three typical utility matrices that are either treated (ash pond overflow and treated chemical metal cleaning wastes) or are once through cooling water (seawater). The computed estimates of the quantitation level (alternative minimum level - AML) for the three matrices are plotted against the corresponding values for reagent water. The parallel lines represent various values of the resulting ratio. As the complexity of the matrix increases (river water to seawater), the resulting AML increases to the point where, in seawater, the matrix AML is well over 10 times the reagent grade water estimate.

5. Performance Based Measurement System

A totally new approach to methods development is on the horizon: the Performance Based Measurement System (PBMS). A performance based method (PBM) is defined as: "an analytical procedure containing the necessary method performance specifications to assure the desired quality of results." In a broad sense PBMS is simply the selection of methods that demonstrate acceptable performance against a set of criteria. The burden of proof is on the user, while the EPA (or another agency) would set the criteria that must be met and how it would be demonstrated. Within the EPA, PBMS development is under the Environmental Monitoring Management Council (EMMC) PBMS Workgroup.

From the EPA's point of view, the PBMS is meant to foster methods development and improve technology insertion, allowing optimization of methods for specific analytes or conditions. The EPA believes that there are too many regulatory hurdles with developing new technology or making major modification to existing approved technology. Under the existing approval process, new analytical methods require scientific, program office and public reviews and then responses by the EPA. The EPA is working on a looser interpretation of the alternative method rules, but they may not be able to make a significant improvement under the current regulatory structure. The EPA (Office of Water/Engineering Analysis Division) has planned a series of public meetings on Streamlining Promulgation of Analytical Methods, the first of which was held in Seattle on September 28, 1995. In the literature provided to the attendees, it was stated that the definitive test criteria against which performance of modified methods would be measured are the QC acceptance criteria in the promulgated method.

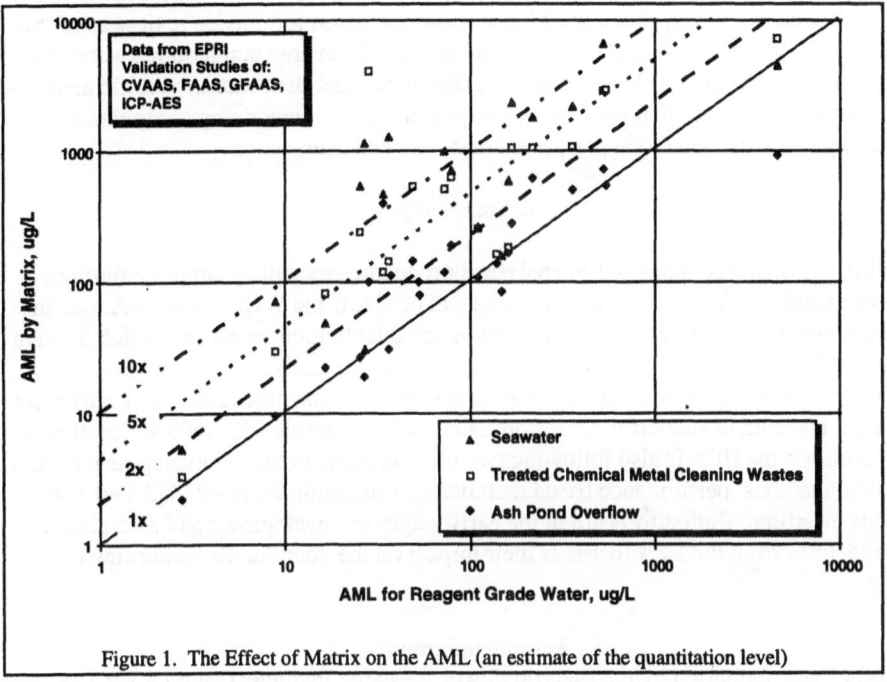

Figure 1. The Effect of Matrix on the AML (an estimate of the quantitation level)

The American Water Works Association (AWWA) sponsored a session on the Performance Based Method System (PBMS) at their Water Quality Technology Conference in San Francisco on November 9, 1994. The session brought together representatives from the EPA, industry and standard setting organizations. The resulting discussion pointed out these problems or concerns:

- If a lab uses a different method, they must develop data on detection limits and interferences, monitor its quality on a routine basis, and have an external (non-governmental) audit. The EPA or another governmental (state) agency will inspect the data collected.
- Lab costs could be 10-20% higher since the user would need a full time QA manager to track the status of the methods and provide confidence in the data collected.

- Due to the cost to develop and maintain new methods under PBMS, there may not be much technology infusion due to the fact that a commercial lab may consider its procedure proprietary and would not be willing to share it.
- The EPA has not resolved the issue of oversight. States would probably have the responsibility for oversight of the program, but they do not have the resources and may not have the skills.
- Implementation in a non-compliance monitoring program will be far easier than in a compliance monitoring program.

The EPA Office of Drinking Water "pilot" program to develop a guideline document for a PBMS has been ongoing, but a draft has not been released for outside review as of this writing. The PBMS is a long way from fruition, but it is an area that requires watching. It offers the user the chance for increased flexibility to meet their matrix specific analysis needs, but it is important that the cost and responsibility of maintaining compliance monitoring methods are not simply passed on to the end user.

6. Summary

The EPA has proposed a new set of analytical procedures including sampling through analysis to address the need to measure metals at or below the WQC. The EPA uses the MDL and the ML as estimates of the detection and quantitation levels and which in themselves are flawed and misrepresent the performance of the analysis method. Given the cost of the implementing these procedures, it is important that procedures have sufficient validation test data to support their performance claims. Besides the 1600 series of proposed methods the EPA is also following two parallel paths to foster development of new analytical methods: performance based methods and streamlining of 40 CFR Part 136 methods insertion. Both efforts are in the early stages of development and await definitive guidelines from the EPA to assess their impact on the compliance monitoring community.

Acknowledgments

This paper was supported by the Electric Power Research Institute, Contract RP 1851.

References

Currie, L.A.: 1968, *Anal. Chem.*, **40**, 586.

Horowitz, A.: 1995, US Geological Survey verbal technical discussion.

Koorse, S. J.: June 21, 1995, "Industry Presentation for EPA's Meeting on the Draft 'National Guidance for the Permitting, Monitoring, and Enforcement of Water Quality-based Effluent Limitations Set Below Analytical Detection/Quantitation Levels'", submitted to the EPA on behalf of a coalition of American Industries by law firm of Hunton & Williams.

Maddalone, R.F., Scott, J.W. and Frank, J.: 1988, "Round -Robin Study of Methods for Trace Metal Analysis," Volumes 1-2, EPRI CS-5910.

National Research Council (NRC) of Canada, Institute for Environmental Research and Technology, Ottawa, Ontario, Canada K1A OR6.

Scott, J.W., N.T. Whiddon, and R.F. Maddalone: 1993, "Compliance Monitoring Detection and Quantitation Levels for Utility Aqueous Discharges," EPRI Report TR-103205.

USEPA: March 22, 1994a, "National Guidance for the Permitting, Monitoring, and Enforcement of Water Quality-based Effluent Limitations Set Below Analytical Detection/Quantitation Levels."

USEPA: May 1994b, "Determination of Trace Elements in Waters and Wastes by ICP-MS," Methods for Determination of Metals in Environmental Samples, Supplement I, EPA/600/R-94/111.

USEPA: October 11, 1994c, "Summary of Engineering and Analysis Division Trace Metals Status."
USEPA: April 1995a, "EPA Method 1638: Determination of Trace Elements In Ambient Waters By Induc-
    tively Coupled Plasma-Mass Spectrometry," EPA 821-R-95-031.
USEPA: April 1995b, "EPA Guidance on the Documentation and Evaluation of Trace Metals Data Collected
    for Clean Water Act Compliance Monitoring," EPA 821-B-95-002
USEPA: April 1995c, "Results of the Validation Study for Determination of Trace Metals at EPA Water Qual-
    ity Criteria Levels."
USEPA: April 1995d, "EPA Method 1669: Sampling Ambient Water for Trace Metals at EPA Water Quality
    Criteria Levels," EPA 821-R-95-034.

TRACE METAL SPECIATION IN NATURAL WATERS: COMPUTATIONAL VS. ANALYTICAL

D. Kirk Nordstrom
U.S. Geological Survey
3215 Marine Street
Boulder, CO 80303

Abstract. Improvements in the field sampling, preservation, and determination of trace metals in natural waters have made many analyses more reliable and less affected by contamination. The speciation of trace metals, however, remains controversial. Chemical model speciation calculations do not necessarily agree with voltammetric, ion exchange, potentiometric, or other analytical speciation techniques. When metal-organic complexes are important, model calculations are not usually helpful and on-site analytical separations are essential. Many analytical speciation techniques have serious interferences and only work well for a limited subset of water types and compositions. A combined approach to the evaluation of speciation could greatly reduce these uncertainties. The approach proposed would be to (1) compare and contrast different analytical techniques with each other and with computed speciation, (2) compare computed trace metal speciation with reliable measurements of solubility, potentiometry, and mean activity coefficients, and (3) compare different model calculations with each other for the same set of water analyses, especially where supplementary data on speciation already exist. A comparison and critique of analytical with chemical model speciation for a range of water samples would delineate the useful range and limitations of these different approaches to speciation. Both model calculations and analytical determinations have useful and different constraints on the range of possible speciation such that they can provide much better insight into speciation when used together. Major discrepancies in the thermodynamic databases of speciation models can be evaluated with the aid of analytical speciation, and when the thermodynamic models are highly consistent and reliable, the sources of error in the analytical speciation can be evaluated. Major thermodynamic discrepancies also can be evaluated by simulating solubility and activity coefficient data and testing various chemical models for their range of applicability. Until a comparative approach such as this is taken, trace metal speciation will remain highly uncertain and controversial.

Keywords: Trace metals, speciation, natural waters, chemical model

1. Introduction

The importance of trace metal speciation in natural waters has often been emphasized because of its relevance to the interpretion of solubility controls, sorption phenomena, bioavailability, and biotoxicity (Butler, 1964; Morrison, 1989; Kushner, 1993). Improvements in the sampling, preservation, and determination of trace metals has not seen concurrent improvements in the speciation of trace metals. In the context of this paper, speciation refers to the partitioning of trace metals among solids, colloids, surfaces, dissolved free ions, complexed with inorganic ligands in the dissolved phase, and complexed with organics in the dissolved phase. The physical separation and determination of these various species forms can, and usually does, disturb the chemical state of the sample so that every technique results in an "operational definition" for the dissolved, colloidal, or complexed metal. Chemical models that calculate aqueous speciation only give meaningful results when all significant complexes are accounted for, the thermodynamic data are reliable, and the input analytical data are reliable. These criteria are often easier to meet for major constituents of natural waters than for trace metals. The purpose of this paper is to outline a systematic comparative approach to trace metal speciation in natural waters whose main goal is to define the limits of

Water, Air and Soil Pollution **90**: 257-267, 1996.

applicability of trace metal speciation methods to natural waters. Some of these studies have been accomplished and will be discussed below.

Rather than discussing all aspects of speciation among aqueous solutions, colloids, and solid phases, I shall focus primarily on aqueous complexation of trace metals by both organics and inorganics. The physical separation of trace metals between dissolved, colloidal, and particulate phases is discussed in numerous books and papers (e.g. Batley, 1989) although some of the problems mentioned in the present paper also apply to colloidal or particulate separation. Furthermore, I shall distinguish three types of aqueous speciation: redox speciation, metal-organic speciation, and metal-inorganic speciation. Reliable redox speciation can only be done analytically, not computationally (except in scenario exercises). Metal-organic speciation can be done either analytically or computationally, but for complex solutions such as natural waters, the analytical approach currently offers the only practical results. For metal-inorganic speciation, analytical techniques are possible but severely limited for several reasons; computational speciation is more practical.

2. Comparison of Different Analytical Techniques with Each Other and with Computed Speciation

Direct comparison of trace metal speciation by analytical determination with computations *via* a chemical model appears to be a useful approach to evaluating and improving speciation. Unfortunately, several problems limit the applicability of both analytical and computational speciation methods:

1. A wider range of water compositions can be computationally speciated than can be analytically speciated.
2. Analytical speciation techniques often have poor sensitivity at the low concentrations of most trace metals.
3. Analytical speciation techniques can disturb the equilibrium speciation of the solution.
4. Numerous interferences limit the applicability of analytical speciation to natural waters (e.g. lead ions will interfere with cadmium electrode measurements).
5. Required thermodynamic and electrolyte properties of trace metals are not always available for computational speciation in natural waters.

Analytical speciation techniques include electrochemical (especially ion-selective electrode (ISE), potentiometry, and anodic stripping voltammetry (ASV)), ion exchange equilibrium (IEE), solvent extraction, UV-visible spectrophotometry, titrimetry, and vapor and hydride generation coupled to atomic-absorption spectrophotometry (AAS) or to inductively-coupled plasma atomic-emission spectroscopy (ICP-AES). Of these techniques, UV-visible spectrophotometry and vapor generation are particularly well-suited for redox species determination such as Fe(II/III) by colorimetry or Se(IV/VI) by hydride generation. Very few species can be determined reliably by more than one method; however, As (III/V) can be done by both colorimetry and hydride generation. Unfortunately, interferences from concomitant species such as phosphate and silica can contribute high errors in the colorimetric technique. ASV and other polarographic techniques can be used for redox speciation but they are not nearly as convenient nor as reliable as most colorimetric and hydride generation techniques. Portable kits for colorimetric techniques are now being used routinely in the field with reasonable success.

At present, comparison of measured electromotive force (EMF) values with those calculated from redox speciation can be done only for Fe(II/III), S(-II/0), and possibly U(IV/VI). The measurement is based on EMF values determined with a Pt electrode and converted to an Eh value (electrode potential caused by electroactive species relative to the standard hydrogen electrode). The calculation is based on the determination of the individual redox species in a pair (like Fe(II/III)). Corrections for ion-pairing and ionic strength are made with a chemical model computation so that an Eh calculation can be made with the Nernst equation to compare with the measured Eh value. Nordstrom *et al.* (1979) did this comparison for Fe (II/III) in acid mine waters using the WATEQ2 program (Ball *et al.*, 1980) and found that the agreement was excellent . Seventy-seven percent of the values compared were within ±30 mV. A later study by Ball and Nordstrom (1985, 1989) for acid mine waters from a different site with some refinement of techniques found that the majority of values agreed to within ±15 mV. The field results of Ball and Nordstrom (1989) are plotted in Figure 1 along with estimated error envelopes of ±30 mV based on field experience (Nordstrom *et al.*, 1979). Furthermore, this later study more clearly defined the limits of Fe(II/III) redox equilibrium at the surface of a noble metal electrode. At dissolved Fe concentrations below about 10^{-6} molar, the measured potential ceases to respond to Fe and the EMF usually drifts, reflecting interference from oxygen adsorbed on the platinum electrode surface. These findings were first observed in laboratory solutions by Morris and Stumm (1967) and later in experiments by Poulson *et al.* (1987) and Macalady *et al.* (1990). Stipp (1990) evaluated the chemical model for the Fe(II/III) system in acid sulfate solutions to determine the sensitivity to errors in the thermodynamic data.

Ahonen *et al.* (1994) found that Eh from Pt electrode measurements compared favorably with Eh based on U(IV/VI) determinations and speciation calculations for uranium-rich groundwaters in crystalline bedrock (gneiss with granitic veins at the Palmottu uranium deposit, Finland). Uranium redox species were carefully separated and measured by alpha spectrometry. Potentially-interfering species such as dissolved iron, oxygen, and sulfide were at such low concentrations that no serious problems were encountered. Speciation computations were done with the PHREEQE program (Parkhurst *et al.*, 1980) using two different databases for uranium species. The results are plotted in Figure 2.

Metal-organic speciation can be determined by several analytical methods. One approach is to determine the complexation capacity of the water sample by titrimetry (Florence, 1989). A heavy metal ion such as Cu is added to the water sample until the complexing capacity is exceeded and free Cu ion appears. Numerous techniques have been used in titrimetry including bioassays, ion exchange, ISE, ASV, and cathodic stripping voltammetry (CSV). Voltammetric methods have been most widely applied. Florence (1989) has pointed out three problems associated with ASV titrimetry for Cu: (1) complexes that are thermodynamically stable become kinetically labile in the presence of an applied current, (2) organic matter can adsorb on the electrode and depress the signal without any compexation taking place, and (3) the formation of the Cu complex may be very slow, requiring several hours between additions of titrant. These same problems apply to most trace metals. The general problem, that of disturbing the solution equilibria by the application of a current and setting up irreversible changes from the original sample speciation, is nearly always present.

Comparisons of computed speciation with ASV-labile speciation have not often been done because it is recognized that these two techniques estimate different quantities. For example, unpolluted seawater containing about 10^{-8} moles/kg$_{H2O}$ of total dissolved Cu may have anywhere from 10^{-9} to 10^{-11} moles/kg$_{H2O}$ of free Cu ions (Nordstrom *et al.*, 1979). In other words, free Cu concentrations may be only a few percent or less of the total dissolved Cu concentrations. In contrast, ASV-labile Cu has been measured at about 45% of the total dissolved Cu (Florence, 1982) because it includes several forms of inorganically but loosely complexed Cu such as the carbonate and bicarbonate ion pairs. ISE determination of free Cu in seawater is deemed unreliable because of strong interference from Cl ions.

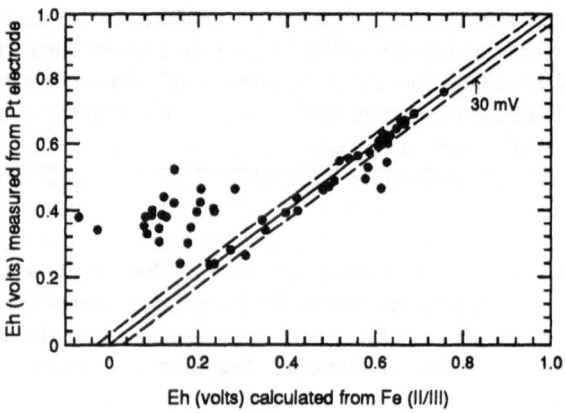

Figure 1. Redox potential calculated from Fe(II/III) analyses compared to Pt electrode measurements.

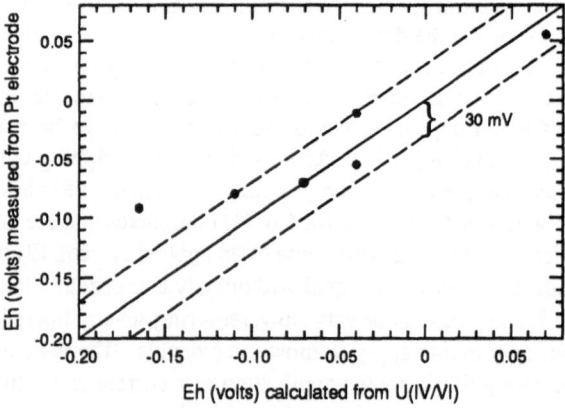

Figure 2. Redox potential calculated from U(IV/VI) analyses compared to Pt electrode measurements.

Combining ASV measurements with ISE measurements, ultrafiltration, and UV irradiation in fresh waters can give a better indication of what the organic complexation of Cu might be (Hart and Jones, 1984). In the study by Hart and Jones (1984) about 10% of the Cu concentration was in the free Cu form and the rest was bound by kinetically-stable organic ligands. The lack of any bound Cu after UV irradiation confirmed the results. Unfortunately, no computed speciation was made for comparison with the analytical inorganic speciation.

There are only five solid-state ISEs that can be used to measure free ion activities for trace metals: Ag, Pb, Cu, Cd, and Hg (Buck, 1979). The Ag electrode is not generally useful because silver is extremely insoluble in most natural waters and many interferences, along with a lack of sensitivity, make it impractical. The other electrodes have been used with varying degrees of success, the most useful being the Cu and Cd electrodes. McGrath *et al.* (1986) compared Cd and Cu free ion concentrations by ISE, IEE, and computation with two different programs for a series of synthetic solutions that included organic acids and a pH range of 3.5-5.5 with good agreement down to about 1 μM in free Cu concentration. In a study of metal complexation by fulvic acids extracted from sewage sludge, Sposito *et al.* (1982) found that free Cd concentrations measured by ISE agreed rather well with calculations from two different models. The results can be seen in Figure 3. Generally, because of problems with interferences and lack of sufficient sensitivity, ISEs do not readily lend themselves to reliable speciation measurements for trace metals in natural waters (Florence, 1989; Lund, 1986).

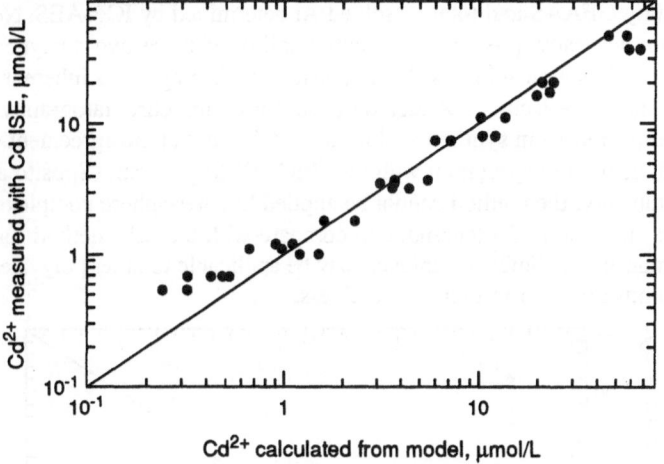

Figure 3. Comparison of Cd^{2+} measured by ISE with Cd^{2+} calculated from model.

Holm and Curtiss (1990) evaluated organic complexation of Cu by comparing determinations made with ISE, fluorescence quenching, and CSV. In their study reasonable values were found for total ligand concentrations and their stability constants when various organic-ligand models were fit to the data for each of the three techniques. This type of comparison gives confidence that for the conditions of the ground-water sample used, these methods can give consistent results. At lower pH values (≤ 6) there was some divergence of the results, indicating that fluorescence quenching is not as reliable a method as the other two. A comparable study of Cd or Pb could provide useful results.

Driscoll (1984) developed an IEE technique for separating organically-bound Al in dilute acidified surface waters. Dissolved Al was assumed to be non-labile, monomeric, organically bound (determined by passage through a strong acid cation exchange column), labile monomeric (acid digestion of a filtered or centrifuged sample), and colloidal (separated by filtration or centrifugation; field procedures were not adequately described). The inorganic monomeric Al was speciated computationally (using a simple computerized chemical equilibrium model, ALCHEMI, Schecher and Driscoll, 1988). Independent measurements by fluoride ISE were done to evaluate the computed inorganic speciation but the results only agreed when the organically-bound Al concentrations were in the very low range of concentration (< 2 µM).

A major U.S. Environmental Protection Agency study was conducted to evaluate five different methods for performing Al speciation in both synthetic and natural water samples (Peden *et al.*, 1989). The five methods were oxine extraction and graphite furnace atomic absorption spectrophotometry (GFAAS), oxine extraction with dialysis and GFAAS, IEE with pyrocatechol violet colorimetry, bound/free fluoride ISE, and lumogallion fluorescence. The results indicated poor agreement between organically-bound Al by IEE (similar to Driscoll (1984), see Kerfoot *et al.*, 1987) and organically-bound Al by dialysis as shown in Figure 4. Unfortunately, considerable differences exist in total dissolved Al depending on the method used, and neither standard additions nor standard reference water samples were analyzed. The dialysis method determined organically-bound Al by difference between dialysed inorganic Al determined by GFAAS and total dissolved Al determined by ICP-AES. No information in the study on accuracy, precision, or comparability of these two analytical methods is available. Hence, it is almost impossible to discuss the discrepancies inherent in the data.

In another study, Bertsch and Anderson (1989) used ion chromatography to determine organically-complexed Al in synthetic solutions and found that the agreement was excellent compared with computed speciation with the GEOCHEM program (Sposito and Mattigod, 1980). Unfortunately, the method cannot be applied to outer-sphere complexes. However, it would appear to be a useful technique to compare with the IEE method for organically-bound Al determinations. Similar techniques may be applicable to other polyvalent metals, but the detection limits may constrain their usefulness.

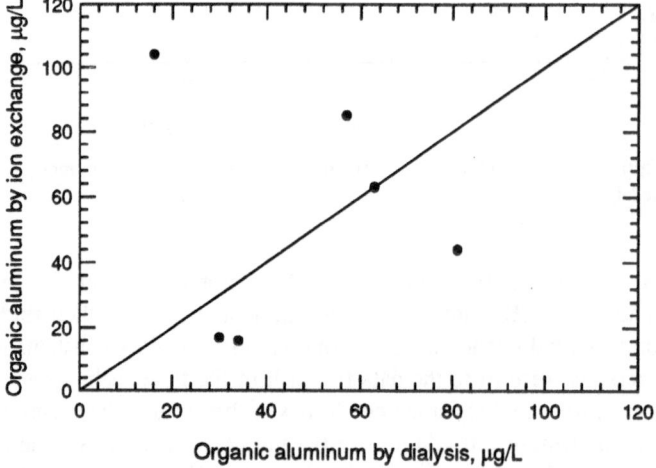

Figure 4. Comparison of organically-bound aluminum by ion exchange with organically-bound aluminum by dialysis.

3. Comparison of Computed Speciation with Experimental Data

Geochemical speciation codes use experimentally-determined values for stability constants and theoretical expressions for activity coefficients to perform iterative numerical approximations for speciation. Examples of popular codes that compute trace metal speciation are MINTEQA2 (Allison et al., 1991; Allison and Brown, 1995), WATEQ4F (Ball and Nordstrom, 1992), GEOCHEM (Sposito and Mattigod, 1980; Parker et al., 1995), and EQ3NR (Wolery, 1992). Comparisons of computational simulations with experimental data are few, and usually emphasize major ions, not trace metals. During the development and application of the Pitzer model (Pitzer, 1992) for calculating activity coefficients in brines, Harvie and Weare (1980) compared the computed solubilities of highly soluble salts (based on single and some mixed electrolyte data) with experimental determinations of their solubilities. Pitzer parameters for electrolyte activity coefficients of $FeSO_4$, $Al_2(SO_4)_3$, $NiSO_4$, and $CuSO_4$ in pure water and sulfuric acid solutions have been obtained by Reardon and Beckie (1987), Reardon (1988), Reardon (1989), and Baes and others (1993), respectively. These studies along with a study on siderite solubility (Ptacek, 1992) have produced data that have proven useful in the geochemical interpretation of acid mine waters. Little else has been done to compare model computations with experimental data for sparingly soluble salts of trace metals. A recent evaluation of the thermodynamic properties for otavite ($CdCO_3$) (Stipp et al., 1993), and earlier reviews of the solubility data for Zn, Cd, Pb, and Hg salts provide excellent evaluations of experimental data, but the comparison with model calculations has not been done.

An insightful evaluation of activity coefficients calculated with different models for both major ions and trace metals was reported by Parkhurst (1990). Comparisons were made in this study between different expressions for mean activity coefficients and experimental data. The results indicated that the ion-association model was in need of modification at higher ionic strengths to include ion triplets and more stability constants to better reproduce the data, although no unique solutions to the activity coefficient expression and related stability constants could be demonstrated.

4. Comparison of speciation by different models

Beginning with the report of Nordstrom et al. (1979), a few intercomparison studies of major speciation codes have been published. In their study, 13 speciation codes were run with the same input water compositions, one a seawater test case and the other a river water test case. All possible major and trace constituents were used in the computations and selected output data such as molalities of some major and minor species, activity coefficients of selected species, ionic strength, and selected saturation indices for minerals were compared in tables. The intercomparison showed that major species compared better than minor and trace species. It also showed that computed species for river water compared better than computed species in seawater. Discrepancies were related to differences in (1) the thermodynamic data for each code, (2) the total number of complexes in a code, (3) the activity coefficients, (4) assumptions regarding the calculation of redox chemistry, (5) the calculation of carbonate chemistry, and (6) the calculation of temperature dependence. In a later report by Broyd et al. (1985), a single ground-water composition was speciated by 10 different codes and the results were similar to those in the previous study except that mineral precipitation was allowed. The mineral precipitation results for non-aluminous minerals gave fair agreement, but the precipitation results for aluminous minerals were inconsistent. Further discussion of these two intercomparisons can be found in Waite (1989). Another comparison of two codes, PHREEQE and EQ3/6, was done by comparing output from five test problems (Noronah and Pearson, 1983). Two of the test problems involved seawater speciation calculations, another involved microcline dissolution, another involved changes in redox chemistry, and the last involved mineral dissolution and precipitation calculations with temperature changes. When these programs were run with the same thermodynamic data, they gave nearly identical results.

Most of the discrepancies in trace metal speciation computations will be caused by differences in the thermodynamic data. This situation will not change until a central clearinghouse or coordination center for chemical modeling databases is established. The International Union of Pure and Applied Chemistry (IUPAC), the Committee on Data for Science and Technology (CODATA), and employees of the Thermodynamic Data Center at the National Institute of Standards and Technology (NIST) have been involved in evaluating and revising basic chemical data (e.g. CODATA key thermodynamic values (Cox *et al.*, 1989)). Unfortunately, there is no liason between the chemical modelers and the international groups who are evaluating data. Modelers who work on natural systems and especially water-rock interactions can do sensitivity analyses on their models that could serve two purposes: (1) to better identify what thermodynamic properties need to be better known or evaluated to address critical water-quality issues and (2) to identify how narrow the uncertainties need to be for certain types of calculations. A few mineral solubilities, such as calcite, are probably known better than they need to be for the temperature range of 0-100°C. A few aqueous species (like Fe^{II} and Fe^{III}), however, are known very poorly for most applied situations. Some sort of central clearinghouse where both thermodynamic experimentalists, evaluators, and geochemical modelers could communicate their data needs and results of recent research should prove useful to the scientific, regulatory, and political communities.

5. Conclusions

Trace metal speciation in natural waters can be accomplished by analytical and computational methods. The reliability of these methods is open to considerable question and debate. Different analytical techniques tend to measure different quantities and although several intercomparisons have been made, improvements can only occur by continued intercomparisons and with more use of combinations of both analytical and computational techniques. When techniques were combined, the information gained was much more useful and quantitative. The limitations and advantages of various methods can be better discerned by such a comparative approach.

Comparison of computed speciation with experimental data on solubilities, potentiometric data, and activity coefficients can provide the range of applicability of speciation models and areas where improvements could be made.

When an intercomparison is made among many different speciation models with the same input water composition, the results show that the trace metal speciation has the greatest discrepancies. Continued improvements in thermodynamic databases with coordination through a central evaluation team or clearinghouse could reduce these problems considerably. There would appear to be an urgent need for a location and a committee to accept the challenge of setting up such a coordinating organization that would benefit both perpetrators and users of basic thermochemical data.

References

Ahonen, L., Ervanne, H., Jaakola, T., and Blomqvist, R.: 1994, *Radiochim.* Acta, **XX**, 1-7.

Allison, J. D. and Brown, D. S.: 1995, "MINTEQA2/PRODEFA2 - A geochemical speciation model and interactive preprocessor," In *Chemical Equilibrium and Reaction Models*, R.H. Loeppert, A.P. Schwab, and S. Goldberg, editors, Soil Science Society of America Special Publication Number 42, 253-270.

Allison, J. D., Brown, D. S., and Novo-Gradac, K. J.: 1991 "MINTEQA2/PRODEFA2, a geochemical assessment model for environmental systems: Version 3.00 user's manual," EPA-600/3-91-021.

Baes,C. F., Jr., Reardon, E. J., and Moyer, B. A.: 1993, *J. Phys. Chem.*, **97**, 12343-12348.

Ball, J. W. and Nordstrom, D. K.: 1985, "Major and trace-element analyses of acid mine waters in the Leviathan mine drainage basin, California/Nevada -- October, 1981 to October, 1982," U.S. Geological Survey Water-Resources Investigations Report 85-4169.

Ball, J. W. and Nordstrom, D. K.: 1989, "Final revised analyses of major and trace elements from acid mine waters in the Leviathan mine drainage basin, California and Nevada - October 1981 to October 1982," U.S. Geological Survey Water-Resources Investigations Report 89-4138.

Ball, J. W. and Nordstrom, D. K.: 1992, "User's manual for WATEQ4F, with revised thermodynamic data base and test cases for calculating speciation of major, trace and redox elements in natural waters," U.S. Geological Survey Open-File Report 91-183, revised and reprinted.

Ball, J. W., Nordstrom, D. K., and Jenne, E. A.: 1980, "Additional and revised thermochemical data and computer code for WATEQ2 - a computerized chemical model for trace and major element speciation and mineral equilibria of natural waters," U.S Geological Survey Water-Resources Invetigations Report 78-116.

Batley, G. E.: 1989, "Physicochemical separation methods for trace element speciation in aquatic samples," In *Trace Element Speciation: Analytical methods and problems*, G.E. Batley, editor, CRC Press, Inc., Boca Raton, Florida, 43-76.

Bertsch, P. A. and Anderson, M. A.: 1989, *Anal. Chem.*, **61**, 535-539.

Broyd, T. W., Grant, M. M., and Cross, J. E.: 1985, *"A report on intercomparison studies of computer programs which respectively model: (i) Radionuclide migration, (ii) Equilibrium chemistry of groundwater,"* EUR 10231 EN, CEC, Luxembourg.

Buck, R. P.: 1979, "Crystalline and pressed powder solid membrane electrodes," In *Ion-Selective Electrode Methodology*, A.K. Covington, editor, CRC Press, Inc., Boca Raton, Florida, 175-250.

Butler, J. N.: 1964, *Ionic Equilibrium: A mathematical approach*, Addison-Wesley, Reading, Massachusetts.

Cox, J. D., Wagman, D. D., and Medvedev, V. A.: 1989, *CODATA key values for thermodynamics*, Hemisphere Publishing Corporation, 271 pp.

Driscoll, C. T.: 1984, *Int. J. Environ. Anal. Chem.*, **16**, 267-283.

Florence, T. M.: 1982, *Talanta*, **29**, 345.

Florence, T. M.: 1989, "Electrochemical techniques for trace element speciation in natural waters," In *Trace Element Speciation: Analytical methods and problems*, G.E. Batley, editor, CRC Press, Inc, Boca Raton, Florida, 77-116.

Hart, B. T. and Jones, M. J.: 1984, "Measurement of the trace metal complexing capacity of Magela Creek waters," In Complexation of Trace Metals in Natural Waters,. C.J.M. Kramer and J.C. Duinker, editors, The Hague: Martinus Nijhoff/Dr. W. Junk Publishers, 201-212.

Harvie, C. E. and Weare, J. H.: 1980, *Geochim. Cosmochim. Acta.*, **44**, 981-997.

Holm, T. R. and Curtiss, C. D., III: 1990, "Copper complexation by natural organic matter in ground water," In *Chemical Modeling in Aqueous Systems II*, D.C. Melchior and R.L. Bassett, editors, American Chemical Society Symposium Series 416, Washington, D.C.: , 508-518.

Kerfoot, H. B., Lewis, T. E., Hillman, D. C., and Faber, M. L.: 1987, *National Surface Water Survey, Eastern Lake Survey (Phase II - Temporal Variability) Analytical Methods Manual*, EPA - 600/X-87/008

Kushner, D. J.: 1993, *Water Pollut. Res. J.Canada*, **28(1)**, 111.

Lund, W.: 1986, "Electrochemical methods and their limitations for the determination of metal species in natural waters," In *The Importance of Chemical "Speciation" in Environmental Processes*, M. Bernhard, editor, Springer-Verlag, Berlin 533-561.

Macalady, D. L., Langmuir, D., Grundl, T., and Elzerman, A.: 1990, "Use of model-generated Fe^{3+} ion activities to compute Eh and ferric oxyhydroxide solubilities in anaerobic systems," In *Chemical Modeling in Aqueous Systems II*, D.C. Melchior and R.L. Bassett, editors, American Chemical Society Symposium Series 416, Washington, D.C., 350-367.

McGrath, S. P., Sanders, J. R., Laurie, S. H., and Tancock, N. P.: 1986, *Analyst (London)*, **111**, 459-465.

Morris, J. C. and Stumm, W.: 1967, "Redox equilibria and measurements of potentials in the aquatic environment," In *Equilibrium Concepts in Natural Water Systems*, W. Stumm, editor, American Chemical Society Symposium Series 67, Washington, D.C., 270-285.

Morrison, G. M. P.: 1989, "Trace element speciation and its relationship to bioavailability and toxicity in natural waters," In *Trace Element Speciation: Analytical methods and problems*, G.E. Batley, editor, CRC Press, Inc., Boca Raton, Florida, 25-42.

Nordstrom, D. K., Jenne, E. A., and Ball, J. W.: 1979, "Redox equilibria of iron in acid mine waters," In *Chemical Modeling in Aqueous Systems*, E.A. Jenne, editor., American Chemical Society Symposium 93, Washington, D.C., 51-80.

Nordstrom, D. K., Plummer, L. N., Wigley, T. M. L., Wolery, T. J., Ball, J. W., Jenne, E. A., Bassett, R. L., Crerar, D. A., Florence, T. M., Fritz, B., Hoffman, M., Holdren, G. R., Jr., Lafon, G. M., Mattigod, S. V., McDuff, R. E., Morel, F., Reddy, M. M., Sposito, G., Thrailkill, J.: 1979, "Comparison of computerized chemical models for equilibrium calculations in aqueous systems," In *Chemical Modeling in Aqueous Systems*, E.A. Jenne, editor, American Chemical Society Symposium Series 93, Washington, D.C., 857-892.

Noronha, C. J., and Pearson, F. J., Jr.: 1983, *Geochemical models suitable for performance assessment of nuclear waste storage: Comparison of PHREEQE and EQ3/6*, Intera Technical Report, ONWI-473.

Parker, D. R., Norvell, W. A., and Chaney, R. L.: 1995, "GEOCHEM-PC - A Chemical speciation program for IBM and compatible personal computers," In *Chemical Equilibrium and Reactrion Models*, R.H. Loeppert, A.P. Schwab, and S. Goldberg, editors, Soil Science Society of America Special Publication Number 42, 253-270.

Parkhurst, D. L.: 1990, "Ion-association models and mean activity coefficients of various salts," In *Chemical Modeling of Aqueous Systems II*, D.C. Melchior and R.L. Bassett, editors, American Chemical Society Symposium Series 416, Washington, D.C., 30-43.

Parkhurst, D.L., Thorstenson, D. C., and Plummer, L. N.: 1980, "PHREEQE - a computer program for geochemical calculations," U.S. Geological Survey Water-Resources Investigations Report 80-96.

Peden, M. E., Amankwah, S. A., Keller, B. J., Krug, E. C., and Peden, J. M.: 1989, *Evaluation of aluminum speciation using synthetic and natural samples: Final report*, Illinois State Water Survey, Champaign, Illinois, USEPA Cooperative Agreement CR813489-01.

Pitzer, K. S.: 1992, "Ion interaction approach: Theory and data correlation," In *Activity Coefficients in Electrolyte Solutions*, K.S. Pitzer, editor, CRC Press, Inc., Boca Raton, Florida, 75-154.

Poulson, R. E., Powers, C. R., and Essington, M. E.: 1987, "Validation of inorganic chemical speciation for geochemical models," Western Res. Inst. DOE/MC/11076 - 2459, DE88 011565.

Ptacek, C. J.: 1992, *Experimental determination of siderite solubility in high ionic-strength solutions*, Ph.D. thesis, University of Waterloo, Waterloo, Canada, 331 pp.

Reardon, E. J.: 1988, *J. Phys. Chem.*, **92**, 6426-6431.

Reardon, E. J.: 1989, *J. Phys. Chem*, **93**, 4630-4636.

Reardon, E. J. and Beckie, R. D.: 1987, *Geochim. Cosmochim. Acta*, **51**, 2355-2368.

Schecher, W. D. and Driscoll, C. T.: 1988, *Water Resour. Res.*, **24**(4), 533-540.

Sposito, G. and Mattigod, S. V.: 1980, GEOCHEM: A computer program for the calculation of chemical equilibria in soil solutions and other natural water systems, Kearney Foundation of Soil Science, University of California, Riverside, CA.

Sposito, G., Bingham, F. T., Yadav, S. S., and Inouye, C. A.: 1982, *Soil Sci. Soc. Am. J.*, **46**(1), 51-56.

Stipp, S. L. : 1990, *Environ. Sci. Technol.*, 24(5), 699-705.

Stipp, S. L. S., Parks, G. A., Nordstrom, D. K., and Leckie, J. O.: 1993, *Geochim. Cosmochim. Acta*, **57**(12), 2699-2713.

Waite, T. D.: 1989, "Mathematical modeling of trace element speciation," In *Trace Element Speciation: Analytical methods and problems*, G.E. Batley, editor, CRC Press, Inc., Boca Raton, Florida, 117-184.

Wolery, T. J.: 1992, "EQ3NR, A computer program for geochemical aqueous speciation-solubility calculations: Theoretical manual, user's guide, and related documentation (Version 7.0)," Lawrence Livermore National Laboratory, UCRL-MA-110662 PT III.

METAL SPECIATION: SURVEY OF ENVIRONMENTAL METHODS OF ANALYSIS

Martin H. Mach[1], Babu Nott[2], Judith W. Scott[1], Raymond F. Maddalone[1], Nina T. Whiddon[1],

[1]*Chemistry Technology Dept., TRW, 1 Space Park, Redondo Beach, CA 90278,* [2]*EPRI, Palo Alto, CA 94303*

Abstract. As part of a recent task under the EPRI Analytical Methods Qualification Program (RP 1851), TRW has surveyed the methods available for monitoring metal species in typical utility aqueous discharge streams. Methods for determining the individual species of these metals can become important in a regulatory sense as the EPA transitions to assessment of environmental risk based on bioavailability. For example, EPA considers methyl mercury and Cr(VI) much more toxic to the aquatic environment than inorganic mercury or Cr(III). The species of a given element can also differ in their transport and bioaccumulation. Methods for speciation generally include a selective separation step followed by standard metals analysis. Speciation, therefore, is mainly derived from the separation step and not from the method of final quantitation. Examples of separation/analysis include: selective extraction followed by graphite furnace atomic absorption or ICP-MS; separation by GC followed by metals detection; chelation and/or direct separation by LC followed by UV measurement or metals detection; and ion chromatography with conductivity, UV, or metals detection. There are a number of sampling issues associated with metal species such as stabilization (maintaining oxidation state), absorption, and filtration that need to be addressed in order to obtain and maintain a representative sample for analysis.

Key Words. Speciation, Environmental Sampling, Instrumental Methods

1. Introduction

1.1 SCOPE

In 1984, TRW conducted a study of pollutants emanating from steam-electric power plants (EPRI, 1984). Pollutants were selected for further study based on the frequency they were found in the aqueous discharges from power plants, and by comparison with intake water concentrations. Criteria and a methodology for identifying and ranking these pollutants were also developed. Of those pollutants identified, this review will concentrate on arsenic, selenium, chromium, mercury, and lead. While not identified as a power plant water emission, mercury is also discussed as it is a good example of an element that forms a variety of both inorganic and organic derivatives, and for which separation and detection schemes have been developed that are applicable to the other pollutants.

This paper summarizes the result of a survey of Chemical Abstracts and NTIS over the last 4 - 5 years and highlights the most widely used and accepted analytical methods for speciation of selected target metals. This survey was not meant to provide a totally inclusive review but to provide the reader with a feel for the methodology and recent trends.

Water, Air and Soil Pollution **90**: 269-279, 1996.
© 1996 *Kluwer Academic Publishers.*

1.2 SPECIATION.

Speciation in the context of this paper is the analysis of metals to determine their chemical forms. For example, chromium can be present as Cr(III) and Cr(VI), two different oxidation states. Other metals can be present as both inorganic (e.g., mercuric ion) and organic (e.g., dimethyl mercury) forms. Lead can be present as a non-volatile ionic species or a volatile tetraalkyl lead.

The speciation process typically involves collection, sample storage, one or more separation steps, and measurement of the metal in the final extract (the analytical "finish"). Problems can occur at any stage of this process, especially when sampling complex, environmental matrices involving dynamic equilibria between different phases (e.g., partitioning between the liquid phase and particulates), adsorption on sampling containers, inadvertent changes in oxidation state, and the effect of high concentrations of dissolved species on the analytical finish. Thus, the challenge of speciation is to separate the different forms and accurately quantitate each species in the particular matrix under study.

2. Analytical Methods

2.1 ARSENIC

2.1.1 Species
Arsenic exists in either its elemental form or the As(III) or As(V) valence states. The (III) and (V) forms can be either inorganic (e.g., As_2O_3) or contain organic groups (e.g., methyl arsine). Commonly reported organic forms include monomethyl arsonic acid (MMAA) and dimethylarsinic acid (DMAA). Monomethyl arsonous acid (MMAA-As(III)) and dimethyl arsonous acid (DMAA-As(III)), formed by the action of hydrogen sulfide on MMAA and DMAA, have also been reported (Hasegawa, 1994). Bioincorporated forms include arsenobetaine (AB, present in marine organisms and used as an indicator of arsenic uptake), and arsenocholine (AC). Certain researchers have determined that the toxicity of arsenic is dependent on its species. The inorganic arsenics are more toxic than the organoarsenicals, and the trivalent forms more toxic than the pentavalents (Fowler, 1983). Therefore much effort has been devoted to methods that will identify the particular molecule in terms of its bonded groups and oxidation state. Reviews of arsenic toxicity, collection, and analysis methods have been published by Clifford (1993) and McLaren (1992).

2.1.2 Sampling and Collection.
Care must be taken to prevent speciation changes during sample collection and storage. Plastic containers should be acid washed with 5% HCl, and traces of oxidizing and reducing agents avoided to preserve the oxidation state and avoid oxidation of organometallic compounds. Freezing samples to -80C in liquid nitrogen has been recommended (Crecelius, 1986). In addition, the trivalent methyl arsenicals are more subject to oxidation than their inorganic As(III) counterparts (Cullen, 1966).

2.1.3 Analytical Methods.

EPA Method 1632 describes the determination of total inorganic As in water: Inorganic arsenic (As(III) + As(V)) is reduced to arsine (AsH$_3$), which is purged and trapped on a cooled adsorbent trap. The trapped arsine is thermally desorbed and detected by flame atomic absorption spectrometry (FAAS). This method quantitates total arsenic, and can also separately quantitate the inorganic arsines (as AsH$_3$) and organic arsines (as higher boiling, volatile arsines) via their boiling points during the thermal desorption step. Method 1632 is reported to have a method detection limit (MDL[1]) of 0.002 µg/L. Inductively Coupled Plasma-Mass Spectrometry (ICP-MS, EPA Method 200.8), stabilized temperature graphite furnace atomic absorption spectrometry (ST/GFAAS, EPA Method 200.9), and ICP - atomic emission spectrometry (ICP/AES, EPA Method 200.15) have also been used for the measurement step, with somewhat higher MDLs of 0.1, 0.5, and 3 µg/L, respectively. For the recent literature methods reported below, detection levels[2] (as quoted in the referenced documents) typically fall in the low to sub- µg/L range using the above-cited analytical finishes.

As(III) can be selectively reduced in the presence of As(V) using NaBH$_4$ buffered at pH 6 (Watras, 1992). This allows differentiation between the inorganic As(III) and As(V). In a research setting, laboratories may be able to achieve detection levels for environmental water samples in the low ng/L range.

For samples containing MMAA and DMAA, a hydride/separation scheme has been reported in which MMAA and DMAA are converted to the hydrides and quantitated by fractional distillation with AAS detection after the inorganic As(III)/(V) is separated and independently quantitated via a carbamate precipitate chelation process (van Elteren, 1994). The authors report detection levels for MMAA and DMAA of 0.2 and 0.5 ng, respectively. This process was applied to sediment interstitial water containing arsenic species at low µg/L levels. Another chelation method using carbamate-based separation followed by hydride generation/heated quartz cell AAS has been applied to the speciation of inorganic and organoarsenicals in natural waters, with reported detection levels of around 0.01 µg/L (Hasegawa, 1994).

Other methods obviate the need for the hydride generation step. Micellar liquid chromatography using surfactants coupled with ICP-MS has been applied to the analysis of arsenic species in biological samples without the need for extensive sample cleanup and deproteinization steps (Ding, 1995). This method was used on urine samples with a reported detection level of 0.9 µg/L for DMAA and 3 µg/L for As(III), As (V), and MMAA. One feature of this method is its freedom from chlorine interfere in the ICP-MS, where the $^{40}Ar^{35}Cl$ ion at mass 75 can interfere with the ^{75}As ion. This increases the applicability of this method for the analysis of seawaters. Capillary zone electrophoresis with on-column preconcentration/field amplified injection can achieve sample enrichment factors of around 150, compensating for the method's small sample size (ca. nanoliters). It has been applied to

---

[1] Method Detection Limit as defined by 40 CFR Part 136, Appendix B, and published in the referenced EPA method. However, MDLs are sample dependent, and may vary as the sample matrix varies.

[2] The detection levels referred to throughout this paper may only be achievable by certain high quality research laboratories, and are not likely to be routinely achieved by commercial laboratories.

the detection of As(V) in tap water (Li, 1995), with a detection level of 25 µg/L; As(III) could not be quantitated by the current separation scheme as it overlaps with a peak arising from an unidentified matrix ion (or ions) in the sample.

2.2 SELENIUM

2.2.1 Species.
Detailed speciation of selenium is complicated by its four possible oxidation states, viz., -II (selenide), 0 (elemental), +IV (selenite) and +VI (selenate), and their selective complexation and/or bonding. Volatile species include dimethyl selenide (DMSe) and dimethyl diselenide (DMDSe). In addition to oxidation state, selenium in water samples is divided into 2 classes: dissolved Se that passes through a 0.45mM filter, and particulate Se (>0.45mM).

Particulate selenium is associated with sediments and other suspended solids. In sediments, selenium may be associated with organic material, iron and manganese oxides, carbonates, or other mineral phases, either adsorbed to or coprecipitated with these phases. Se(II) can be covalently bound to the organic portion of the sediment and other materials to give organoselenides. Two recent reviews of selenium analysis and speciation have been published (Cappon, 1994; Dauchy, 1994).

2.2.2 Sampling and Collection.
Selenium loss by adsorption on the container can be minimized by acidifying to pH 2 with a non-oxidizing acid such as HCl and using 1 liter or larger high density polyethylene or borosilicate glass containers. Se(IV) and Se(VI) in natural waters can be preserved at the 1 µg/L level in pyrex or polyethylene bottles at pH 1.5 (H_2SO_4); acidification with HNO_3 can interfere with the hydride generation process. Determinations should be performed immediately after collection, because organic selenium compounds such as dimethyl selenide are lost within 24 hours. Conversely, organoselenium compounds have been found to increase in concentration due to biological activity; refrigeration can mitigate this process, as well as prevent loss of volatile species.

2.2.3 Analytical Methods.
The basic analytical method for total selenium is generation of the volatile hydride (H_2Se) from Se(IV) using $NaBH_4$, followed by graphite furnace AAS (GFAAS, e.g., EPA Method 200.9, MDL 0.6 µg/L). Watras (1992) describes a scheme for quantitating the different valence states using a differential reduction method followed by a GFAAS finish.

A review of analytical techniques for the speciation of selenium (Olivas, 1994) lists a number of methods for the speciation of different forms of selenium (primarily Se(IV) and Se(VI)), along with the matrices examined and detection levels. Matrices include Milli-Q deionized water, natural waters, ground waters, sediments, and biological materials, with detection levels in the low to sub-µg/L range. Methods include differential pulsed cathodic scanning voltammetry (DPCSV, for selenium in water, 0.04 µg/L), ion chromatography-liquid chromatography (IC-LC) with conductometric detection (in soil extracts, ca. 100 µg/L), molecular fluorescence spectrometry and isotope dilution MS (in natural waters, 5 - 10 ng/L), and liquid chromatography-ICP-AES (in Milli-Q deionized water, 14 - 54 ng/L). A schematic diagram outlining the preferred analytical separation and analysis methods for selenium speciation is also presented in the above reference.

2.3 CHROMIUM

2.3.1.Species.
There are significant differences in the toxicities of Cr(III) and Cr(VI), with Cr(VI) classified by EPA as a human carcinogen based on studies showing an increase in lung cancer. The adverse health effects of chromium compounds have been reviewed (Cohen, 1993). Cr(III) is considered to be essential for the maintenance of lipid, protein, and glucose metabolism, but Cr(VI) is reported to be toxic due to its facile penetration of biological membranes and its oxidizing potential.

2.3.2 Sampling and Collection.
A number of methods have been used to avoid alteration of the oxidation state during sampling and storage. These include separation of the two forms by solvent extraction or by ion exchange and separate analysis of the fractions, or by coprecipitation techniques. As with other polyvalent species, the use of oxidizing acids or oxidizing/reducing agents in the treatment of collection vessels, etc., is to be avoided.

2.3.3 Analytical Methods
EPA Method 1636 for Cr(VI) consists of isolation of Cr(VI) by ion chromatography, post-column derivatization with diphenylcarbazide, and detection of the colored complex at 530nm, with a claimed MDL of <1 µg/L. Total chromium can be analyzed by EPA Methods 200.8 (ICP-MS, MDL 0.08 µg/L), 200.9 (ST/GFAAS, MDL 0.1 µg/L), and 200.15 (ICP-AES, MDL 2 µg/L).

An automated high performance liquid chromatography (HPLC) method has been reported (Posta, 1993) in which Cr(III) and Cr(VI) are separated by ion pair chromatography followed by AAS detection with high pressure nebulization. This method has been applied to chromium speciation in drinking water, waste water, and soil extracts, with a detection level of 20 - 30 µg/L.

Cr(III) has been detected at the low µg/L levels in estuarine and sea waters via preconcentration on iminodiacetate resin prior to determination by flame AAS (Pasullean, 1995). Concentration on cellulose adsorbents followed by flame AAS detection has been used to speciate Cr (III) and Cr(VI) in tap and surface waters at 0.8 and 1.4 µg/L, respectively (Naghmush, 1994).

Ion chromatography has been used to separate Cr(III) from Cr(VI) using a post-column derivatization method based on the catalytic oxidation of luminol (Beere, 1994), with detection levels of 0.05 µg/L (Cr(III)) and 0.1 µg/L (Cr(VI)) in a simulated fresh water standard reference material. Direct methods that bypass the pre-separation step include fluorimetry of Cr(VI) in the presence of Cr(III) after complexation with a crystal violet-iodine reagent (Hayashi, 1993), and electrochemical techniques. These methods typically allow for detection in the low ppb range for laboratory (i.e., analytical) samples.

2.4 MERCURY

2.4.1 Species
Common forms include inorganic mercuric (II), methyl mercury cation (MeHgX), and dimethyl mercury (Me$_2$Hg). Humic matter can methylate mercury (Weber, 1993), reduce Hg(II) to Hg$^{\circ}$, and form complexes that EPA considers to be of greater toxicity than the inorganic starting form. Approximately 50 to 90% of total mercury in coastal waters and estuaries is bound to humic matter (Weber, 1988). Because methyl mercury appears to bioaccumulate in organisms, it is of greater concern than other forms of this metal. Several reviews have been published in recent years covering sampling and collection methods (Baeyens, 1992), analytical methods (Puk, 1994), and human health risks (Galli, 1993).

2.4.2 Sampling and Collection
Samples collected for total Hg analysis are usually acidified to a 0.1% HCl concentration to complex Hg and prevent loss through volatilization. The integrity of samples for organomercury speciation can be preserved by freezing without the addition of HCl (Watras, 1992). EPA Method 1631 specifies 125 ml to 1L Fluoropolymer bottles deactivated by a multi-step cleaning process with HCl. After cleaning, the bottles are tightly capped (with a wrench) and double-bagged in new, polyethylene zip-lock bags.

2.4.3 Analytical Methods
EPA claims its Methods 200.8 (ICP-MS) and 200.15 (ICP-AES) have MDLs for mercury of 0.2 and 3 ppb, respectively. EPA Method 1631, using cold vapor atomic fluorescence spectrometry (CVAFS) in combination with stricter contamination control protocols, claims an MDL of <0.0002 µg/L. Again, these MDLs provide only a rough indication of instrumental sensitivity. A typical scheme for the species separation and quantitation of four different mercury fractions has been reported by Watras (1992). Volatile mercury species (Hg$^{\circ}$, Me$_2$Hg) are determined by purging an untreated water sample through stacked Carbotrap and Au-coated sand. The volatile organomercurials are trapped on the Carbotrap for subsequent analysis of individual compounds by GC and CVAFS detection. Hg$^{\circ}$ is trapped on the gold, with quantitation by thermal desorption CVAFS. Total mercury is analyzed by oxidation of all species to Hg(II) by BrCl, reduction to Hg$^{\circ}$ by SnCl$_2$, and the resulting Hg$^{\circ}$ quantitated by Au adsorption followed by thermal desorption and CVAFS. Methylmercury and Hg(II) are ethylated, the compounds trapped on Carbotrap, and the Hg(II) (as diethyl mercury) and methylmercury (as methylethyl mercury) separated by GC and quantitated by CVAFS.

Hydride generation methods have also been applied to mercury speciation. Hg(II), monomethyl mercury cation (MeHg), dimethyl- and diethyl mercury were determined at levels of from 50 to 110 pg in estuarine marsh grass samples by hydride generation/volatilization, trapping and separation on a GC column, and detection by quartz furnace AAS (Puk, 1994a)

Another scheme for the separation of mercury compounds present in soil and sediments into five different classes has been proposed (Miller, 1995) in which the sample is first divided into toluene- and water soluble fractions, plus insoluble residues. The insoluble fraction is further differentiated into dilute HNO$_3$ soluble (e.g., HgO), HNO$_3$ soluble (Hg$^{\circ}$), and

HNO$_3$/HCl soluble (e.g., HgS) fractions.   Mercury-containing fractions in soil were quantitated by ICP-MS at the ppm level.

Inorganic, methyl- and ethylmercury have been analyzed in natural waters (Emteborg, 1993) and biological fluids (Bulska, 1992) at low and sub ng/L levels via preconcentration on a dithiocarbamate resin, followed by extraction, butlylation, and detection by GC-AES.

Speciation of inorganic and methyl mercury in seawater at 0.1 - 0.2 μg/L has been accomplished using HPLC-CVAFS (Aizpun, 1994). Determination of total mercury requires conversion of all forms to Hg(II), and this may involve decomplexation from organic ligands and alkylated forms. Sn(II) or NaBH$_4$ is then used to reduce Hg(II) to Hg$^{\circ}$ which can be quantitated by GC/electron capture (Rubi, 1994), or atomic fluorescence (Ritsema, 1994). Speciation of MeHg-X as the hydride (MeHgH) has been accomplished by an acidification/extraction scheme followed by reduction with cysteine (Westwoo, 1967).

2.5 LEAD

2.5.1 Species
The most important organic forms are the tetraalkyl leads (TALs). Researchers have found that these compounds can be readily absorbed by the lungs and also penetrate the skin and biological membranes. They therefore are considered more harmful than inorganic lead. TALs are degraded by sunlight and ozone to trialkyl lead compounds, which are further degraded to inorganic leads via a dialkyllead intermediate.

2.5.2 Sampling and Collection
Adsorption on sample bottles can be avoided by using containers of Teflon or polyethylene that have been acid treated and thoroughly rinsed. In some cases, acid treatment may activate surface sites and lead to adsorption of lead species. Taking these factors into consideration, lead samples should be stored in suitably conditioned containers and analyzed as soon as possible after collection. Samples themselves should not be acidified prior to speciation, as this might change the distribution of lead species.

In addition to the sample containers, another issue is whether or not to filter the samples prior to analysis. Adsorption on particulates is a dynamic process and the distribution between dissolved and adsorbed lead may change with time after the sample is collected. Filtration itself may be problematical, as filtration efficiency for colloidal particles changes depending on the filter load. On the other hand, inhomogeneity of unfiltered samples may lead to errors if care is not taken to assure a uniform sample.

While acidification of samples can lead to speciation changes, samples containing low levels of alkyllead species should be acidified to minimize wall adsorption (Blaszkewicz, 1987). TALs can be extracted from water using organic solvents, with extraction efficiencies significantly higher if the sample does not contain suspended solids. This problem is more severe with filtered samples, since some of the TALs remain on the filter residue.

2.5.3 Analytical Methods

EPA Method 1640 preconcentrates acid-soluble lead using an iminodiacetate functionalized chelating resin to achieve a reported MDL of 0.0081 µg/L. The complex is eluted from the resin using an ammonium acetate buffer at pH 5.5, and determined using on-line ICP-MS. EPA Methods 200.8 (ICP-MS, 0.02 µg/L), 200.9 (ST/GFAAS, 0.7 µg/L), and 200.15 (ICP-AES, 4 µg/L), do not involve preconcentration, and have somewhat higher reported MDLs. The EPA methods are applicable to acid-soluble lead species, and are not suitable for the direct determination of volatile TALs and other acid-insoluble species.

Trapping and concentrating lead species on adsorbent columns allows for its detection at lower levels than if the samples were run directly. Speciation of lead in fresh waters has been accomplished by trapping on three different adsorbents: Chelex-100 (for labile complexes), Sep-Pak C-18 (for non-polar organics) and Fractogel DEAE (for ion exchangeable materials), followed by ICP-MS. Detection levels are in the low ng/L range, while the lead concentration range in fresh waters is 100 - 950 ng/L (Haraldsson, 1993).

A review of the speciation of organolead compound analysis (Lobinski, 1994) details current separation and analysis methods. Ionic organolead compounds in environmental water samples have been extracted as diethyldithiocarbamate complexes into an organic solvent, butylated with a Grignard reagent, and the volatile products detected by GC-quartz furnace AAS or GC-AES . Detection levels of 15 ng/g have been reported for river sediments, and 0.6 - 10 ng/g for soils.

Volatile alkyl lead compounds can be concentrated directly on Porapak and Tenax cartridges and analyzed by GC/MS with selective ion monitoring (Nerin, 1994). Lead-containing particles (Wonders, 1994) have been analyzed for total lead by differential pulsed anodic scanning voltammetry (DPASV), X-ray fluorescence, graphite furnace AAS, and laser microprobe mass analysis.

3. Detection Limits

In Table 1, we have summarized the MDLs stated for EPA's 200 Series and the new 1600-Series methods, plus detection levels for representative methods cited in the literature. The numbers indicate that at least for determining the total metals in a sample, the 1600-series of methods appear at or below the current water quality criteria (WQC) standards. However, it must be emphasized that these new methods have been tested in only 1 to 3 laboratories and may not provide the same low detection levels or reliable results for actual environmental samples with complex matrices. As the EPA transitions toward bioaccumulation risk models which are based on the bioavailability of a particular metal species, these detection levels will require validation as part of a complete metal speciation approach which includes the species-specific separation step.

The literature contains new methods and separation/detection schemes that are being investigated for specific substrates and matrices. While many methods are reported, it is not always possible to meaningfully compare the detection levels for the different analytical techniques because of the different ways detection levels are defined and the different sample matrices examined, e.g., acidic solutions, uncharacterized "natural waters", tapwater,

Table 1
Representative EPA and Literature Cited Analytical Methods

	As	Se	Cr(III)/(VI)	Hg	Pb
EPA WQC[1] (μg/L)	0.018	5.0	57/10.5	0.012	0.14
EPA "200-Series"					
MDL (μg/L)	0.1	0.5	0.08/0.08	0.2	0.02
Method No. [2]	200.8	200.8	200.8	200.8	200.8
MDL (μg/L)	0.5	0.6	0.1/0.1	n/a	0.7
Method No. [3]	200.9	200.9	200.9	200.9	200.9
MDL (μg/L)	3	5	2/2	3	4
Method No. [4]	200.15	200.15	200.15	200.15	200.15
EPA "1600-Series"					
Technique	Hyd/FAAS[5]	ST-GFAAS	ST-GFAAS /IC	CVAFS	ICP/MS
MDL (μg/L)	0.002	0.83	0.10/0.23	0.0002	0.0081
Method No.	1632	1639	1639/1636	1631	1640
Literature-Cited Methods					
Technique	Hydride/ AAS	Molec Fluor.	Flame AAS	Hydride/At. fluorescence	Trap/GC-MS
Matrix	Nat'l waters	Nat'l waters	Seawater	Seawater	Air
DL (μg/L)[6]	0.005	0.005	5	0.001	0.7 - 2.6 pg (TALs)
Reference	Watras	Olivas	Pasullean	Ritsema	Nerin
Technique	Carbamate/ hydride	CZE	Adsorp/ Flame AAS	Adsorp/ GC-ICP-AES	Adsorp/ ICP-MS
Matrix	Nat'l. waters	Tap, spring water	Tap/surface water	Natural. waters	Fresh water
DL (μg/L)	0.01	2	1	0.001	0.005
Reference	Hasegawa	Li	Naghmush	Emteborg	Haraldsson
Technique	LC/ ICP-MS		IC/Luminol	HPLC/ CVAFS	Extract/ GC-QFAAS
Matrix	Seawater		Fresh water	Seawater	Environ. water
DL (μg/L)	1		0.1	0.1	1
Reference	Ding		Beere	Aizpun	Lobinski

---

[1] Lowest of the freshwater, marine, and human health WQC [Water Quality Criteria] promulgated by EPA for 14 states at 40 CFR Part 131 (57 FR 60848), with hardness-dependent freshwater aquatic life criteria adjusted in accordance with 57 FR 60848 to reflect the worst case hardness of 25 mg/L $CaCO_3$ and all aquatic life criteria adjusted in accordance with the Oct 1, 1993 Office of Water guidance to reflect dissolved metals criteria.
[2] Inductively coupled plasma mass spectrometry in the selective ion monitoring mode
[3] Stabilized temperature graphite furnace atomic absorption spectrometry
[4] Inductively coupled plasma - atomic emission spectrometry
[5] Hydride/flame atomic absorption spectrometry
[6] Detection Level from the reference cited

seawater, biological samples, etc. In addition, some authors report sensitivity in absolute amounts rather than concentrations, making comparative data more difficult to interpret.

4. Conclusions

The EPA has proposed a variety of unvalidated (i.e., no interlaboratory confirmation) analytical methods for total metals that reportedly have detection at or below the current EPA Water Quality Criteria (WQC) levels. However, it is recognized that the biotoxicities of all metal species are not the same, and that future WQC requirements may require speciation in addition to total metals. The literature cites a variety of separation steps combined with different detection techniques to speciate metals at low concentrations. At present, neither the EPA nor the literature methods have been validated in different matrices of environmental concern or in interlaboratory tests, so their performance in real-world matrices is unknown. Factors that must be taken into account in a validation program include sampling, sample preservation, general laboratory cleanliness, specificity, detection levels, ease of use, potential interfering compounds or substances, and cost effectiveness. The candidate methods should be evaluated by a single laboratory using typical utility matrices to screen out those methods not robust enough for the utility environment. A formal procedure can then be proposed for the selected method(s), followed by validation by multiple laboratories using several utility matrices.

References

Aizpun, B. J., Fernandez, M. L., Blanco, E., Sanz-Medel, A., 1994, *Anal. At. Spectrom.*, **9**, 1279 - 1284.

Baeyens, W., 1992, *Trends Anal. Chem.*, **11**, 245 - 254.

Beere, H., Jones, P., 1994, *Anal. Chim. Acta.*, **293**, 237-243.

Blaszkewicz, M., 1987, *Int. J. Environ. Anal. Chem.*, **28**, 207.

Bulska, E, Emteborg, H., Baxter, D. C., Frech, W., Ellingsen, D., Thomassen, Y., 1992, *Analyst*, **117**, 657-663.

Cappon, C.J., 1994, *Anal. Contam. Edible Aquat. Resour.*, 205-224, VCH Pub., New York, NY.

Clifford, D., Zhang, Z., 1993, *Proc. Water Qual. Technol. Conf.*, pt. 2,1955-1968.

Cohen, M., Kargacin, B., Klein, C. B., Costa, M., 1993, *Crit. Rev. Toxicol.*, **23**, 255 - 281.

Crecelius, E., 1986, *"Speciation of Selenium and Arsenic in Natural Waters and Sediments"*, **2**: Arsenic Speciation, EPRI Final Report, EA-4641.

Cullen, W., 1966, *Adv. Organomet. Chem.*, **4**, 145.

Dauchy, X., Potin-Gautier, M., Astruc, A, Astruc, M., 1994, *Fresenius' J. Anal. Chem.*, **348**, 792 - 805.

Ding, H., Wang, J., Dorsey, J. G., Caruso, J. A., 1995, *J. Chromatogr.*, A, **694**, 425 - 431.

Emteborg, H., Baxter, D. C., Sharp, M., Frech, W., 1993, *Analyst*, vol. **118**(8), 1007 - 1013.

EPA Method 200.8, 1994, *"Determination of Metals and Trace Elements in Water and Wastes by Inductively Coupled Plasma - Mass Spectrometry"*, Rev. 5.3.

EPA Method 200.9, 1994, *"Determination of Trace Elements by Stabilized Temperature Graphite Furnace Atomic Absorption Spectrometry"*, Rev. 2.2.

EPA Method 200.15, 1994, *"Determination of Metals and Trace Elements in Water by Ultrasonic Nebulization Inductively Coupled Plasma - Atomic Emission Spectrometry"*, rev. 1.2.

EPA Method 1631, 1995, *"Mercury in Water by Oxidation, Purge and trap, and Cold Vapor Atomic Fluorescence Spectrometry"*, EPA-R-95-027.

EPA Method 1632, 1995, *"Determination of Inorganic Arsenic in water by Hydride Generation Flame Atomic Absorption"*, EPA 821-R-95-028.

EPA Method 1636, 1995, *"Determination of Hexavalent Chromium by Ion Chromatography"*, EPA 821-R-95-029.

EPA Method 1640, 1995, *"Determination of Trace Elements in Ambient Waters by On-Line Chelation Preconcentration and Inductively Coupled Plasma-Mass Spectrometry"* EPA 821-R-95-033.

EPRI Report CS-3741, 1984, *"Aqueous Discharges from Steam-Electric Power Plants: Data Evaluation."*

Fowler, B., 1983, *"Biological and Environmental Effects of Arsenic"*, Elsevier, Amsterdam, New York.

Galli, C., Restani, P., 1993, *Pharmacol. Res.*, **27**, 115 - 127.

Haraldsson, C., Lyven, B., Pollak, M., Skoog, A., 1993, *Anal. Chim. Acta.*, **284**, 327 - 335.

Hasegawa, H., Sohrin, Y., Matsui, M., Hojo, M, Kawashima, M., 1994, *Anal. Chem.*, **66**, 3247 - 3252.

Hayashi, H., Segami, Y., Koizumi, S., 1993, *Hyomen Gijutsu,*. **44**, 697 - 698.

Li, K., Li, S. F. Y., 1995, *Analyst*, **120**, 361 - 366.

Lobinski, R., Dirkx, W. M. R., Szpunar - Lobinska, J., Adams, F. C., 1994, *Anal. Chim. Acta.*, **286**, 381 - 390.

McLaren, J., 1992, *Instrum. Trace Org. Monit.*, pp. 195-208, Clement, R.E., Ed., Pub. Lewis, Chelsea, MA.

Miller, E., Dobb, D.E., Heithmar, E.M., 1995, *"Speciation of Mercury in Soils by Sequential Extraction"*, presented at the USEPA Metal Speciation and Contamination of Surface Water Workshop, Jekyll Island, GA.

Naghmush, A. M., Pyrzynska, K., Trojanowicz, M., 1994, *Anal. Chim. Acta*, **288**, 247 - 257.

Nerin, C., 1994, *Appl. Organomet. Chem.*, **8**, 607.

Olivas, R., Donard, O., Camara, C, Quevauviller, P, 1994, *Anal. Chim. Acta.*, **286**, 357 - 370.

Pasullean, B., Davidson, C. M., Littlejohn, D., 1995, *J. Anal. At. Spectrom.*, **10**, 241 - 246.

Posta, J., Berndt, H., Luo, S-K, Schaldach, G., 1993, *Anal. Chem.*, **65**, 2590 - 2595:

Puk, R., Weber, J. H., 1994, *Appl. Organomet. Chem.*, **8**, 293 - 302.

Puk, R., Weber, J. H., 1994a, *Anal. Chim. Acta.*, **292**, 175 - 183.

Ritsema, R., Donard, O., 1994, *Appl. Organomet. Chem*, **8**, 571.

Rubi, E., Lorenzo, R. A., Casais, M., C., Cela, R., 1994, *Appl. Organomet. Chem.*, **8**, 677.

Van Elteren, J. T., Das, H. A., Deligny, C. L., Agterdenbos, J., Bax, D., 1994, *J. Radioanal. Nucl. Chem.*, **179**, 211 - 219.

Watras, C, 1992, *"Quantification of Mercury, Selenium, and Arsenic in Aquatic Environments"*, EPRI Report TR 101141.

Weber, J., 1993, *Chemosphere*, **26**, 2063 - 2077.

Weber, J., 1988, *"Humic Substances and Their Role in the Environment"*, J. Wiley and Sons, New York, N.Y., 165.

Westwoo, G., 1967, *Acta. Chem. Scand.*, **21**, 1790.

Wonders, J., 1994, *Int. J. Environ. Anal. Chem.*, **56**, 193.

THE EFFECT OF MEMBRANE FILTRATION ON DISSOLVED TRACE ELEMENT CONCENTRATIONS

ARTHUR J. HOROWITZ[1], KEN R. LUM[2],* JOHN R. GARBARINO[3], GWENDY E. M. HALL[4], CLAIRE LEMIEUX[2+] AND CHARLES R. DEMAS[5]

[1]U.S. Geological Survey, Peachtree Business Center, 3039 Amwiler Road, Atlanta, GA 30360, U.S.A., [2]Centre Saint-Laurent, Environment Canada, Montreal, Quebec H2Y 2E7, Canada, [3]U.S. Geological Survey, Branch of Analytical Services, 5293 Ward Road, Arvada, CO 80002, U.S.A., [4]Geological Survey of Canada, 601 Booth Street, Ottawa, Ontario K1A 0E8, Canada, [5]U.S. Geological Survey, 3535 S. Sherwood Forest Blvd., Suite 120, Baton, Rouge, LA 70816, U.S.A.

*Current address: I.U.C.N.-World Conservation Union, 380 St. Antoine, W., Suite 3200, Montreal, Quebec H2Y 3X7, Canada
+Current address: Multisources, 2875 Rue Holt, Montreal, Quebec H1Y 1P7, Canada

Keywords: membrane filters, filtration, filtration artifacts, dissolved, major elements, trace elements

Abstract: The almost universally accepted operational definition for dissolved constituents is based on processing whole-water samples through a 0.45-µm membrane filter. Results from field and laboratory experiments indicate that a number of factors associated with filtration, other than just pore size (e.g., diameter, manufacturer, volume of sample processed, amount of suspended sediment in the sample), can produce substantial variations in the 'dissolved' concentrations of such elements as Fe, Al, Cu, Zn, Pb, Co, and Ni. These variations result from the inclusion/exclusion of colloidally-associated trace elements. Thus, 'dissolved' concentrations quantitated by analyzing filtrates generated by processing whole-water through similar pore-sized membrane filters may not be equal/comparable. As such, simple filtration through a 0.45-µm membrane filter may no longer represent an acceptable operational definition for dissolved chemical constituents. This conclusion may have important implications for environmental studies and regulatory agencies.

1. Introduction

Examination of a variety of standard methods compendia dealing with water samples indicates that among environmental scientists, the current and almost universally accepted/mandated definition of dissolved constituents is an operational one—only materials which pass through a 0.45-µm membrane filter are considered to be dissolved (U.S. EPA, 1983; OWDC, 1984; APHA-AWWA-WPCF, 1989; ASTM, 1995) Obviously, this definition concerns only the pore size and type of filter, **no other** aspect of filtration is involved.

About 25 years ago, marine chemists began to recognize that prior dissolved sea water trace element data were elevated due to contamination introduced by improper handling (e.g., sampling, processing, preservation, analyses; Brewer, 1970; Jones, 1978; Bruland, 1983). This led to the introduction of so-called 'clean/ultraclean' sampling, processing, preservation, and analytical techniques to obtain reliable dissolved trace element data (Windom, et al., 1991; Benoit, 1994). During the past 10 years, freshwater chemists have begun to employ similar techniques to those used by the oceanographic community (Shiller and Boyle, 1987; Flegal and Coale, 1989; Windom, et al., 1991; Nriagu, et al., 1993; Benoit, 1994; Horowitz, et al., 1994; Taylor and Shiller, 1995). These changes have led to marked reductions in sample contamination, and concomitant decreases in reported levels of ambient dissolved trace elements in aquatic systems. This pattern of decreasing trace element

Water, Air and Soil Pollution 90: 281–294, 1996.
© 1996 *Kluwer Academic Publishers.*

concentrations due to contaminant reduction, has led many water chemists to believe ax-
iomatically, that the lowest dissolved concentration determined for a sample, or a site, is
probably the most representative concentration.

During this same period, the importance of colloids to the concentration, transport, and
redistribution of trace elements in aquatic systems has been demonstrated (Martin and
Meybeck, 1979; Hoffman, et al., 1981; Beckett, et al., 1988; Rees and Ranville, 1990).
Colloids are capable of sorbing large concentrations of trace elements and fall on the con-
tinuum between suspended sediment and dissolved constituents; as such, there is some controversy as to when a solid-phase material changes from a suspended sediment to a colloid,
and when a colloid changes to a dissolved form (Martin and Meybeck, 1979; Hoffman, et
al., 1981). However, colloidal material typically is considered to be finer than 1-μm.

Prior to the use of 'clean/ultraclean' techniques, and major improvements in analytical instrumentation (e.g., ICP-MS), the differences in 'dissolved' concentrations caused by the
presence of varying amounts of colloidally-associated trace elements probably were
masked by relatively high levels of contamination and/or relatively insensitive analytical
techniques (Nriagu, et al., 1993). However, as the reported ambient levels of dissolved
trace elements have declined from the tens of a part per billion (ppb, μg/L), into the single
digit ppb range, and now well into the part-per-trillion (ng/L, ppt) range (Shiller and Boyle,
1987; Flegal and Coale, 1989; Windom, et al., 1991; Nriagu, et al., 1993; Benoit, 1994;
Taylor and Shiller, 1995), the inclusion of varying amounts of colloids and their affect on
'dissolved' (filtered) trace element concentrations has become significant (Wagemann and
Brunskill, 1975; Ledin, 1993; Karlsson, et al., 1994).

Anyone filtering a water sample has observed flow rate reductions as water is processed.
Typically, these reductions are the result of physical clogging by both organic and inorganic material. These reductions in flow rate indicate a decrease in the nominal pore size of
the filter and are likely to affect the quantity of trace element-rich colloidal material contained in the filtrate. Different filters clog at different rates because of their specific physical characteristics (e.g., surface area) as well as external conditions (e.g., amount of organic or inorganic suspended sediment in the samples, amount of water processed). The underlying assumption behind processing samples through 0.45-μm membrane filters is that
all the chemical data from the filtrates are comparable because they all are based on the
same operational definition. If this assumption is invalid as a result of filtration or processing artifacts, it could lead to changes in how environmental scientists and regulatory
agencies operate. To evaluate the effects of these artifacts on a variety of dissolved trace
elements, a series of field and laboratory studies were undertaken by the U.S. Geological
Survey, the Geological Survey of Canada (GSC), and Environment Canada. The results
from these studies are reported herein.

2. Methodology

This study was divided into two segments—each with a laboratory and a field phase. The
laboratory portions were used to evaluate various aspects of filtration under controlled conditions which could not be maintained in a field setting (e.g., trace element concentrations,
suspended sediment concentration and composition, presence of colloidal material) or with
actual environmental samples.

2.1. SEGMENT 1 - LABORATORY STUDIES

Throughout this segment of the study, 3 types of filter were used. Two were tortuous path-filters (TP, manufactureby MicroFiltration Systems* and Millipore) and one was a sieve-filter (S, manufactured by Nuclepore). The TP filters had nominal pore sizes of 0.45-μm whereas the S filters had an actual pore size of 0.40-μm . Both 47 mm and 142 mm versions of each filter were tested and filtrations with each brand and diameter of filter were carried out using both vacuum (VF) and pressure methods (PTF).

The laboratory phase of Segment 1 evaluated the effect of filter type (TP vs. S), filter diameter (47 mm vs 142 mm), and filtration method (VF vs. PTF) on 'dissolved' Fe and Al concentrations in a laboratory-prepared dissolved/colloidal solution with and without varying concentrations of a laboratory-prepared 'suspended sediment'. Fe and Al were selected because they represent excellent proxies for colloidal material as well as many colloidally-associated trace elements such as Pb, Zn, and Mn (Wagemann and Brunskill, 1975; Laxen and Chandler, 1982).

Temporal changes in the efficiency of the filters, and the chemistry of the filtrates were evaluated by analyzing sequentially-collected 50 or 100-mL filtrate aliquots. Each aliquot was stored in separate acid-rinsed bottles for subsequent chemical analysis. Quantitation of Fe and Al in the various laboratory- and field-derived filtrates was carried out using standard flame atomic absorption spectrophotometric (AAS) techniques.

2.2. SEGMENT 1 - FIELD STUDIES

At the conclusion of the laboratory phase, field samples were collected from differing environments in Georgia to further evaluate the filtration effects on dissolved trace element concentrations. The three sites were: Chattahoochee River at Fairburn , Ohoopee River on Highway 1 near Oak Park, and Keg Creek near Deepstep . The suspended sediment concentrations (material >0.45-μm) for the sites were, respectively: 21, 9, and 11 mg/L.

2.3. SEGMENT 2 - FIELD STUDIES

The purpose of these studies was to evaluate the effect(s) of using different (e.g., manufacturer, diameter) 0.45/0.40-μm membrane filters to field-process whole-water samples for subsequent 'dissolved' trace element quantitation. Three filters were used in this study: 1) a 47 mm, 0.40-μm polycarbonate Nuclepore plate filter (#111107) with a surface area of 17.3 cm^2, 2) a 142 mm, 0.45-μm cellulose nitrate MicroFiltration Systems (MFS) plate filter (#A045A142C) with a surface area of 158 cm^2, and 3) a 47 mm, 0.45-μm polyethersulfone Gelman capsule filter (#12176) with a surface area of 600 cm^2. The Nuclepore filter was selected because it is the filter of choice for numerous aquatic chemists (Shiller and Boyle, 1987; Nriagu, 1993; Benoit, 1994; Taylor and Shiller, 1995); the MFS filter was selected because it was used extensively by the USGS and other monitoring agencies until the development of a new sampling and processing protocol; and the Gelman capsule filter was selected because of its very large surface area and because it is currently used by the USGS (Horowitz, et al., 1994).

Field samples were collected from the Mississippi River at St. Francisville and the Tangipahoa River at Robert, Louisiana. A field blank was passed through all the sampling and compositing equipment, as well as through each separate filtration system, prior to collect-

---

* The use of brand names is for identification purposes only and does not represent an endorsement by the U.S. Geological Survey, the Geological Survey of Canada, or Environmrent Canada.

compositing equipment, as well as through each separate filtration system, prior to collect-ing and processing actual samples (field blank volumes were the same as sample volumes). Whole-water samples were then processed through each filter type, and sequential aliquots (250 mL) collected, until there was an obvious decrease in filtration rate. Data from the field blanks were either below the method detection limit and/or were sufficiently low, relative to the measured concentrations, as to be insignificant.

Filtrate aliquots from the field studies were analyzed by the USGS using inductively-coupled plasma mass spectrometry (ICP-MS) and inductively-coupled plasma optical emis-sion spectrometry (ICP-OES). Al, Sb, Ba, Be, Cd, Cr, Co, Cu, Pb, Mn, Mo, Ni, Ag, Sr, Tl, U, and Zn concentrations were measured by ICP-MS whereas B, Ca, Fe, Li, Mg, Si (as SiO_2), Na, and V were measured by ICP-OES . Analytical blanks were generally less than the detection limit for all analytes.

2.4. SEGMENT 2 - LABORATORY STUDIES

The laboratory portion of this segment was designed to measure dissolved trace element concentrations in samples processed under controlled conditions through four different fil-ters. All the filtrations were conducted in a Class 100 clean room. In addition to the filters used in the field study, a 47 mm, 0.45-μm mixed cellulose acetate/nitrate Millipore filter (#HAWP-04700) was added. This membrane has the same surface area as the Nuclepore, but is a tortuous path filter rather than a sieve filter (Horowitz, et al., 1992). The effects of two different pretreatments (precentrifugation and prefiltration) also were evaluated. Precentrifugation entailed spinning 1.5 L of whole water at 3000 rpm (~2000 g) for 30 min-utes. Prefiltration entailed passing sample aliquots through a 10-μm polyester screen in a 142 mm polycarbonate filter holder; a separate screen was used to process each liter of whole-water. After pretreatment, the filtration experiments proceeded in the same manner as for the untreated aliquots. Water samples collected from the Mississippi River at St. Francisville and the Tangipahoa River at Robert were used for this part of the study.

The experiment was designed so that the total volume of whole- or pretreated water pro-cessed through a particular filter was proportional to its surface area. Thus, the experiment attempted to normalize the effects of filter loading (clogging). Initial and final filtrate aliquots were collected during sample processing to evaluate any chemical effects due to changes in effective filter pore-size. A total of 100 mL was processed through each 47 mm membrane with 50 mL collected for the first and the last aliquots; a total of 900 mL was processed through the 142 mm MFS with the first and last 125 mL collected; and a total of 3500 mL was processed through the Gelman filters with the first and last 125 mL collected. Five separate aliquots were filtered, using new membranes each time, for all four filter types. Prior to processing each sample, the filters were purged/conditioned with deionized water (500 mL for the plate filters; 1000 mL for the capsule filters); then, system blanks (100 mL for the 47 mm membranes, 250 mL for the 142 mm and capsule filters) were pro-cessed prior to actual sample filtration. As in the field study, blank concentrations were either below the method detection/reporting limit or were sufficiently low relative to the measured concentrations as to be insignificant.

Analyses for the laboratory phase of this segment were performed by both the USGS in Denver, CO, and the GSC in Ottawa, Ontario on sample splits. The data from both facili-ties indicate that the *trends* shown in the various samples were the same for the USGS and the GSC data; however, there appears to be a bias in *absolute concentrations* between the sets for some elements. Considering the independent methodologies/techniques used to generate the two data sets, the agreement between the laboratories is very good.

3. Results and Discussion

3.1. SEGMENT 1

During the laboratory phase, it became apparent that the filtration step used to operationally define dissolved constituents had a major effect on the quantity of colloidally-associated trace elements incorporated in a filtered sample (Horowitz, et al., 1992). The colloidal concentration was affected by the rate at which the membrane became clogged, thus reducing its nominal pore size. Clogging could be affected by such diverse variables as: 1) filter type, 2) filter diameter, 3) method of filtration (vacuum or pressure), 4) suspended sediment concentration, 5) suspended sediment grain size distribution, 6) concentration of colloids, and 7) volume of sample processed because of 4, 5, and 6, and 9). As cited previously, the pattern of decreasing trace element concentrations during the past 25 years, due to contaminant reduction, has led many water chemists to believe that the lowest concentration determined for a sample, or a particular site, is probably the most representative concentration . The results from these laboratory experiments implied that this might not be true.

The field studies supported the results/conclusions reached during the laboratory phase, using natural, as opposed to laboratory-manufactured whole water samples. In addition to the filtration procedures used during the laboratory phase, an additional technique was examined: the effect of centrifuging a sample prior to filtration. This procedure was added because it appeared to be a logical outgrowth of the laboratory results. That is, the differences in filtrate Fe and Al appeared to be caused by differential clogging rates. Precentrifugation could reduce or eliminate clogging. Centrifugation conditions were established to remove all particles >0.45-μm.

Despite a wide concentration range for Fe and Al among the three natural samples, the pattern of results for the 0.45/0.40-μm filtrates was extremely consistent. Invariably, the concentrations, in decreasing order were: 142 mm TP > 142 mm S > 47 mm TP > 47 mm S (see Fig.1 for Fe). The chemical concentrations differed according to both filter diameter and type, with the former appearing to be more important than the latter (e.g., both 142 mm filtrates were higher than both 47 mm filtrates, while the TP filtrate was higher than the S filtrate for each size). A similar result can be inferred from a prior comparison of 47 mm metal and 142 mm TP filters used on natural samples (Wagemann and Brunskill, 1975). For example, the differences in Fe between the various filtrates ranged from about a factor of 2 (Ohoopee River) to a factor of 9 (Chattahoochee River). Centrifugation eliminated differences due to type for the 142 mm filters, as the results for both were similar (Fig. 1); however, it only marginally altered the differences for the 47 mm filters.

The chemical trend in the 0.45/0.40-μm filtrates that received no pretreatment was one of decreased concentrations with increased processed volumes. This probably is attributable to filter clogging. Follow-on processing with the same type and diameter 0.10-μm filters did not appear to eliminate or reduce the differences noted in the larger pore-sized filtrates; however, it did lower the absolute concentrations. Of the 4 tested filters, the 142 mm TP filters appeared to be the most resistant to clogging. Precentrifugation reduced clogging in the 142 mm TP filters; further, it reduced clogging in the 142 mm S filters to the point where these filtrates were equivalent to their TP counterparts. However, precentrifugation did not substantially reduce clogging in the 47 mm filters. Based on temporal changes in filtrate Fe and Al concentrations in both the laboratory and field trials, it appears that a 142 mm TP filter alone, or any type of 142 mm S filter used after precentrifugation, is likely to provide the most representative filtrate for the subsequent quantitation of dissolved trace elements. On the other hand, the use of any type of 47mm diameter filter appears to pro-

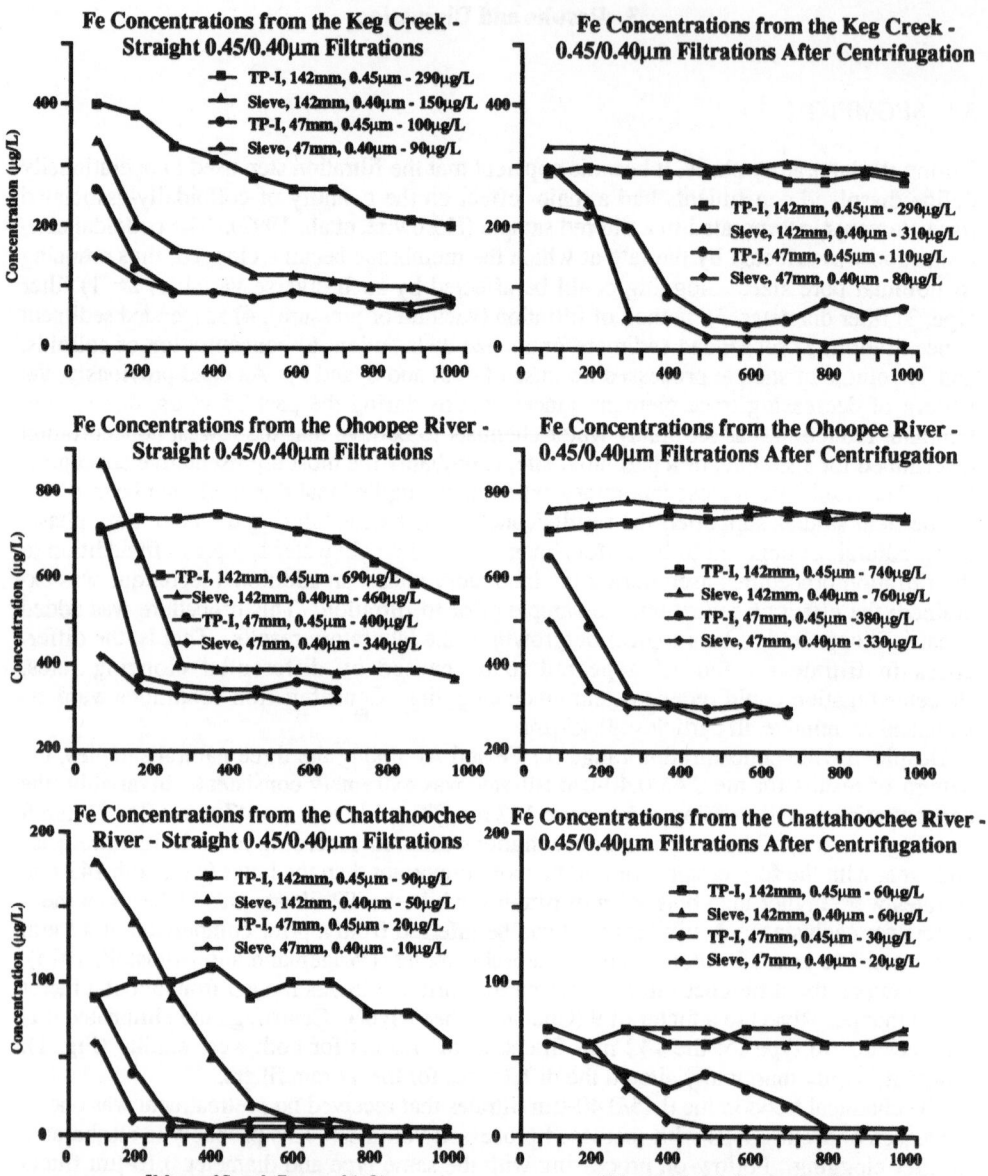

Figure 1. Iron concentrations in 0.45/0.40-μm filtrates from various rivers using different filters. Each pair of graphs is for a single river; the graph on the left is for straight filtrates whereas the graph on the right is for filtrates after the sampe had been precentri- fuged. The concentrations appearing in the legends are for the Fe concentrations for a 1 liter sample for each type of filter.

duce filtrates with concentrations which are biased low, relative to 142 mm diameter filtrates, and to actual environmental levels.

3.2. SEGMENT 2

The detectable Mississippi dissolved trace element data fell into one of three categories: 1)
affected by filtration artifacts, 2) possibly affected by filtration artifacts and/or dilution, and
3) not affected by either filtration artifacts or dilution (Table 1). The first group includes
Al, Fe, Ni, Cu, and Zn. In each case, the general pattern was the same; the lowest concentrations occurred in the Nuclepore filtrates, the next highest concentrations occurred in the
MFS filtrates, and the highest concentrations occurred in the Gelman filtrates. This pattern
is proportional to the surface areas of the filters. This also may imply a correlation with
filtration rate and/or clogging.

The effects on Al and Fe are particularly obvious (Table 1). The 8.2µg/L Al concentration in the 100 mL Nuclepore filtrate was not reached in the MFS filtrate until some 500
mL had been processed, and was not reached in the Gelman filtrate until over 2000 mL had
been processed. In fact, the Gelman filtrates eventually contained lower Al levels than the
Nuclepore filtrates, but only after 2600 mL had been processed. The patterns for Ni, Cu,
and Zn are not as pronounced as for the Al and Fe, but are detectable. The Cu concentration (50 µg/L) in the first 250 mL Gelman aliquot was 'verified' (it was quantitated in
both halves of a split field sample) but almost certainly represents some form of contamination introduced during sample handling. Even so, the second Gelman aliquot (250-500
mL) still contains more Cu than either the first Nuclepore or MFS aliquots. There appears
to be a direct correlation between filter surface area and the Fe and Al concentrations in the
filtrates. On the other hand, although filtration artifacts appear to have affected Ni, Cu, and
Zn, and despite the known sorptive capacity of Fe and Al oxides and hydroxides for these
elements, there does not appear to be a direct correlation between Ni, Cu, and Zn concentrations and filter surface area. These results and conclusions are similar to those reported
previously (Martin and Meybeck, 1979; Horowitz, et al., 1992; Karlsson, et al., 1994).

Co, Mo, Pb, and possibly Cr concentrations may have been affected by a different type of
artifact (Table 2). This effect only seems to be associated with the Gelman filter; and may
be a function of its very high surface area and/or of its composition (polyethersulfone). The
concentrations of these elements increased slightly after some 1500 mL had been processed. This may indicate that some proportion of these elements had sorbed to the filter during
initial processing, then, once all the potential sorption sites had been filled, higher concentrations began to appear in the filtrate.

The concentrations of several elements, notably Sr, Ba, and Ca appear to have been affected by dilution due to the presence of some entrained deioninzed water that was used to
condition the filters. Note the concentration increases in the second and third filtrate aliquots, relative to the first aliquot, for these elements in the Gelman and the MFS filtrates
(Table 1). This effect has been noted before (Horowitz, et al., 1994). If dilution is affecting trace element concentrations, it should affect all of them equally; however, due to low
concentrations and/or limited analytical sensitivity, it does not appear to be significant
below 50 µg/L. On the other hand, this pattern may indicate that some proportion of these
elements initially had been sorbed to the filter, then, once all the potential sorption sites had
been filled, higher concentrations begin to appear in the filtrate. Although the data were not
included, the concentrations of a variety of dissolved nutrients (nitrate, nitrite, ammonium,
orthophosphate, and total phosphorus) did not appear to be affected by filtration artifacts or
dilution. In fact, constituents occurring at mg/L (ppm) concentrations do not appear to be
affected by filtration artifacts at all.

Table 1. Comparison of Selected Dissolved Trace Element Concentrations in Sample Filtrates From the Mississippi River at St. Francisville and the Tangipahoa River at Robert, Louisiana, [all concentrations in μg/L except Ca, Mg, Na, and Si (mg/L)].

Analytical Method[1] and Constituent

Mississippi River[2,3] Filter[4] and Aliquot Volume	MS Al	OES Fe	MS Cr	MS Mn	MS Co	MS Ni	MS Cu	MS Zn	MS Mo	MS Ba	MS Pb	MS U	OES B	OES Sr	OES Ca	OES Mg	OES Na	OES Si
Nuclepore																		
0-100mL	8.2	7	0.5	14	0.4	<0.8	1.8	2.1	1.3	57	<0.2	0.9	33	160	39	11.1	19.5	7.0
MicroFiltration Systems																		
0-250mL	26	27	1.3	14	0.4	<0.8	2.2	4.6	1.3	54	<0.2	0.9	31	153	37	10.7	18.2	6.7
250-500mL	8.6	14	0.8	14	0.5	<0.8	2.1	3.8	1.5	58	<0.2	1.0	34	164	39	11.3	19.8	7.1
500-750mL	4.9	4	0.9	14	0.6	<0.8	2.4	4.1	1.5	58	<0.2	0.9	31	161	39	11.3	19.4	7.1
Gelman Capsule																		
0-250mL	43	58	<0.3	13	0.2	2.2	50	6.1	1.2	52	<0.2	0.9	31	152	37	10.6	18.6	6.9
250-500mL	41	56	0.3	14	0.4	2.0	3.9	2.7	1.2	57	<0.2	1.0	32	160	39	11.2	19.6	7.2
500-750mL	32	50	<0.3	14	0.3	1.7	2.2	2.5	1.1	59	<0.2	1.1	34	163	39	11.3	19.7	7.2
750-1000mL	27	43	0.3	14	0.3	2.0	1.9	1.5	1.3	55	0.6	1.1	37	155	39	11.3	19.7	7.2
1125-1375mL	22	34	0.4	14	0.4	2.1	2.2	2.4	1.5	56	0.6	1.1	31	161	39	11.2	19.3	7.1
1375-1625mL	18	36	0.4	13	0.4	2.5	2.2	1.9	1.7	56	0.8	1.1	34	163	40	11.3	19.2	7.2
1625-1875mL	15	19	0.5	14	0.4	3.0	2.5	2.2	1.8	57	0.7	1.1	32	161	39	11.2	19.5	7.0
1875-2125mL	11	24	0.6	13	0.5	2.3	2.7	2.8	2.0	58	0.7	1.1	32	161	39	11.2	19.5	7.1
2125-2375mL	8.6	20	0.6	14	0.5	2.6	2.5	2.1	1.8	57	0.9	1.0	32	162	39	11.2	19.4	7.1
2375-2625mL	8.2	20	0.7	13	0.5	2.8	3.0	2.7	1.9	58	1.0	1.0	34	164	40	11.4	19.6	7.1
2625-2875mL	7.4	17	0.6	14	0.5	2.6	2.7	2.5	1.9	59	0.9	1.1	32	161	39	11.2	19.9	7.0
2875-3125mL	6.8	15	0.6	14	0.5	2.9	2.7	2.6	2.0	59	1.0	1.1	31	161	39	11.1	19	7.0
3125-3375mL	3.4	13	<0.3	14	0.3	1.9	2.1	2.4	1.2	58	<0.2	1.1	33	162	39	11.2	19.6	7.0

[1]Analytical methods included inductively coupled plasma-mass spectrometry (MS) and iductively coupled plasma optical emission spectrometry (OES).

[2]The concentrations for Be (0.6μg/L), Ag (0.2μg/L), Cd (0.2μg/L), Sb (0.2μg/L), Li (1μg/L) and V (3μg/L) were excluded because they were all below the method detection limit (given in parentheses).

[3]The concentration of suspended sediment was 157 mg/L.

[4]Filters were a 47mm, 0.40-μm Nuclepore, a 142mm, 0.45-μm MicroFiltration Systems, and a 47mm, 0.45-μm Gelman capsule.

Analytical Method[1] and Constituent

Tangipahoa River[6,7] Filter[4] and Aliquot Volume	MS Al	OES Fe	MS Mn	MS Co	MS Ni	MS Cu	MS Zn	MS Ba	MS Pb	OES B	OES Sr	OES Ca	OES Mg	OES Na	OES Si
Nuclepore															
0-100mL	36	55	75	<0.2	<0.8	1.0	2.2	24	<0.2	10	13	1.3	0.7	2.3	4.2
MicroFiltration Systems															
0-250mL	185	252	81	0.3	<0.8	1.4	4.1	28	0.5	13	13	1.4	0.7	2.3	4.7
250-500mL	66	109	81	<0.2	<0.8	0.7	2.5	27	<0.2	11	14	1.4	0.7	2.4	4.4
625-875mL	42	74	77	<0.2	<0.8	0.8	1.8	28	<0.2	11	14	1.4	0.7	2.4	4.4
Gelman Capsule															
0-250mL	320	302	72	0.3	1.4	4.8	5.6	27	0.8	11	12	1.5	0.6	2.2	4.4
250-500mL	308	289	89	0.3	<0.8	1.6	3.1	33	0.5	10	14	1.5	0.7	2.4	4.7
500-750mL	153	248	81	0.3	<0.8	2.0	3.0	30	0.4	11	14	1.5	0.7	2.4	4.7
750-1000mL	145	228	85	0.4	<0.8	1.5	2.6	31	0.3	12	14	1.4	0.7	2.4	4.6
1125-1375mL	64	169	82	0.2	<0.8	1.6	2.2	30	<0.2	13	14	1.5	0.7	2.4	4.6
1375-1625mL	75	143	83	0.3	<0.8	1.4	2.2	30	<0.2	12	14	1.5	0.7	2.4	4.5
1625-1875mL	50	142	81	<0.2	<0.8	0.9	2.3	28	<0.2	11	13	1.5	0.7	2.4	4.3
1875-2125mL	54	105	80	<0.2	<0.8	0.8	0.9	27	<0.2	12	14	1.2	0.7	2.4	4.4
2250-2500mL	54	98	79	<0.2	<0.8	0.8	0.8	28	<0.2	12	14	1.5	0.7	2.4	4.4
2500-2750mL	45	95	79	<0.2	<0.8	0.7	0.8	28	<0.2	12	14	1.4	0.7	2.4	4.4
2750-3000mL	44	93	79	<0.2	<0.8	0.3	0.6	27	<0.2	10	14	1.4	0.7	2.3	4.4
3000-3250mL	44	87	80	0.2	<0.8	0.6	1.0	28	<0.2	11	14	1.4	0.7	2.4	4.3

[5]Analytical methods included inductively coupled plasma-mass spectrometry (MS) and iductively coupled plasma optical emission spectrometry (OES).

[6] The concentrations for Be (0.6μg/L), Cr (0.3μg/l), Mo (0.6μg/l), Ag (0.2μg/l), Cd (0.2μg/l), Sb (0.2μg/l), U (0.1μg/L), Li (1μg/L), and V (3μg/L) were excluded because they were all below the method detection limit (given in parentheses).

[7]The concentration of suspended sediment was 39 mg/L.

[4]Filters were a 47mm, 0.40-μm Nuclepore, a 142mm, 0.45-μm MicroFiltration Systems, and a 47mm, 0.45-μm Gelman capsule.

Data from the Tangipahoa River are similar to those from the Mississippi River (Table 1). The elements affected by filtration artifacts in the Mississippi sample also were affected in the Tangipahoa sample, with the exception of Pb. Pb concentrations in the Tangipahoa sample appear to be affected by filtration artifacts in the same way as Al, Fe, Cu, and Zn. In the Mississippi sample, Pb did not appear to be affected by these artifacts, but may have been affected by initial sorption to the Gelman filter. The sorption artifacts noted in the Mississippi sample for Cr, Co, and Mo, could not be evaluated in the Tangipahoa sample due to lower concentrations. All the unaffected elements, or those possibly affected by dilution, in the Mississippi sample also displayed similar patterns in the Tangipahoa sample.

The laboratory study confirmed many of the the field study results. The concentrations in the initial filtrate aliquots for the untreated Mississippi sample indicate that Al, Fe, Mn, Co, Ni, Cu, and Zn can vary significantly, even though all four filtering procedures fall within the currently accepted operational definition for dissolved constituents (Fig.2). As in the field study, the variations appear to be proportional to the surface areas of some of the membranes. On the other hand, substantial concentration differences occur between the Millipore and Nuclepore filtrates despite their similar surface areas (Fig. 2). This probably indicates that at least some of the concentration differences between the various filters may be caused by filter type (tortuous path vs sieve) (Horowitz, et al., 1992). Surprisingly, the concentrations in the Millipore filtrates were higher than those in the Gelman filtrates, despite the fact that the latter has nearly 35 times more surface area than the former (Fig. 2). The higher Millipore filtrate concentrations likely are the result of the smaller sample volumes processed (50 mL vs 250 mL). The elements unaffected by filtration artifacts in the field study (Sr, Ba, Ca, Mg, Na, and SiO_2) also were unaffected in the laboratory study. The dilution/sorption effect detected in the field study also occurred in the laboratory study (e.g., Sr, Fig. 2).

Comparison of the data for the first and last filtrate aliquots for the Mississippi sample further emphasizes the effect of filtration artifacts on dissolved concentrations (Fig. 3). Concomitantly, it also indicates that the volume of processed sample can affect filtrate chemical concentrations. These concentration differences occurred even though all four filters met the currently accepted operational definition for processing whole-water samples for the subsequent quantitation of dissolved constituents. The concentrations in the Nuclepore filtrates, for those elements affected by filtration artifacts, typically were the lowest in the group, usually by factors ranging from three- to fivefold. This rather consistent negative bias (relative to the other filters) was neither the result of contamination nor analytical insensitivity, and casts doubt on the view that the lowest concentration determined for a sample, or a particular site, is the most representative concentration given the current operational definition of a dissolved constituent (U.S. EPA, 1977; OWDC, 1984; APHA-AWWA-WPCF, 1989; ASTM, 1995; Horowitz, et al., 1992; Nriagu, et al., 1993; Taylor and Shiller, 1995).

For the artifact-affected trace elements, precentrifugation and prefiltration reduce variability in the filtrate replicates as well as the concentration differences between them, (Figs.2 and 3). The pretreatments also reduce the effects associated with processing varying sample volumes (Figs. 2 and 3). Of the two pretreatments, precentrifugation appears to be the more effective procedure for reducing these differences. However, both techniques substantially lower trace element concentrations in pretreated filtrates, relative to untreated filtrates (Fig. 2). If the chemical differences between the various filtrates are the result of the inclusion/exclusion of colloidally-associated trace elements, then it appears as if both pretreatments remove some proportion of colloidal material from the treated samples.

The data for the Tangipahoa sample are similar to those for the Mississippi sample. For most samples, the elements affected by filtration artifacts (Al, Fe, Mn, Co, Ni, Cu, and Zn),

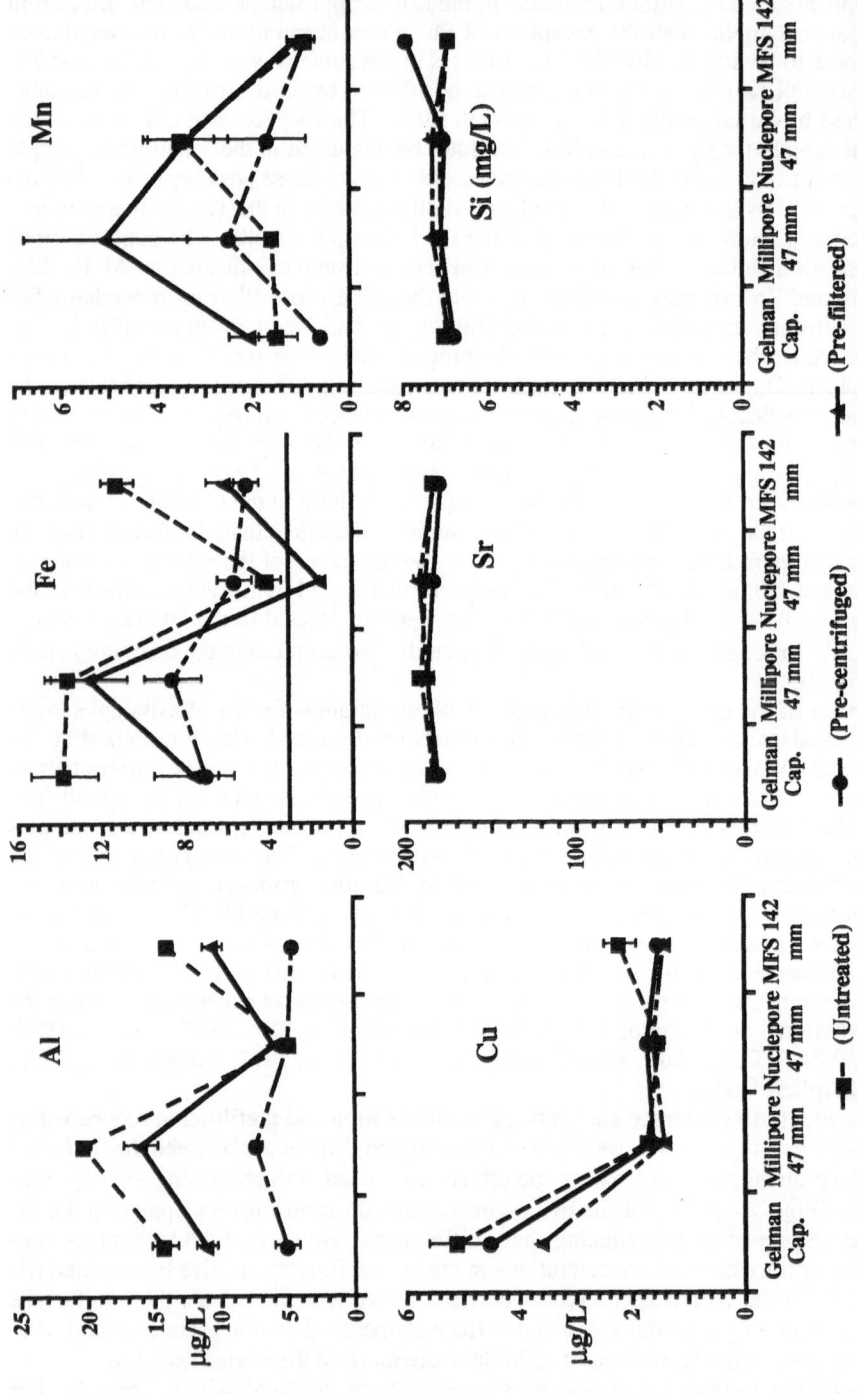

Figure 2. Concentrations of selected dissolved trace elements in a Mississippi River water sample in the first aliquots processed through four different types of filters. The error bars represent the standard error of the mean at one standard deviation; the solid horizontal line in some of the graphs indicates the method detection/reporting limit.

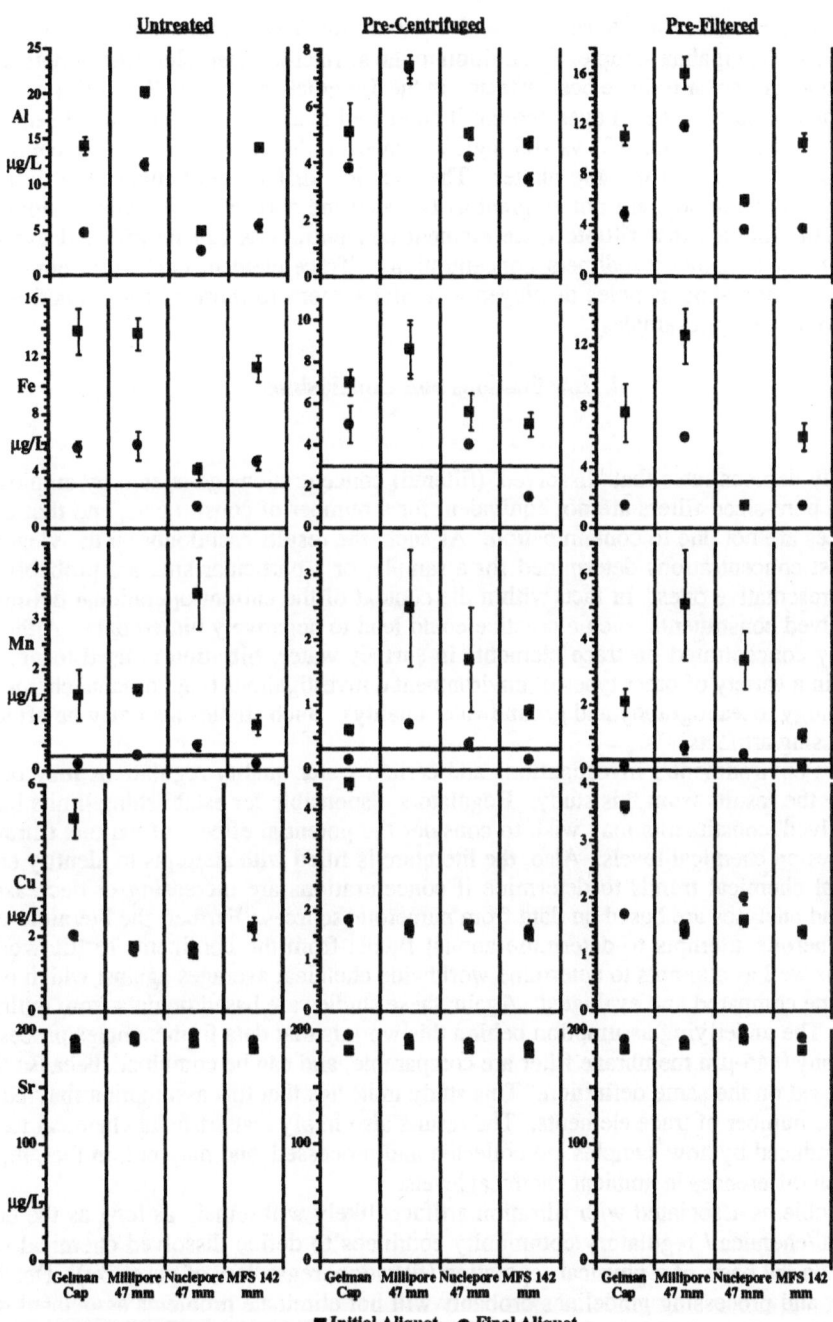

Figure 3. Selected trace element concentrations in the first and last filtrate aliquots of four different types of filters for a Mississippi River water sample (the error bars represent the standard error of the mean at one standard deviation, the solid horizontal line in some of the graphs indicates the method detection/reporting limit).

or dilution effects (Sr, Ba, Ca, Mg, Na, and SiO_2), in the Mississippi sample also were affected in the Tangipahoa sample. In addition to the artifact-affected elements in both samples, Pb and Cr appear to have been affected in the Tangipahoa sample. The artifacts in the Tangipahoa sample occurred even though it contained markedly less suspended sediment than the Mississippi sample (22 vs 263 mg/L). However, the Tangipahoa sample contained higher concentrations of organic matter. The chemical differences between the first and last Tangipahoa aliquots are not as great as between the corresponding Misssissippi aliquots. This implies that filtrate trace element concentrations can be affected even by relatively low suspended sediment concentrations. Trace element concentrations in the pretreated Tangipahoa samples displayed a similar pattern to those already noted in the pretreated Mississippi samples.

4. Implications and Conclusions

This study demonstrates that 'dissolved' (filtered) concentrations generated by employing the same pore-sized filters are not equivalent for a number of constituents, and that these differences are not due to contamination. As such, the results cast doubt on the view that the lowest concentrations determined for a sample, or a particular site, are probably the most representative ones. In fact, within the context of the current operational definition for dissolved constituents, such a practice could lead to negatively biased data. Although this study concentrated on trace elements in surface water, filtration is used to process samples in a variety of other types of environmental investigations (e.g., organic chemistry, microbiology, oceanography and groundwater quality). Such studies also may be affected by processing artifacts.

Various environmental investigations and certain water quality regulations may be affected by the results from this study. Regulators responsible for establishing limits based on 'dissolved' constituents may wish to consider the potential effects of various filtration procedures on chemical levels. Also, the literature is filled with attempts to identify environmental chemical trends to determine if concentrations are increasing or decreasing. Such trend analyses are based on data from numerous sources. Further, the literature contains numerous attempts to determine annual fluxes from the continents to the world's oceans, as well as attempts to determine worldwide chemical averages against which other data can be compared and evaluated. Again, these studies are based on data from multiple sources. The underlying assumption behind this work is that data from samples processed through any 0.45-μm membrane filter are comparable, and can be combined, because they all are based on the same definition. This study indicates that this assumption may be invalid for a number of trace elements. The results also imply that artificial chemical trends may be induced by how samples are collected and processed, and may not, in fact, represent actual differences in ambient chemical levels.

The problems associated with filtration artifacts likely will remain as long as the environmental/chemical/ regulatory community continues to define dissolved chemical constituents on the basis of a physical separation (filtration, regardless of pore size). Detailed sampling and processing guidelines probably will not eliminate problems associated with these artifacts because of randomly changing environmental factors. Finally, although possible, it is unlikely that correction factors can be developed and applied to filtered water chemical data to eliminate the effects of processing artifacts.

There appear to be three viable options for dealing with filtration artifacts; two entail substantial changes in the current definition of dissolved constituents. The artifacts can be reduced/eliminated by employing filters with very high surface areas (e.g., capsule filters) and collecting the initial aliquots for the subsequent quantitation of artifact-affected constituents. This choice does not entail a substantive change in the current operational definition of dissolved constituents and is the philosophical basis for the new USGS protocols (Horowitz, et al., 1994), as well as the procedures employed by Windom, et al. (1991).

A second option uses pretreatment, notably filtration or centrifugation, to limit artifact-induced differences in trace element concentrations. This option could permit the continued use of any 0.45-μm membrane filter. However, the pretreatments enhance the chances for random contamination because of increased sample handling, and more importantly, markedly lower a number of trace element concentrations in the subsequent filtrates. This option represents a substantive change in the current operational definition of dissolved constituents, and may be a source of negative bias in dissolved chemical data. However, this may be viewed as an acceptable tradeoff if it permits the continued use of a relatively simple and inexpensive procedure (filtration through any 0.45-μm membrane filter) for defining 'dissolved' concentrations.

The last option assumes colloids represent a contaminant which should be excluded from 'dissolved' samples. Although there is much controversy over what constitutes a colloid, current data indicate that material coarser than 0.015 to 0.005-μm would have to be removed [some 30 to 90 times finer than at present (Martin and Meybeck, 1979; Hoffman, et al., 1981; Taylor and Shiller, 1995)] This approach represents the most substantive departure from the current definition of dissolved constituents, and would require the use of multiple filters Martin and Meybeck, 1979; Hoffman, et al., 1981; Taylor and Shiller, 1995), much more expensive equipment (e.g., tangential-flow filtration systems (Taylor and Shiller, 1995), or the use of such subjective procedures as 'exhaustive filtration (Shiller and Boyle, 1987; Taylor and Shiller, 1995).

Of the three options, it is probably far simpler and more effective to specify the use of large surface area membrane filters, than to either pretreat samples, or to try to remove a majority of the 'colloidal' material, prior to filtration . Although all three methods can limit the effects of filtration artifacts, they obviously do not produce comparable data. Therefore, until there is consensus on the most appropriate procedure for dealing with these artifacts, a detailed description of all sampling and processing procedures is needed with the publication or storage of any chemical data from filtered water. Regardless of which option is employed to limit artifacts, it is apparent that for many dissolved constituents, the current operational definition (processing whole-water through an unspecified 0.45-μm membrane filter) is no longer acceptable.

References

APHA-AWWA-WPCF: 1989, *Standard Methods for the Examination of Water and Wastewater*, 17th Ed., 1193p.

ASTM: 1995, *Annual Book of ASTM Standards*, Vol 11.01, Water (I), 814p.

Beckett, R.; Nicholson, G.; Hart, B.T.; *et al.* : 1988, *Water Research*, **22**, p. 1535 -1545

Benoit, G.: 1994, *Environmental Science & Technology*, **22**, 1987-1991

Brewer, P.G. and Spencer, D.W.: 1970, *Trace Metal Intercalibration Study.* , Woods Hole Oceanographic Institute, Report No. 70-62.

Bruland K.: 1983, Trace Elements in Sea-Water, In *Chemical Oceanography*, **5**, (Edited by Riley, J., Chester, R.), 157-220.

Flegal A. and Coale K.: 1989, *Water Resources Bulletin*, **25**, 1275.

Hoffman, M.R., Yost, E.C., Eisenreich, S.P., and Maier, W.P.: 1981, *Environmental Science and Technology*, **15**, 655-661.

Horowitz, A.J., Demas, C.R., Fitzgerald, K.K., *et al.*, :1994, *U.S. Geological Survey Open-File Report No. 94-539*, 56p.

Horowitz, A.J., Elrick, K.A., and Colberg, M.R., 1992, *Water Research*, **26**, 753-763.

Jones, P.G.W.: 1978, *International Council for the Exploration of the Seas*. Copenhagen, ICES, CM 19787/E:16.

Karlsson, S., Peterson, A.;,Hakansson, K., and Ledin, A.: 1994, *The Science of the Total Environment*, **194**, 215-223.

Laxen D. and Chandler, I.: 1982, *Analytical Chemistry*, **54**, 1350-1355.
Ledin, A.: 1993, *Colloidal Carrier Substances — Properties and Impact on Trace Metal Distribution in Natural Waters.*, Linkoping Studies in Arts and Science No. 91.
Martin, J.M.. and Meybeck, M.: 1979, *Marine Chemistry*, **7**, 173-206.
Nriagu, J.O., Lawson, G., Wong, H.K.T., and Azcue, J.M.:1993, *Journal of Great Lakes Research*, **19**, 175-182.
Office of Water Data Coordination: 1984, *National Handbook of Recommended Methods for Water-Data Acquisition.*, Chapter 5, 5-14-5-15.
Rees, T.F. and Ranville, J.F.: 1990, *Journal of Contaminant Hydrology*, **6**, 241-250.
Shiller A. and Boyle E.:1987, *Geochimica et Cosmochimica Acta*, **51**, 3273-3277.
Taylor, H.R. and Shiller, A.M.:1995, *Environmental Science and Technology*, **29**, 1313-1318.
U.S. Environmental Protection Agency: 1983, *Methods for Chemical Analysis of Water and Wastes* EPA-600/4-79-020, 375 p.
Wagemann, R. and Brunskill G.:1975, *International Journal of Environmental Analytical Chemistry*, **4**, 75-84.
Windom, H., Byrd, H., Smith, Jr. H., and Huan, F.: 1991, *Environmental Science & Technology*, **25**, 1137-1142.

RESiDUAL SULFITE AFTER DECHLORINATION OF WATER

N. V. EKKAD and C. O. HUBER

Department of Chemistry and Center for Great Lakes Studies, University of Wisconsin-Milwaukee, P O Box 413 Milwaukee, WI 53201

Abstract. The kinetics of dechlorination show that chlorinated organic amines are not completely dechlorinated during typical contact times. Analytical techniques for measuring residual sulfite must maintain pH neutrality in order to represent the actual extent of dechlorination and must allow for rapid and convenient operation near the sampling site in order to minimize errors due to air oxidation during the procedure. A portable, analog circuitry-based instrument using constant current coulometry with amperometric end point detection was developed and evaluated. Laboratory and field operation of the instrument showed an analytical range of 0.015-to-25.0 mg sulfite/l. Relative standard deviation was typically 1-2 %.

key words: dechlorination, coulometry, sulfite, water, rate, chloramine, residual, kinetics

1. Introduction

The objective of dechlorination of process effluents is to minimize the toxicity and mutagenicity associated with residual chlorine species in water. Addition of a stoichiometric excess of sulfur dioxide is the usual method of dechlorination. On dissolution the sulfur dioxide forms sulfurous acid. The pK_a values for sulfurous acid are 1.8 and 7.2, respectively. Thus, under the usual conditions for dechlorination at pH 7-8, the predominant species are bisulfite and sulfite. These species will be referred to subsequently as 'sulfite'. . The measurement of residual sulfite is important because in addition to cost considerations, residual sulfites released to the environment can deplete dissolved oxygen and threaten aquatic life.

Chlorinated water typically contains inorganic monochloramine and organic chloramines, All dechlorination reactions are rapid at low pH, but in neutral, more realistic, pH waters, the reaction of sulfite with with organic chloramines is much slower. A summary of the literature for the reaction kinetics of dechlorination for some typical chlorination species is given in Table I. Reaction half times were obtained by applying second order rate theory and assuming initial sulfite and chlorine species concentrations of 10 midromolar, which corresponds to about one part per million of sulfite.

Water, Air and Soil Pollution **90**: 295-300, 1996.
© 1996 *Kluwer Academic Publishers.*

Table 1

Reaction times assuming 10 micromolar concentrations

Chorine Species	pH	Time	Reference
Free Chlorine	<11	13 ms	Fogelman et al., 1989
	>11	4.3 s	Fogelman et al., 1989
Monochloramine	4	1.8 s	Fogelman et al., 1987
	8	2.0 min.	Fogelman et al., 1987
Organic chloramine	7	1-20 min.	Stanbro and Lenkevich,, 1982,4
	8	2-200 min.	Ekkad, 1995
Chlorine Dioxide	9	0.1 s	Gordon and Suzuki, 1978
	11	0.1 s	Gordon and Suzuki, 1978

Typical contact times for dechlorination processes are less than twenty minutes. These data show that a portion of the organic chloramines survive the usual dechlorination process and enter the receiving water.

Any analytical method for residual sulfite determination based on acidification of the sample causes rapid completion of dechlorination reactions and indicates the sulfite concentration if the dechlorination process had been complete, rather than the extent of actual dechlorination.. Thus, to measure actual sulfite concentration the sampling and measurement must maintain the sample pH above seven. These considerations preclude use of methods for sulfite based upon tiration with iodine produced by the reaction of acidified potassium iodide and iodate solutions. (APHA, Laila, 1989).

Sulfite is oxidized to sulfate by dissolved oxygen from air in sample solutions. Although the kinetics of this reaction have been studied, it is not possible to predict the rate quantitatively in actual effluents. Reports of coexisting oxygen and sulfite in dechlorinated effluents (Helz, 1984) indicate that the reaction is probably slow. Air oxidation of the sulfite in samples dictates minimal time between sampling and measurement. This can best be achieved if the apparatus is portable and features ease of operation. Commercially available instruments with these capabilities were not found. The design and development of a new method for the determination of sulfite was therefore necessary.

Objectives for the new method included maintaining neutral pH, portability, convenience of operation, and concentration measurements down to the parts per billion range.

Constant current coulometric generation of iodine accompanied by amperometric equivalence point detection was the analytical mode selected. .

2. Instrumentation

The block diagram of the instrument is shown in Figure 1. The titration cell contains a set of two generator electrodes and a set of three amperometric detection electrodes.
At the generator anode the constant current process yielding iodine is accompanied by

$$2I^- \ -----> \ I_2 + 2e^- \tag{1}$$

oxidation of the sulfite by the iodine in the stirred solution. The high overvoltage for sulfate cathodic reduction prevents sulfite formation at the generator cathode. At the equivalence point accumulation of excess iodine begins. The first excess iodine yields a cathodic current at the indicator electrode of the amperometricdetector. The current-follower output, which is the endpoint signal is transfered through the optocoupler and triggers lateral solid state switches which terminate the constant generator current and the count on the digital display counter. The generation time indicated on the digital display is proportional to the sulfite content in the sample. The instrument is ordinarily calibrated so that the digital display reads directly in mg SO_3 / l.

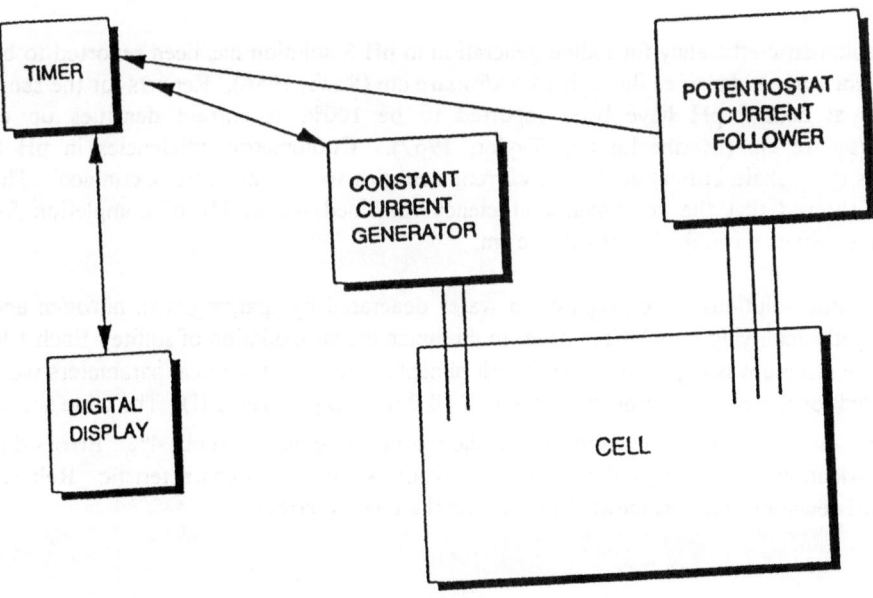

Fig. 1. Block diagram of the coulometer instrument.

Circuitry design and construction details (Ekkad, 1995) are based on analog components in order to permit a compact and portable instrument. For the amperometric endpoint detection circuit, a conventional in-house built potentiostat and current follower based on a quad JFET-input operational amplifier chip was used. This detector circuitry was electrically isolated from the generator circuitry by using a battery as the power supply and by propogating it's output signal to the generator/timer circuit via an optocoupler. The titration cell was a 200 ml beaker. The iodine generator electrodes are a commercially available dual platinum electrode assembly of area 1.4 cm^2 each. For the detector, both the indicator and the counter electrodes are platinum wires. The indicator electrode area is 0.45cm^2. The applied voltage is +0.1 V vs. a commercially available saturated calomel electrode. A magnetic stirrer is used to continuously stir the cell contents.

In the procedure for operation of the instrument, 100 ml of sample is introduced into the titration cell and about 2g of potassium iodide crystals are added. Then the electrode assembly is inserted, and the titrator is turned on. The red LED "on" indicates completion of the measurement, and the count is read.

3. Reagents and Chemicals

0.1M Phosphate buffer solutions at pH 8.0 were used unless otherwise noted. Reagent grade chemicals were used as supplied.

4. Results and Discussion

The coulometric efficiency for iodine generation in pH 5 solution has been reported to be 100 % for current densities through 15mA/square cm (Raab, 1984), Reports for the same process at neutral pH have been reported to be 100% at current densities up to 5.0mA/square cm (Marinenko and Taylor, 1967). Coulometric efficiencies in pH 8 samples (phosphate buffer) at higher current densities were , therefore, examined. The results showed that the coulometric efficiency remained within 1% of completion for current densities through 8.3mA/square cm.

Stock sulfite solutions were prepared in water deaerated by sparging with nitrogen and also slightly acidifying with HCl in order to minimize the air oxidation of sulfite. Each 100 ml sulfite sample was adjusted to pH 8 with phosphate buffer. Analytical parameters were evaluated for the concentration range 0.015 - 10.0 mg SO_3/L (Table II). The table shows that four out of the six amounts measured show a negative bias of about 4%. Errors due to air oxidation of the sample during titration would show such a characteristic. Relative standard deviations were somewhat lower than the relative errors.

Table II

Analytical Results

present	measured	rel. std. dev., N = 5	mA/cm^2
0.015 mg SO$_3^=$ / l	0.015 mg SO3= / l	0.8 %	0.19
0.100	0.098	0.9	"
0.300	0.320	1.0	0.82
1.00	0.960	1.1	"
5.00	4.80	2.0	8.3
10.0	9.60	2.2	"

Field testing of the new instrument was performed in the Valley Plant of the Wisconsin Electric Power Company near downtown Milwaukee, . One application in the plant was monitoring dechlorinated city tap water for residual sulfite. The method normally used was the standard method by titration with iodine generated by addition of acidified iodate and iodide with starch as indicator.(APHA, 1989) The analytical results for the two methods were found to be in agreement at the 1.0 mg SO$_3$/l level using a 90 % confidence t-test. Power plant monitoring of sulfite content in dechlorinated cooling water was not regularily performed, hence no comparison method was available. Cooling water samples on standing after collection show depletion of sulfite due to air oxidation. The half time for the reaction is approximately 30 minutes, thus minimizing the sampling time by having the instrument close to the sampling site is essential..

In combination, the above results show that the instrument described provides for the determination of residual sulfite under dechlorination conditions. Those conditiions include maintaining neutral pH and accommodating the air oxidation of sulfite in the samples..

Acknowlegement

The support and cooperation of The Wisconsin Electric Power Company were essential to this work.

References

APHA, AWWA, WPCF,; 1989, *Standard Methods for Examination of Water and Wastewater, 17th Edition;* 4199-4203..

Ekkad, N. V.; 1995,   PhD. Thesis, University of Wisconsin-Milwaukee., 55-61,

Fogelman, K. D.; Walker, D. M.; Margerum, D. W.; 1989, *Inorganic Chemistry,* **28,** 986-93.

Gordon, G., and Suzuki, K.; 1978 *Inorganic Chemistry,* **17,** 3115-3118.

Helz, G. R.; 1984, *Environmental Science and Technology,* **18,** 48A-53A.

Laila, A.H.; 1989, *Talanta,* **36,** 1145-1146.

Marinenko, G. and Taylor, J.K.; 1967. *Analytical Chemistry,* **39,** 1568-1571.

Raab, D.H., and Huber, C.O.; 1984, in *Water Chlorination Environmental and Health Effects,* R. Jolley ed.,, Lewis Publishers, Chelsea MI, USA, 5, 1073-1079..

Stanbro, W. D., and Lenkevich, M. J.; 1982, *Science,* **215, 19,** 967-968.

Stanbro, W.D., and Lenkevich, M. J.; 1984, *International Journal of Chemical Kinetics,* **14,** 251-258.

Yin, B. S., Walker, D. M., Margerum, D. W.; 1987, *Inorganic Chemistry,* **26,** 3435-3441.

PART VI

WATER RESOURCES

THE MODULAR MODELING SYSTEM (MMS) -- THE PHYSICAL PROCESS MODELING COMPONENT OF A DATABASE-CENTERED DECISION SUPPORT SYSTEM FOR WATER AND POWER MANAGEMENT

G.H. Leavesley, S.L. Markstrom, M.S.Brewer, and R.J. Viger

U.S. Geological Survey, WRD, Box 25046, MS 412, DFC, Denver, CO 80225

Abstract . The Modular Modeling System (MMS) is an integrated system of computer software that is being developed to provide the research and operational framework needed to support development, testing, and evaluation of physical-process algorithms, and to facilitate integration of user-selected sets of algorithms into operational physical-process models. MMS uses a module library that contains compatible modules for simulating a variety of water, energy, and biogeochemical processes. A model is created by selectively linking modules from the library using MMS model-building tools. A geographic information system (GIS) interface also is being developed for MMS to support a variety of GIS tools for use in characterizing and parameterizing topographic, hydrologic, and ecosystem features, visualizing spatially and temporally distributed model parameters and variables, and analyzing and validating model results. MMS is being coupled with the Power Reservoir System Model (PRSYM) to provide a database-centered decision support system for making complex operational decisions on multipurpose reservoir systems and watersheds. The U.S. Geological Survey and the Bureau of Reclamation are working collaboratively on a project titled the Watershed Modeling Systems Initiative to develop and apply the coupled MMS - PRSYM models to the San Juan River basin in Colorado, New Mexico, Arizona, and Utah.

Keywords: hydrologic models, watershed models, decision support system, water management, reservoir management, electric power production, water resources planning, geographic information system.

1. Introduction

The interdisciplinary nature and increasing complexity of environmental and water-resource problems require the use of modeling approaches that can incorporate knowledge from a broad range of scientific disciplines. Selection of a model to address these problems is difficult given the large number of available models and the potentially wide range of study objectives, data constraints, and spatial and temporal scales of application. Coupled with this are the problems of characterizing and parameterizing the study area once the model is selected. Guidelines for estimating parameters are few and the user commonly has to make decisions based on an incomplete understanding of the model developer's intent.

To address the problems of model selection, application, and analysis, a set of modular modeling tools, termed the Modular Modeling System (MMS) (Leavesley *et al.*, 1996) is being developed. MMS uses a master library that contains compatible modules for simulating a variety of water, energy, and biogeochemical processes. A model is created by selectively coupling the most appropriate process algorithms from the library to create an "optimal" model for the desired application. Where existing algorithms are not appropriate, new algorithms can be developed.

MMS provides a flexible framework in which to develop a variety of physical process models that can be coupled with resource-management models for use in addressing a wide range of

Water, Air and Soil Pollution **90**: 303-311, 1996.
© 1996 *Kluwer Academic Publishers.*

management issues. The operation and management of water resource systems is one such application being addressed jointly by the U.S. Geological Survey (USGS) and the Bureau of Reclamation (USBR) under a program called the Watershed Modeling Systems Initiative (WMSI). MMS is being coupled with the Power Reservoir System Model (PRSYM), a general purpose reservoir simulation and optimization model, to provide a set of decision support tools for use on any river basin. The purpose of this paper is to provide an overview of MMS, the database-centered approach to linking MMS and PRSYM, and the initial application of this work under the WMSI program.

2. MMS Overview

The conceptual framework for MMS has three major components: pre-process, model, and post-process (Figure 1). The pre-process component includes the tools used to prepare, analyze, and input spatial and time-series data for use in model applications. The model component includes the tools to develop and apply models. The post-process component includes tools to display and analyze model results, and to pass results to management models or other types of software. The model component currently is the most fully developed of the three components. However, a number of the pre-processing and post-processing tools are being developed, tested, and made available for linkage to the model component.

A graphical user interface (GUI) is being developed to provide user access to all the components and features of MMS. The present framework has been developed for UNIX-based workstations and uses X-windows and Motif for the GUI. The GUI provides an interactive environment for users to access model-component features, apply selected options, and graphically display simulation and analysis results. Completion and public release of MMS is proposed for mid 1996.

2.1 PRE-PROCESS COMPONENT

The pre-process component includes all data preparation and analysis functions needed to meet the data and parameterization requirements of a user-selected model. A goal in development of the pre-process component is to take advantage of existing data-preparation and analysis tools and to provide the ability to add new tools as they become available. Spatial data analysis is accomplished using GIS tools that have been developed and tested in both the Arc/Info system (ESRI, 1992) and the Geographical Resources Analysis Support System (GRASS) (U.S. ARMY Corps of Engineers, 1991). Functions developed include the ability to [1] delineate and characterize watershed subbasin areas for distributed-parameter modeling applications, [2] estimate selected model parameters for these subbasins using digital elevation model (DEM) data and digital databases that include information on soils, vegetation, geology, and other pertinent physical features, and [3] generate an MMS input parameter file from these estimates.

Time-series data from existing databases as well as from field instrumentation are prepared for use in selected model applications by generating and combining these data into a single flat ASCII file. Additional tools are being developed to detect and replace erroneous or missing values, aggregate data to longer time steps, disaggregate data to shorter time steps, and apply

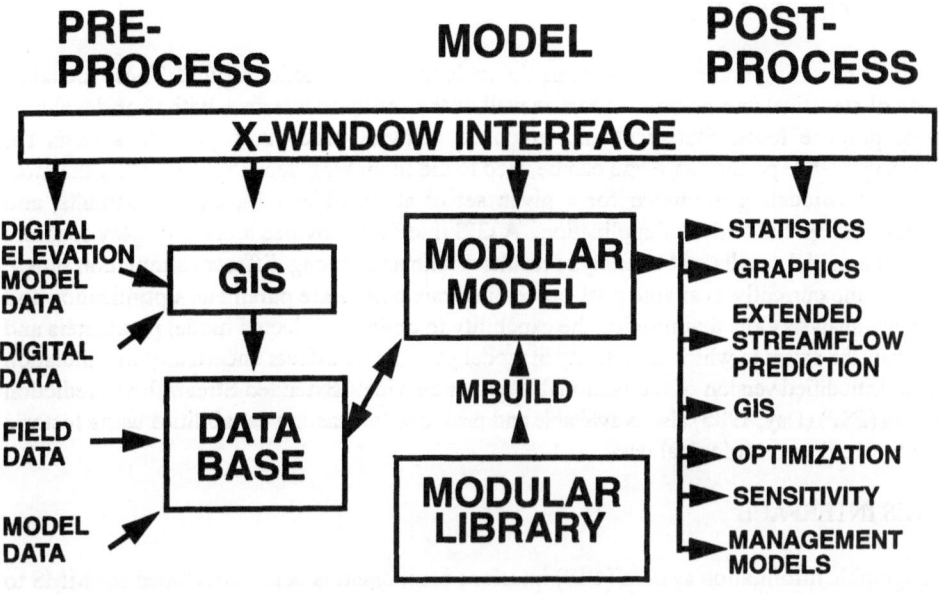

Fig. 1. A schematic diagram of the components of the Modular Modeling System (MMS).

transform functions to produce a new time series. Methods to create simulated time series from model output or from the analysis and extrapolation of measured data to unmeasured points or gridded fields are also being developed. The databases used to store the spatial and time-series data provide the interface between the pre-process and model components. Interface to the model component is provided by the use of database-manipulation scripts that access the database.

2.2 MODEL COMPONENT

The model component is the core of the system and includes the tools to build a model by selectively linking process modules from the module library and to interact with this model to perform a variety of simulation and analysis tasks. The library can contain several modules for a given process, each representing an alternative conceptualization or approach to simulating that process. The user, through an interactive model builder interface (MBUILD), selects and links modules to create a specific model. Once a model has been built, it may be saved for future use without repeating the MBUILD step. This capability allows "canned" versions of models to be provided to end users. User interactions with MBUILD and the model are provided using a variety of X-window and graphical techniques.

2.3 POST-PROCESS COMPONENT

The post-processing component contains the tools to analyze model results. These include a variety of statistical and graphical tools as well as the ability to interface with user-developed special purpose tools. Statistical and graphical analysis procedures provide a basis for comparing module performance and can be used to aid in making decisions regarding the most appropriate modeling approach for a given set of study objectives, data constraints, and temporal and spatial scales of application. A GIS interface provides tools to display spatially distributed model results and to analyze results within and among different simulation runs.

Two of the currently available post-processing capabilities are parameter-optimization and sensitivity-analysis tools that provide the capability to optimize selected model parameters and to evaluate the extent to which uncertainty in model parameters affects uncertainty in simulation results. A modified version of the National Weather Service's Extended Streamflow Prediction Program (ESP) (Day, 1985) also is available and provides forecasting capabilities using historic or synthesized meteorological data.

2.4 GIS INTERFACE

A geographic information system (GIS) interface component is being developed for MMS to facilitate model development, parameterization, application, and analysis. This interface permits application of a variety of GIS tools to lumped- and distributed-parameter modeling approaches. These tools permit development and testing of a variety of objective characterization and parameterization techniques. They also permit visualization and analysis of the spatial distribution of model parameters and simulated state variables at a variety of spatial and temporal scales.

Within the model component, the GIS interface provides an animation tool to enable the visualization of the spatial and temporal variation of simulated state variables during a model run. Selected images from this animation for user-defined time periods can be stored and used in a post-modeling analysis to compare simulated and measured spatial and temporal variations in the selected state variable. Remotely sensed snow-covered area and soil moisture are examples of variables that can provide important additional independent measures of distributed-parameter model performance.

3. PRSYM

PRSYM (Fulp et al., 1995) is a general-purpose reservoir simulation and optimization modeling system developed by the University of Colorado's Center for Advanced Decision Support (CADSWES) through a cooperative program with the Electric Power Research Institute (EPRI), USBR, Western Area Power Administration (WAPA), and Tennessee Valley Authority (TVA). PRSYM is being developed using object-oriented programming methodology. Within PRSYM, the basic network topology of a river system is constructed by the user. Reservoir, river reach, confluence, and other objects are selected from a GUI palette and linked in the appropriate sequence to form the river and infrastructure network of a given basin. Data associated with each object can be entered into the modeling system through the GUI or imported from a database or ASCII files. This methodology allows the user to customize the modeling process for any basin.

The operations policy and rules associated with a given basin are added to PRSYM through an existing constraint editor. Current modeling methods within PRSYM include basic reservoir simulation and the distribution of water throughout a network using linear programming, goal programming, and rule-based simulation approaches. Water right allocation is being developed as an additional modeling method. Computational time steps supported include 6-hour, daily, weekly, and monthly time steps.

4. Database-Centered Decision Support System

The physical process models in MMS are being linked with the reservoir and river management models of PRSYM via a common database, thus providing a database-centered decision support system (Figure 2). A number of ancillary tools provide GIS, statistical analysis, and data query and display capabilities that are shared by MMS and PRSYM. In the current WMSI project, the database is an INGRES (INGRES Corp.,1991) based system called the Hydrologic Data Base (HDB). However, this approach is not limited to INGRES but can be used with any relational database system.

MMS and PRSYM access the database through user-written data management interfaces (DMIs). Users can use a variety of standard DMIs, or write customized DMIs in any standard programming language that has database bindings, to access data from a variety of data repositories, including other relational databases. Changing the central database requires only that the DMIs be modified to support the selected database. No changes need to be made in MMS or PRSYM.

Communication between MMS and PRSYM is being designed to use a scenario file in the database. A scenario is defined as a sequential list of modeling operations to be run using MMS and PRSYM. Different scenarios can be developed to address a range of resource-management decisions. The scenario file is accessed by both modeling systems and the specified models are executed in the order listed. A given resource-management decision may require the results of one or several models in both systems. Model results from MMS are written to the database for use as inputs to PRSYM and vice versa. This exchange of data and model results is an iterative procedure, the magnitude of which is dependent on the complexity of the decision process.

An example of such a procedure would be the management of a multi-reservoir river system within the constraints of competing water users and selected environmental constraints such as water temperature limits or fisheries habitat needs. Here a scenario of MMS and PRSYM runs might begin with the execution of a watershed model in MMS to provide estimates of the daily time series of water inflows to all reservoirs in the system. Then PRSYM would be executed to use these inflows to develop a number of different management options and produce a time series of reservoir releases associated with each. These options might reflect different mixes of water use for power generation, agriculture, and municipal water supply.

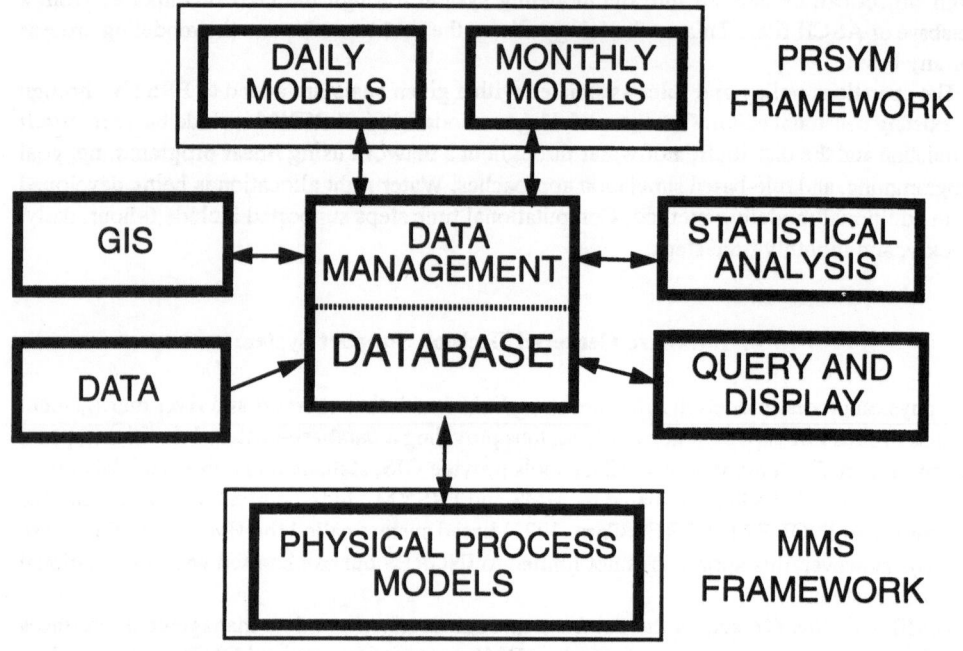

Fig. 2. Schematic of the database-centered decision support system.

Each release option has implications with regard to the environmental constraints on system operation and would need to be examined for those river reaches where these constraints apply. Thus, a one- or two-dimensional river hydraulic model would be run in MMS to assess the effects of the different reservoir-release options on the critical river reaches. Then PRSYM would be run again using the additional reach information to further refine the management options for these specific sets of conditions and constraints.

The generation of management options is not limited to PRSYM but they can be developed within MMS as well. Using the ESP capability of MMS, alternative time series of water inflow to reservoirs can be developed, each with an associated estimate of a probability of occurrence. The manager could then select various levels of probability of occurrence to assess the effects of uncertainty on water management options. Alternatively, new time series of meteorological variables could be developed to reflect the potential effects of global climate change. These time series could be used as input to the watershed model to provide estimates of reservoir inflows for use in assessing the effects of climate change on basin management strategies.

Given that different river basins may have different management objectives and operational constraints, the flexibility of MMS enables the development and application of those models determined to be most appropriate for a specific basin. Alternative modeling approaches can be evaluated to determine the optimal set of models for various management decisions. Coupled with PRSYM, the result is a very flexible set of water-resource management tools.

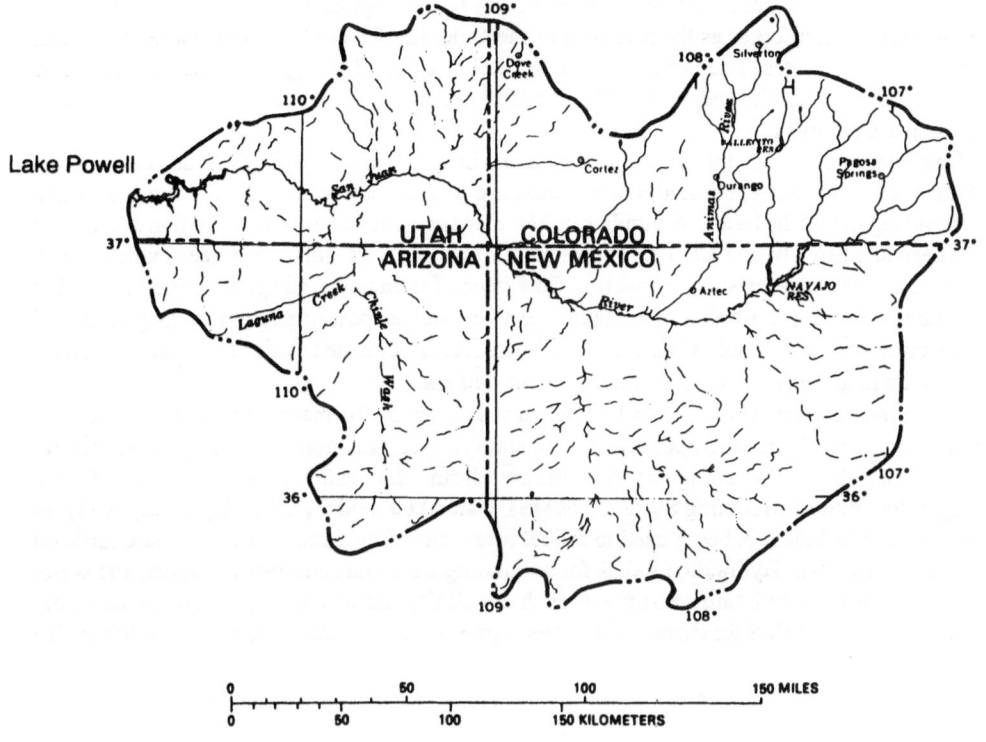

Fig 3. Map of the San Juan River basin showing principal streams.

5. The Watershed Modeling Systems Initiative

The Watershed Modeling Systems Initiative (WMSI) is a cooperative effort between the USGS and USBR to develop and implement water-resource models and integrated database management systems in the western United States. The objectives of WMSI are to [1] improve water management so as to increase the benefits associated with water uses such as hydropower, agriculture, water supply, and recreation, [2] improve environmental quality of water resources, and [3] increase the supply of water by more efficient water management and use. The database-centered decision support system approach using MMS and PRSYM is being developed, tested, and implemented under this initiative. The San Juan River basin has been selected for the initial development and application of this approach. The decision support system is proposed to be operational in the San Juan River basin by the end of 1996, at which time another basin will be selected for the development and testing of additional system tools.

The San Juan River basin is a tributary to the Colorado River and has a drainage area slightly in excess of 23,000 mi 2 at Bluff, Utah (Figure 3). USBR operates four reservoirs within the San Juan River basin, and six other reservoirs with storage capacities greater than 5,000 acre-feet also exist within the basin. The San Juan River empties directly into Lake Powell, a USBR reservoir that serves as the regulator of the Colorado River Compact between the lower and upper basin States. The San Juan River basin has a varied topography and climate which results in a varied quantity and quality of streamflow. As a result, the management issues within the basin are complex.

Headwater drainages in the San Juan Mountains in the Colorado portion of the basin are the major sources of water, but major water uses are located downstream in the more arid and semi-arid regions of the basin. A number of key management issues within the San Juan basin have been identified by the USBR. These include efficiency of water-resources management, environmental concerns such as meeting flow needs for endangered species, and optimizing operations within the constraints of multiple objectives such as power generation, irrigation, and water conservation. In addition, the San Juan River is regulated under the Colorado River Compact which further constrains management options.

The initial application in the San Juan basin will focus on the management of the Vallecito and Navajo reservoirs so as to optimize their operation in conjunction with unregulated inflows to the San Juan, in particular the inflow from the Animas River. The USGS Precipitation-Runoff Modeling System (PRMS) watershed model (Leavesley et al., 1983), as implemented in MMS, is being used to model reservoir inflows and unregulated streamflows for use by PRSYM. Hydraulic models for simulating flow and sediment transport, and water quality models for simulating constituents such as salinity and water temperature are currently being developed in MMS for channel-reach-level applications to address these issues in the San Juan basin.

6. Summary

MMS is an integrated system of computer software that has been developed to provide the research and operational framework needed to support the development, testing, and evaluation of water, energy, and biogeochemical process algorithms and to facilitate the integration of user-selected sets of algorithms into operational models. MMS provides a common framework in which to develop and apply models that are designed for basin- and problem-specific needs. MMS is being coupled with the reservoir simulation and optimization model PRSYM using a common database interface to provide a database-centered decision support system for use in water-resources and power management. As the physical process modeling component of this decision support system, MMS provides a flexible framework in which to integrate these current activities and to easily incorporate any future advances in science and technology.

Aknowledgement

The use of trade, product, industry, or firm names is for descriptive purposes only and does not imply endorsement by the U.S. Government.

References

Day, G.N.: 1985, *Journal of Water Resources Planning and Management*, American Society of Civil Engineers, **111**, pp.157-170.

Environmental systems Reasearch Institute: 1992, *ARC/INFO 6.1 user's guide*. Redlands, CA.

Fulp, T.J., Vickers,W.B., Williams, B., and King, D.L.: 1995, *in* L. Ahuja, J. Leppert, K. Rojas, and E. Seely, (eds.), *Workshop on computer applications in water management*. Colorado Water Resources Research Institute, Fort Collins, CO., Information Series No. 79, pp. 24-27.

INGRES Corporation: 1991, *INGRES/SQL Reference Guide for the UNIX and VMS Operating Systems*. Alameda, CA.

Leavesley, G. H., Lichty, R.W., Troutman, B. M., and Saindon, L.G.: 1983, "Precipitation-Runoff Modeling System: User's Manual." *U.S. Geological Survey Water Resources Investigations Report* **83-4238**.

Leavesley, G.H., Restrepo, P.J., Stannard, L.G., Frankoski, L.A., and Sautins A.M.: 1996, *in*   M. Goodchild, L. Steyaert, B. Parks, M. Crane, M. Johnston, D. Maidment, and S. Glendinning, (eds.),   *GIS and Environmental Modeling: Progress and Research Issues*, GIS World Books, Ft. Collins, CO.

U.S. Army Corps of Engineers: 1991, *GRASS Version 4.0 User's Reference Manual*, USACERL, Champagne, IL.

VALIDATION AND SENSITIVITY OF A CONVECTIVE PRECIPITATION MODEL FOR MOUNTAINOUS AREAS

J. H. Copeland

U. S. Geological Survey, Water Resources Division, Denver Federal Center, MS412, Lakewood, Colorado 80225, USA

Abstract. A convective precipitation model for use in regions of complex terrain has been developed and applied to the Gunnison River Basin in southwestern Colorado. Spring snowfall in the Rocky Mountain region often has a significant convective component which orographic precipitation models are unable to simulate. Additionally, summertime precipitation is predominately convective in this area and is responsible for a large portion of summer streamflow variability. Streamflow typically increases by 50 to 100 percent of baseflow for moderate rainfall events for periods of up to one week. Larger precipitation episodes can produce peak discharges that exceed the spring snowmelt peaks. Convective precipitation also is important for plant growth, minimum streamflows and fire hazard conditions. In addition, an accurate assessment of the response of hydrologic systems to climate variability and change requires an accurate estimate of convective precipitation in mountainous areas. The convective model accurately reproduced the trend and amount of observed precipitation for the test period August 14-20, 1989. The convective model has applicability for downscaling large-scale model precipitation to smaller scales for use in water quality and quantity assessments.

Keywords: Colorado, convection, downscaling, modeling, orographic, precipitation, summer, sensitivity.

1. Introduction

Water quality and quantity are both dependent upon convective precipitation to some extent. The additional volume of water from surface runoff of convective precipitation into rivers, streams, and lakes can dilute pre-existing concentrations of impurities leading to an improvement in water quality. In other instances the convective precipitation may degrade water quality through runoff over mine tailings or agricultural fields, or through increased turbulence in streams and rivers disturbing pollutant laden sediments. All of these situations may impact water quality trend and or background monitoring.

In the Rocky Mountain region of the western United States much of the available water comes from snowmelt runoff during the late spring and early summer. An accurate assessment of snowpack is important for determining water availability in the coming year. Snowpack accumulation has been modeled with reasonable success using an orographic precipitation model (Rhea, 1977). Spring snowfall in the Rocky Mountain region often has a significant convective component which orographic precipitation models are unable to simulate. Additionally, summertime precipitation is predominately convective in this area and is responsible for a large portion of summer streamflow variability. Streamflow typically increases by 50 to 100 percent of baseflow for moderate rainfall events for periods of up to one week. Larger precipitation episodes can produce peak discharges that exceed the spring snowmelt peaks. The operation of hydroelectric

Water, Air and Soil Pollution 90: 313-320, 1996.
© 1996 *Kluwer Academic Publishers.*

and agriculture supply reservoirs, maintenance of minimum streamflows, plant growth, and summer fire hazard conditions are all affected by convective precipitation.

To overcome the deficiencies in orographic precipitation models during predominantly convective periods of spring and summer a convective precipitation model for use in regions of complex terrain has been developed and applied to the Gunnison River Basin in southwestern Colorado. A description of the convective model will follow. The results of a verification and sensitivity to input data study will also be presented.

2. Model Description

Convective precipitation has traditionally been estimated by using 3-dimensional atmospheric models. These models are computationally expensive to use, especially for studies of long time periods (10 years or more) at fine spatial scales (on the order of 10 kilometers). Therefore, a convective precipitation model for mountainous regions has been developed. The convective model is designed to be flexible and therefore can be run using either observed atmospheric soundings or outputs from other atmospheric models. This flexibility makes the convective model useful for examining hydrologic sensitivities and water availability under both current and altered climatic conditions.

There are several components to the convective model: initialization, convective initiation and the cloud model. The convective model is a one-dimensional vertical column model. It can be used spatially but no communication between adjacent columns occurs. The convective model can be used spatially on grids with a cell size of 20 to 120 kilometers. These limits exist due to the basic assumption that only a single storm exists in a grid cell when convection is occurring, and that the convective life cycle is the same as the model time step of one hour.

The initialization component of the convective model takes an input profile and interpolates it to the model grid. The convective model requires vertical profiles of temperature, pressure, humidity, and wind vector components. If input profiles are not available on a hourly basis then a time interpolation is done to obtain a valid profile. The input profile is then vertically interpolated to the model grid which has a grid separation that varies from 200 to 500 m. The grid extends from the ground surface to the top of the input profile, approximately 20 km.

The convective initiation component performs a vertical analysis on the model grid to determine if convection will occur. Several criteria must be met in order to evaluate the initiation of convection. The first is the presence of conditional instability, this is done by checking for model levels at which the moist static energy decreases with height. If conditional instability exists, a source level for the convective air is determined as the level of maximum equivalent potential temperature near the surface. The source air is lifted dry adiabatically to find the lifting condensation level (cloud base) and then moist adiabatically to establish the free convection and equilibrium (cloud top) levels. Other necessary convective initiation criteria are net positive convective available potential energy (CAPE), upward motion at cloud base, minimum depth of cloud, and minimum altitude of cloud top.

If it is determined that convection should occur then the cloud model component is called to compute the total condensate and precipitation efficiency used in determining the precipitation rate due to convection.

The cloud updraft model is based on the convective parameterization of Kuo (1974), in which the vertically integrated horizontal moisture convergence plus surface evaporation is equal to atmospheric moisture storage plus precipitation. This allows the precipitation to be calculated as a percentage of the total condensate in the model sounding, assuming negligible surface evaporation during the convection time period (1 hour). The percentage is the precipitation efficiency, which is an empirical function of wind shear (Marwitz, 1972).
The updraft ascent is assumed non-entraining and conserves the equivalent potential temperature of the source air. Only vapor and liquid phases are accounted for.

3. Model Verification

Verification of precipitation models is a very difficult process complicated by factors relating to the discontinuous nature of precipitation. These factors include the density and quality of the observing precipitation gauges, and the resolution of and processes simulated by the model. Precipitation measured by a gauge and simulated by a model are not the same. The first is a realization at a point through a storm while the second is a spatial average. It might be more appropriate to compare model output with spatial averages of radar observations but difficulties also exist with this method. Initiation of atmospheric models also has its difficulties due to the low density of upper air sounding stations, which has a typical spacing of 300 km.

To avoid the problems inherent in model initiation and verification with observed data the method of Observing Systems Simulation Experiments (OSSEs) (Charney et al., 1969; Kasahara, 1972) is used. The OSSE method involves the use of a control model to generate "observations" to be used as both input and verification for a second model. This method is very useful since in addition to providing a complete dataset for model verification, the sensitivity of the model to observational error and temporal and spatial resolution of observations can also be examined. Comparisons with observed precipitation data from Grand Junction, Colorado will also be done in this study.

The control model used is the Regional Atmospheric Modeling System (RAMS) (Pielke et al., 1992). It is a non-hydrostatic 3-dimensional numerical weather prediction model capable of simulating scales from meters to hundreds of kilometers. In this work it is used to produce control observations on a 160,000 km^2 grid with a 20 km horizontal grid separation over southwestern Colorado. The RAMS model was run for one week, from 00Z August 14 1989 through 00Z 21 August 1989, to generate hourly 3-dimensional fields of pressure, temperature, mixing ratio, horizontal and vertical winds, and 2-

dimensional surface distributions of hourly accumulated precipitation. Hourly vertical profiles were extracted from the RAMS grid cell corresponding to the location of Grand Junction, Colorado to initiate the convective precipitation model.

The convective model and RAMS both produce precipitation averages due to a single storm in a 400 km^2 grid cell. To directly compare model precipitation with observed would be in error. To make a valid comparison the assumption of a stationary convective storm of 80 km^2 has been made. This is based on a typical spatial scale for individual thunderstorm cells of ~10 km (Byers and Braham, 1949). This yields a scaling factor of 5 which has been applied to the output of RAMS and the convective model.

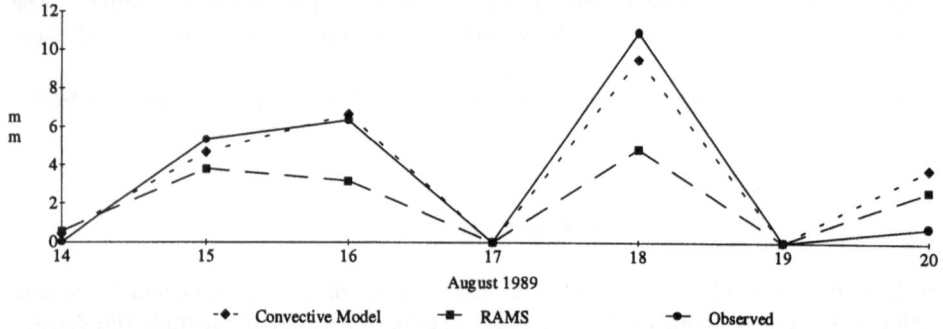

Fig. 1. Simulated and observed daily precipitation for the period August 14 through August 15, 1989.

Daily precipitation amounts from both models reflect the pattern of observed daily precipitation at Grand Junction (Figure 1). The convective model produces more precipitation than RAMS. This is a result of various physical processes which are not accounted for in the convective model such as: updraft entrainment, horizontal advection, and downdraft evaporation. The convective model does not appear to have a consistent bias while RAMS tends to underpredict the daily precipitation amount. A quantitative assessment of the two models is shown in Table I. Percent errors in daily precipitation amount for the convective model tend to be below 15% with no obvious bias. RAMS daily errors are significantly higher in the range from 30 to 50%, and consistently fall below the observed amount. August 14, is a day when precipitation was simulated but not observed. This situation can occur when the environment around Grand Junction is conducive to convective development but no storms pass over the observation rain gauge. August 20 is a similar day. In this case both models simulated significantly more precipitation than was observed. The small amount of observed precipitation indicates a grazing pass over the observation rain gauge.

Accumulated precipitation errors over the week for the convective model are within 10% of observed. The lack of bias error in the convective model results in a lower accumulated precipitation error than daily precipitation error due to errors on one day canceling out errors from another day. RAMS with its consistent negative bias exhibits error growth over the week, up to 45% for accumulated precipitation.

It should be noted that the validation study is not a test to determine the better model. The scaling factor could have been tuned a posteriori to favor either model. The important results of the validation study are that each model reproduces the temporal trend of observed precipitation and each model also produces precipitation amounts close to observed with reasonable scaling.

TABLE I

Percent differences between observed and simulated precipitation

	Daily Precipitation		Accumulated Precipitation	
Date	Convective Model	RAMS	Convective Model	RAMS
14	+	+	+	+
15	-12	-29	-4	-18
16	5	-50	1	-36
17	0	0	1	-36
18	-13	-56	-6	-45
19	0	0	-6	-45
20	400	249	7	-36

+ precipitation simulated but not observed

4. Model Sensitivity

For validation simulation the convective model required hourly input data. Observed atmospheric soundings are available 12-hourly and archived model datasets may not be available hourly. There are two types of sensitivity to time that can be explored. The first is the period of the input data, or the number of hours between available atmospheric sounding. The second is the actual time that the sounding is valid for. In both cases the convective model still requires hourly sounding data. The missing sounding in each case will be interpolated from adjacent valid data.

The effect of varying the time between valid soundings is shown in Table II. In each case the initial sounding of the day is valid at 00 GMT. The control scenario is the hourly validation convective model simulation. The convective model is highly sensitive to the sounding update period. Accumulated precipitation errors range from a low of 5.3% for a 2-hour update period to a high of 61.6% more precipitation in the 8-hour update scenario, and 94.3% less precipitation with a 6-hour update period.

TABLE II

Convective model sensitivity to input sounding update period

Update period (h)	Accumulated precipitation (mm)	% difference from control
1	25.08	control
2	26.41	5.3
3	4.22	-83.2
4	31.91	27.2
6	1.44	-94.3
8	40.53	61.6
12	2.60	-89.6
24	6.57	-73.8

The reason for the sensitivity in the convective model relates to how the available soundings resolve the structure of the atmosphere over time. Average hourly precipitation from the control scenario is shown in Figure 2. The pattern is typical of summertime convection in the Rocky Mountain region. The peak near 02 GMT occurs around local sunset and is a result of increased convective instability due to radiative cooling at cloud top. The peak near 17 GMT is a late morning instability release due to daytime heating and slope circulations.

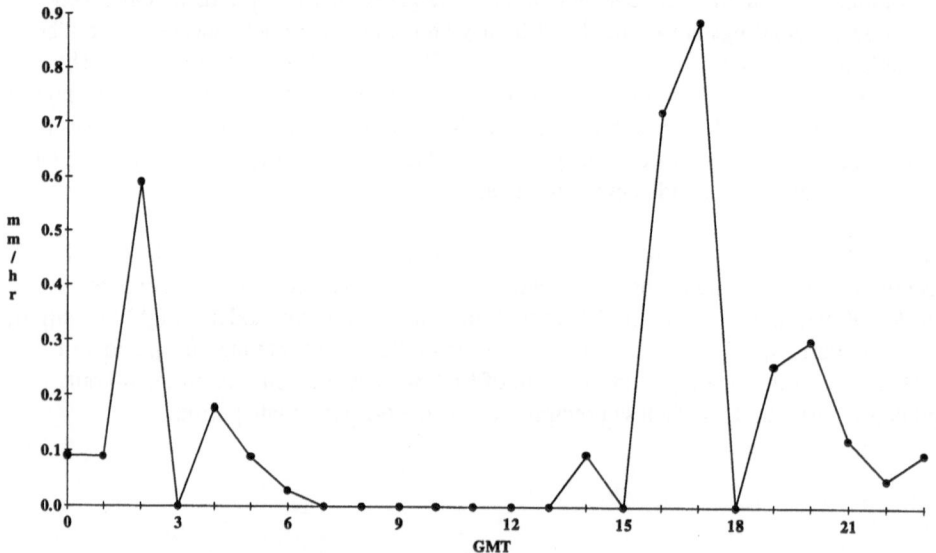

Fig. 2. Averaged hourly precipitation from the convective model.

As the sounding update period changes, the time evolution of atmospheric conditions that lead to the morning and evening precipitation peaks will get resolved differently. In the case of the 8-hour update period the valid soundings are at 00, 08, and 16 GMT. The evening peak is missed but the morning peak is captured. Due to the linear time interpolation used to obtain soundings at the remaining times of the day, the convectively conducive 16 GMT sounding has a longer time influence than for an hourly update period. This causes the increase in precipitation of 61.6%. This situation is similar for the 6-hour update period scenario, but in this case both peaks are missed and the minima in convective activity are resolved leading to a reduction of precipitation by 94.3%.

The final sensitivity is to examine the effect of varying the valid times of input soundings for a given update period on accumulated precipitation. Since the 6-hour update period scenario had the lowest accumulated precipitation it was chosen to see if by judicious choice of sounding times it is possible to replicate the accumulated precipitation of the control scenario. The start time of each 6-hour update period was varied from 00 to 05 GMT. The accumulated precipitation and percent errors from the hourly control scenario are shown in Table III. The results indicate that it is possible to improve the simulated accumulated precipitation for a given update period by changing the times at which the input soundings are valid. By moving the time of the first sounding of the day from 00 GMT to 02 GMT the percent error went from -94.3% to only -0.1%.

TABLE III

Convective model sensitivity to input sounding times for a 6 hour update period		
Sounding times (GMT)	Accumulated precipitation (mm)	% difference from control
5, 11, 17, 23	22.66	-9.7
4, 10, 16, 22	44.26	76.5
3, 9, 15, 21	1.43	-94.3
2, 8, 14, 20	25.06	-0.1
1, 7, 13, 19	11.07	-56.9
0, 6, 12, 18	1.44	-94.3

5. Conclusions

The results of the validation study indicate that the convective model if provided quality hourly input soundings can accurately simulate both the trend and amount of observed precipitation. The sensitivity study demonstrated the need for the input data to properly resolve the phenomena under study. In this case it was possible for 6-hourly input soundings to resolve the dual peak nature of summertime convective precipitation in the Rocky Mountain region. It is not possible for the convective model to use 12-hourly input soundings to simulate accumulated precipitation since the interpolated time evolution of atmospheric structure would not resolve the two precipitation peaks. The consequence of this is that the convective model is unable to be run from the synoptic soundings which are available every 12 hours. The convective model is therefore better suited to downscaling precipitation from large-scale models, such as GCMs, to finer scales though additional tests still need to be performed.

References

Byers, H. R., and Braham, R. R. Jr.: 1949, *The Thunderstorm*, U.S. Department of Commerce, Weather Bureau, Washington D.C., 287 pp.

Charney, J. G., Halem, M., and Jastrow, R.: 1969, *J. Atmos. Sci.* **26**, 1160-1163.

Kasahara, A.: 1972, *Bull. Amer. Meteorol. Soc.* **53**, 252-264.

Kuo, H. L.: 1974, *J. Atmos. Sci.* **31**, 1232-1240.

Marwitz, J. D.: 1972, *J. Res. Atmos.* **6**, 367-370.

Pielke, R. A., Cotton, W. R., Walko, R. L., Tremback, C. J., Nicholls, M. E., Moran, M. D., Wesley, D. A., Lee, T. J., and Copeland, J. H.: 1992, *Meteorol. Atmos. Phys.* **49**, 69-91.

Rhea, J. O.: 1977, *Orographic precipitation model for hydrometeorological use.*, Colorado State University, Department of Atmospheric Science, Ft. Collins, Colorado.

Predictive Techniques for River Channel Evolution and Maintenance

J. M. NELSON

U.S. Geological Survey, Water Resources Division- National Research Program, Box 25046, Lakewood, Colorado 80225

Abstract. Predicting changes in alluvial channel morphology associated with anthropogenic and natural changes in flow and/or sediment supply is a critical part of the management of riverine systems. Over the past few years, advances in the understanding of the physics of sediment transport in conjunction with rapidly increasing capabilities in computational fluid dynamics have yielded new approaches to problems in river mechanics. Techniques appropriate for length scales ranging from reaches to bars and bedforms are described here. Examples of the use of these computational approaches are discussed for three cases: (1) the design of diversion scenarios that maintain channel morphology in steep cobble-bedded channels in Colorado, (2) determination of channel maintenance flows for the preservation of channel islands in the Snake River in Idaho, and (3) prediction of the temporal evolution of deposits in lateral separation zones for future assessment of the impacts of various dam release scenarios on lateral separation deposits in the Colorado River in Grand Canyon. With continued development of their scientific and technical components, the methodologies described here can provide powerful tools for the management of river environments in the future.

Keywords: Sediment transport, geomorphology, channel maintenance, instream flows, fluid mechanics

1. Introduction

Predicting the response of river channel morphology to alterations in flow hydrograph or sediment supply is one of the most difficult and important tasks carried out by engineers and geomorphologists providing information to resource managers. The difficulty of making predictions of this type is a direct result of the complex, nonlinear nature of flow and sediment transport in channels. With the exception of a few simple, easily characterized situations, there is no general theoretical solution for predicting the interaction between turbulent flows and an erodible bed. In view of this, the most common method for developing even semi-quantitative assessments of channel response in the past has been almost entirely empirical, requiring the acquisition of large data sets over significant periods of time. For example, to assess the effects of flow diversion from a channel, hydrologists typically develop both flood frequency and duration curves and sediment-transport rating curves to predict how changes in the natural flood frequency distribution as a result of diversion might affect the sediment load in the channel over some relatively long period of time. Developing these empirical relations requires the acquisition of stage, discharge, and sediment load data over a long enough period of time (often many years) to acquire statistically adequate characterizations of the flow and transport. Unfortunately, this kind of long-term data is not always available, so predictions must be made based upon extrapolations of data from other (hopefully similar) basins to the basin of interest. Furthermore, in many more complicated situations, even data for extrapolation is unavailable, and predictions must be made based upon rules of thumb or engineering judgement. When adequate basic data are not available, empirical relations cannot be developed and, generally, predictions of channel response cannot be made in any quantitative form. This represents a serious shortcoming in our ability to manage

Water, Air and Soil Pollution **90**: 321-333, 1996.
© 1996 *Kluwer Academic Publishers.*

riverine environments in a rational, scientifically based manner, and points to the need for the development of new tools for addressing this problem.

Over the past decade or so, developments in computational approaches in fluid mechanics coupled with increasing understanding of the basic mechanics of sediment transport in turbulent flows have started to produce tools for predicting channel response in situations where basic data are unavailable. In this paper, three examples of this kind of approach are described both in order to outline how these methodologies work and to provide some perspective on the kinds of problems to which these techniques can be applied. Each of the three examples represents a case where the use of empirical techniques would be difficult, if not impossible. The basic structure of the three approaches is the same: an appropriate model for flow in the channel of interest is coupled to a sediment-transport algorithm to predict the role of various flows in maintaining channel morphology. Due to space limitations, the flow models are described only briefly with reference to other publications. The three approaches here are comprise situations where one, two, and three-dimensional flow models, respectively, are required to characterize the sediment transport in the reach of interest.

2. Applications

I. STEEP MOUNTAIN CHANNELS

The first situation involves making predictions of flow and sediment transport in steep, mountain channels where the size of sediment clasts making up the channel bed may be comparable to the flow depth. There are two essential features of this kind of channel for the purposes of the discussion here. First, the vertical structure of the flow in channels of this type is significantly different than that found in typical open channel flows where the flow depth is much greater that the sizes of sediment grains on the bed (Jarrett, 1989). In consequence, the relation between flow velocity and bed stress in mountain streams varies strongly with flow depth or, in other words, the effective friction changes with flow depth in a relatively complex manner (Limerinos, 1970; Jarrett, 1989; Wiberg and Smith, 1991). Second, when sediment moves, it usually does so in a marginal transport mode (Parker et al., 1982; Andrews, 1994) meaning that the boundary shear stress value is only slightly above the critical value for the initiation of sediment motion. In channels with marginal transport conditions, significant sediment transport tends to occur only for relatively large flow events, often at or near the bankfull flow condition of the channel. In at least some significant fraction of these channels, a majority of the transport may take place during flows that occur as infrequently as once every decade or even longer.

One important result of marginal transport conditions is that it may take a relatively long period of time to obtain sediment transport data over a wide enough range of flow conditions to define an empirical relation between flow and transport. In general, defining such a relation accurately requires that measurements be concentrated during rare occurrences of high flow, which is often difficult both logistically and in terms of measurement techniques. Thus, this situation represents an ideal case for application of predictive modeling techniques.

To address the flow and sediment transport in steep, rough channels, the flow model described by Nelson et al. (1991) is coupled to the marginal bedload sediment-transport relation developed by Parker et al. (1982). The flow model, which is conceptually similar to that described by Wiberg and Smith (1991), differing only in the prescription of turbulence structure near the bed and in the method whereby the protrusion of sediment grains on the bed is determined, predicts the spatially averaged vertical distributions of velocity and stress given detailed grain geometry information. Essentially, by treating the drag on grains making up the bed, the approach predicts the relation between depth and velocity that can be used to develop a roughness relation that depends on depth in a manner that explicitly incorporates the variation in flow structure resulting from effects of large relative roughness. The approach yields depth-discharge and stress-discharge relations based only on inputs of discharge and grain geometry. In conjunction with measured cross-sectional channel geometry, the first of these two relations is used to develop a stage-discharge relation and the second, along with Parker's marginal transport bedload equation, is used to predict a relation between discharge and bedload transport rate.

Thus, using this approach, it is possible to generate stage-discharge and bedload-discharge relations based only on measured cross-sectional geometry and grain geometry on the bed, which is done using the pebble-counting methodology described by Nelson et al. (1991). This approach has been applied on several channels in Colorado both to test validity and to explore the impact of various diversion and climate change scenarios on mountain channels (Parker et al., 1992). In Figure 1, the predicted stage-discharge relation for the East River, a tributary of the Gunnison River located in Southwestern Colorado, is shown along with the measured stage discharge data for the site. In general, the predictions of the model are in reasonable agreement with the measured relation between stage and discharge. However, there is some discrepancy at the highest flows, with the measurements showing higher stage at a given discharge than the model predicts. Inspection of field notes and the site itself indicates that there is some backwatering effect during high flows from the confluence with the Taylor River immediately downstream of the experimental site. Thus, the principal source of error appears to be associated with streamwise convective accelerations produced by channel nonuniformity in the streamwise direction. As the approach assumes that the flow is both steady and at least

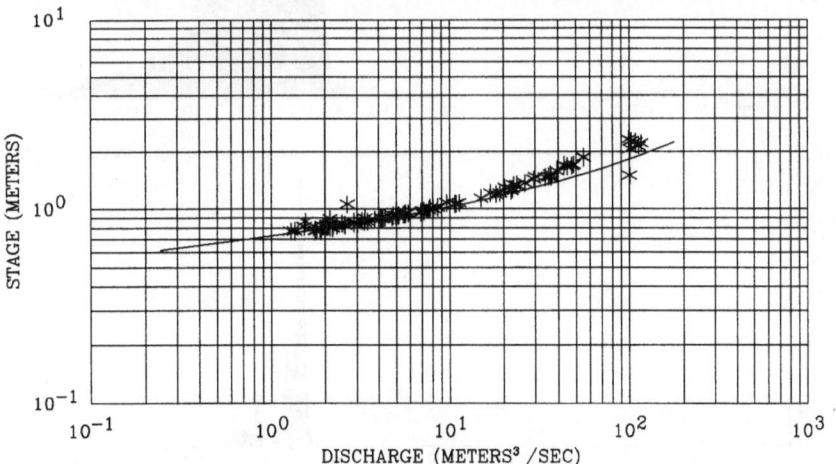

Figure 1. Predicted (solid line) versus measured (asterisk points) stage-discharge relation for the East River immediately upstream of its confluence with the Taylor River.

uniform in an averaged sense, meaning that channel slopes are in equilibrium with local roughness and are unaffected by backwater effects, this is not surprising.

In Figure 2a, daily discharge records for 1984 are shown. Using this record as input, the model predicts the associated sediment load, as shown in Figure 2b. Note that by far the majority of the transport in this channel occurs during relatively rare large flows, as is typical of channels of this type. This behavior is one of the reasons that it is critical to have some predictive capability for assessing instream or channel maintenance flows. Restricting this discussion to the physical characteristics of the channel, the precise definition of channel maintenance flows must be all those flows that move sediment in the channel. In other words, if we assume that sediment supply from the hillslopes and tributaries is a given quantity, the channel flows that must be preserved in a given diversion scenario such that the channel will experience no aggradation or degradation relative to the natural, undiverted flow situation are all flows that move sediment in the channel. In general, this strict definition can be relaxed because finer material in the channel, which may be transported at relatively low flows, is limited by supply to the channel, rather than by the transport capacity of the channel, so that all of the fine material supplied to the channel can easily be moved by a few short high flows even in the absence of lower flows that may move this finer material. Thus, it is common to restrict attention to the relatively coarse gravel and cobbles making up the majority of the bed material in channels of the type considered here when considering channel maintenance flows. With this in mind, the definition of channel maintenance flows may be relaxed to include only those flows that move the relatively coarse material making up most of the channel bed.

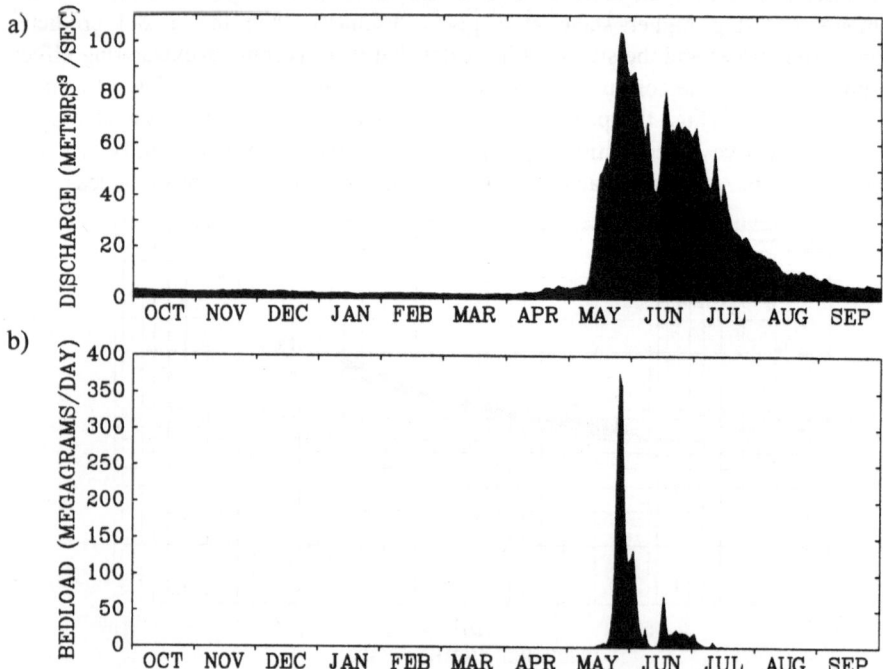

Figure 2. Measured mean daily discharge (a) and computed bedload transport (b) for the East River site during the 1984 water year.

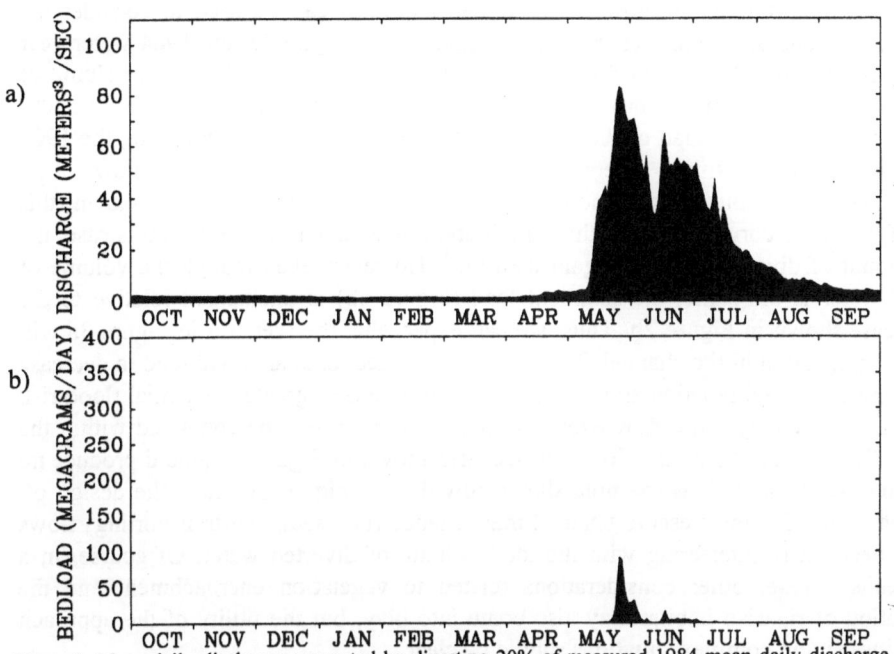

Figure 3. Mean daily discharge computed by diverting 20% of measured 1984 mean daily discharge (a)   and computed bedload transport (b) for the East River site.

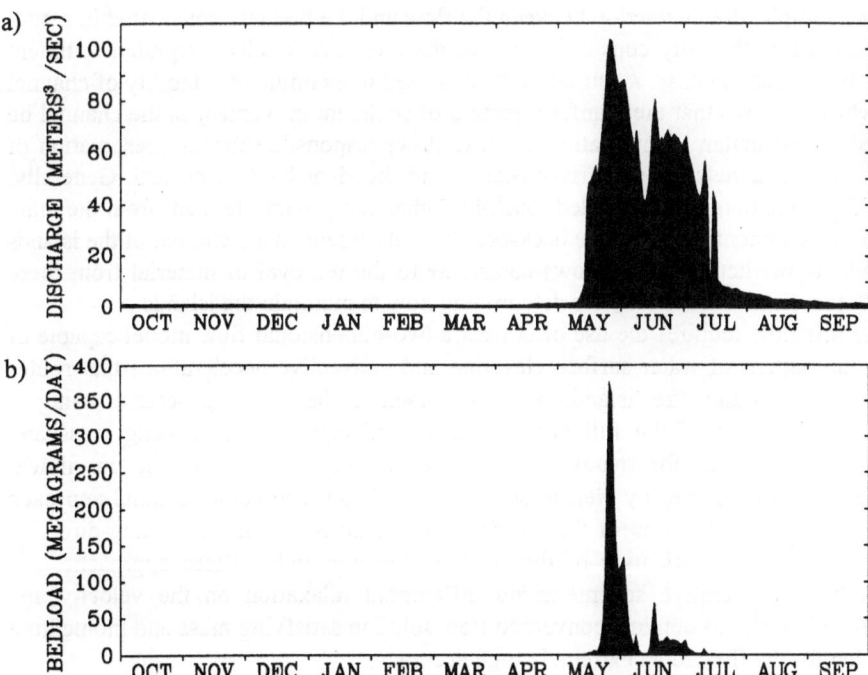

Figure 4. Mean daily discharge computed by diverting 70% of all flows below 40 m³/s and nothing at greater flows of 1984 mean daily discharge (a) and computed bedload transport (b) for the East River site. Note that the volume of water diverted is the same in Figures 3 and 4.

Figures 3 and 4 illustrate how the described approach can be used to consider the impacts of various diversion scenarios on the channel. In Figure 3a, the 1984 water year hydrograph modified by a 20% diversion of daily flows is shown. The total volume of water diverted is 0.38 km^3. Figure 3b shows that this diversion results in a more than 80% decrease in the sediment load of the channel over this water year. In Figure 4a, the 1984 hydrograph modified using a diversion scenario that removes 70% of daily flows at all flows below 40 m^3/s but removes nothing when the flow exceeds 40 m^3/s is shown (this value of discharge corresponds roughly to initiation of sediment motion). In this case, the total amount of diverted water is again 0.38 km^3. However, even though the volume of diverted water is the same, the sediment load (Figure 4b) is unchanged relative to the natural case shown in Figure 2b. Thus, we can expect that the scenario in Figure 3a will result in aggradation in the channel. The channel cross-sectional area will tend to decrease in response to this aggradation and, as a result, there will be a greater potential flood risk during rare, extremely large flow events, because less flow will be contained within the channel. On the other hand, the diversion scenario shown in Figure 4a should produce no impact on the channel. This example shows how the technique can allow the design of diversion scenarios that preserve channel maintenance (i.e., sediment-transporting) flows without necessarily interfering with the total volume of diverted water. Of course, in a more realistic case, other considerations related to vegetation encroachment and the preservation of riparian habitat may also come into play, but the utility of the approach even in these more complex situations should be clear.

II. CHANNEL ISLANDS

In the first example, the interaction between the flow and the bed was considered in a one-dimensional sense- the only consideration was the cross-sectionally integrated sediment transport. In this second case, a similar method is used to examine the stability of channel islands, which requires that the planform pattern of sediment movement in the channel be considered. In particular, identification of those flows responsible for the preservation of relatively small channels between river islands and the river bank is critical. Generally, channel islands are bars with so-called 'back-bar' channels separating them from the main channel bank; sedimentation in these back-bar channels results in attachment of the islands to the bank, so predicting which flows contribute to the removal of material from these regions of low flow is necessary for determining how to maintain the islands.

Treating this flow requires the use of at least a two-dimensional flow model capable of resolving the pattern of water surface elevation and convective accelerations that routes the river flow around the island. Results presented here are generated using a computational solution of the full vertically averaged equations expressing mass and momentum conservation for steady flow. The equations are cast in the curvilinear coordinate system described by Nelson and Smith (1989a). The computational approach consists of an explicit solution of the momentum equations and an alternating-direction implicit scheme for modification of the surface elevation field. These approaches are combined with an iterative scheme using differential relaxation on the velocity and pressure field in order to obtain a converged flow solution satisfying mass and momentum conservation for the flow to an extremely high accuracy.

Results of this computational approach for a reach of the Snake River in Idaho are shown in Figure 5a-c for three different discharges. The figures show both the bathymetry of the channel reach and the pattern of bottom stress predicted by the flow model. The critical difference between this kind of approach and a one-dimensional approach is the prediction of local flow routing in the channel. For the case of channel island maintenance and the preservation of back-bar channels, determining the patterns of flow and bottom stress around the island and how those patterns vary with discharge is critical. As shown in the topographic contours in Figure 5, the channel used as an example here consists of a relatively deep, high-velocity channel on the right side of the island (looking downstream) and a shallower low-velocity channel on the left, which is the so-called back-bar channel. In general, the bottom stress and hence the sediment transport capacity of the left channel is much lower than that on the right. During periods of relatively low discharge, fine sediment accumulates in this channel and, over time and in the absence of any higher flows, this channel may narrow, become shallower and, in some cases, become completely plugged with sediment.

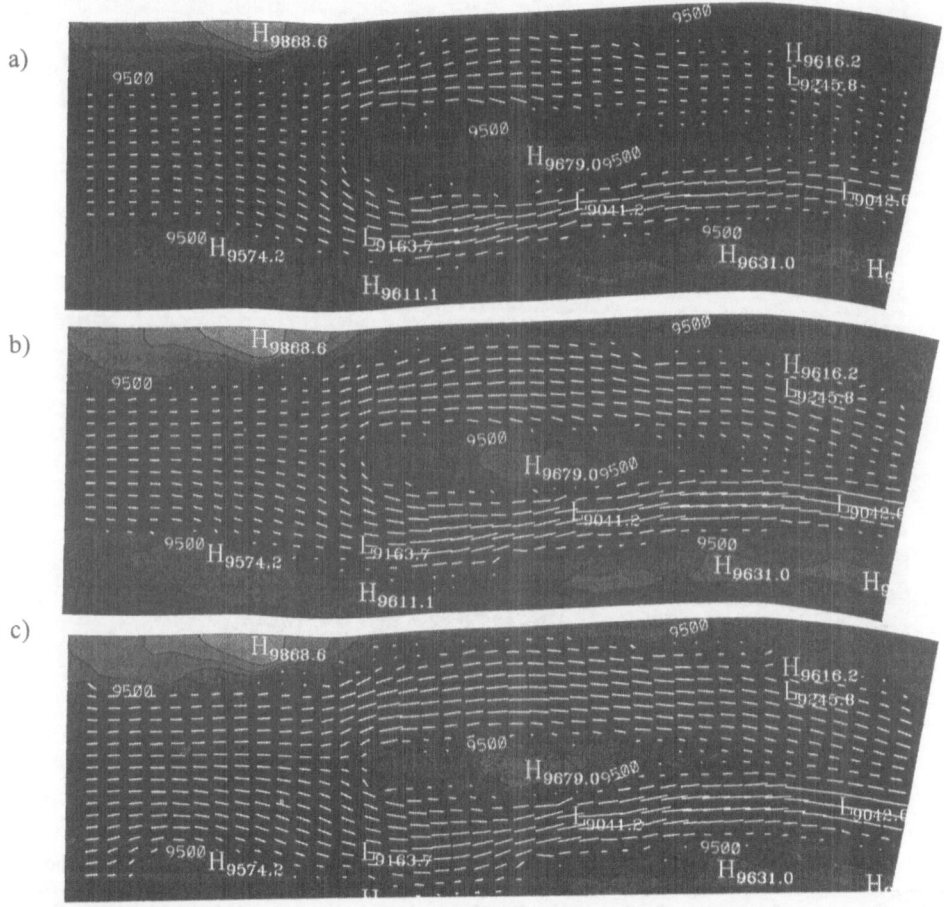

Figure 5. Bed contours and bottom stress vectors for the Silo Island reach of the Snake River in Idaho for discharges of 200 m³/s (a), 425 m³/s (b), and 700 m³/s (c). The stress vectors are normalized using 145 dy/cm², 197 dy/cm², and 232 dy/cm², respectively. Contours are in meters.

To identify the flows that can prevent long-term aggradation in the back bar channel requires consideration of the geomorphology of the reach. Most of the bed material in the reach studied here is made up of coarse (> 1 cm) gravel. Most of this material moves infrequently at high flows. In addition, there is fine sediment supplied to the channel via tributaries and bank failures. This finer material is deposited in regions of low flow and in the interstices of the gravel deposits. In general, removal of this finer material ('flushing') requires flows that are able to move the finer material out of relatively low-flow regions and flows that produce some very low transport rate of the coarser material in the channel. Generally, this latter constraint is a far more stringent one (requiring higher flows) so the principal condition for keeping the back-bar channel free of finer material is that the gravel in this channel be moved at least sporadically.

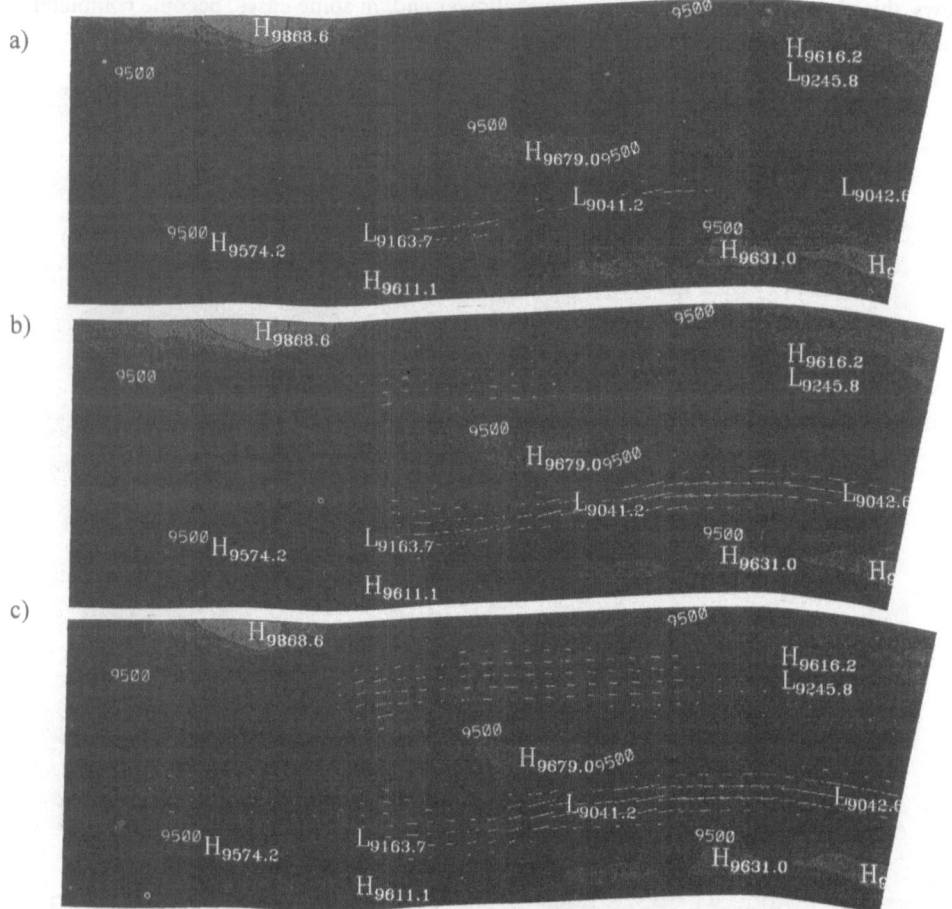

Figure 6. Bed contours and excess shear stress for the Silo Island Reach of the Snake River in Idaho for discharges of 200 m³/s (a), 425 m³/s (b), and 700 m³/s (c). Contours are in meters.

For the case of interest, the low-velocity channel contains gravel deposits made up of material with grain sizes of roughly 2 cm, although there is some spatial variability. For the example here, only this size will be considered. The initiation of motion for this size can be expected to occur at around a bottom stress of 100 dynes/cm², at least for the case where we consider true initial motion, rather than significant motion (Wiberg and Smith, 1987; Parker,1982). Using this value, it is possible to compute the excess stress, defined as zero if the bottom stress is less than the critical value, and defined as the difference between the bottom stress and the critical stress normalized by the critical stress for cases where the stress exceeds the critical value. Thus, this quantity is zero when the bottom stress equals the critical value, is equal to one when the bottom stress is twice the critical value, and so forth. Results for the example computation are shown in Figure 6a-c. At a flow of 200 m^3/s, 2 cm material can only be moved in the right-hand channel. At 425 m^3/s, there is some limited motion in the left-hand channel, and at 700 m^3/s there is motion of this size throughout most of the back-bar channel. Thus, using the computational flow solution, it is possible to conclude that flows around 700 m^3/s are required to produce gravel movement and the associated flushing of fines out of this channel.

Thus, as in the first example, it is possible to draw meaningful conclusion about the roles of various flows in producing channel stability even in the absence of the desired comprehensive data set. Furthermore, it is clear that, as the question to be answered grows in specificity and complexity, the difficulty in addressing the problem empirically increases enormously and the value of the computational approach increases commensurately. This trend continues in the last and most complex example.

III. LATERAL SEPARATION DEPOSITS

In the last example, the formation of sediment deposits in lateral separation zones is considered. Lateral separation zones occur in rivers and streams where bank curvature produces separation of the downstream flow from the bank, producing a region bordered by slow upstream flow near the bank and by strong lateral shear along its riverward margin. These regions, also referred to a lateral separation eddies, are efficient traps for sediment and organic material, and play important roles for riparian habitat and, in some rivers, for recreational use (Schmidt and Graf, 1990). One aspect of these regions is the nearly ubiquitous presence of characteristic sediment deposits within them, as described by Schmidt et al. (1990). Observations of these deposits show that variations in river flow can have a significant effect on the their spatial extent and morphology. Specifically, it appears that the removal of relatively high flows from the hydrograph (e.g., via flow regulation) results in a decrease in the frequency and extent of these deposits. In general, this can be explained by noting that rivers cannot deposit sediment above the surface of the water, so that if the highest flows are removed, sand removed from existing eddy deposits by wind or beach-like processes will not be replaced by subsequent high flows. But what flows, in terms of magnitude, duration, and frequency, are required to maintain these deposits? This is a very difficult question to answer using an entirely empirical approach, because the answer depends, at least to some extent, on local channel geometry and the efficacy of the processes acting to move sediment into the eddies.

In the past, the mechanism responsible for sediment deposition in the eddies has been assumed to consist of a purely diffusive process whereby sediment from the relatively high-concentration main river flow is diffused into the eddy across the shear layer separating the eddy from the main flow. However, laboratory observations (Nelson, 1991) show that this is not correct; the magnitude of the transfer between the eddy and the main river flow cannot be explained using physically reasonable values of diffusivity. These studies show that there is a strongly three-dimensional effect producing direct advection of fluid and sediment into the eddy near the bed. This effect is principally a product of secondary flows generated at the riverward margin of the eddy, where secondary flow is defined as flow with no net discharge directed orthogonally to the vertically averaged velocity vector. Thus, in order to assess the role of various flows in producing eddy deposits, a three-dimensional flow model must be used.

To construct an appropriate model, the vertically averaged flow solution described above is combined with an algorithm for computing flow vertical structure both along the streamlines of the vertically averaged flow and perpendicular to those streamlines using the technique described by Nelson and Smith (1989b). This method permits computation of secondary flows arising from streamline curvature of the vertically averaged flow. In addition, as this approach is intended for investigating the time rate of change of the bed morphology, the flow model is coupled to algorithms for computing both bed and suspended load. Using the computed sediment fluxes, it is possible to predict the change in bathymetry and then, by iteratively computing the flow and the sediment-transport field in succession, the temporal change in the bed elevation is predicted.

For this example, laboratory data provides a clear example of the method and, due to the ease of comprehensive data collection in the laboratory, also provides an opportunity to test the accuracy of model predictions. In Figure 7, laser-Doppler measurements of streamwise flow in a simple, flat-bedded laboratory lateral separation zone are shown with the predictions of the flow model for this case. The computational model yields reasonably accurate predictions of the mean flow in the eddy. In Figure 8, the temporal evolution of the initially flat laboratory bed is shown over a period of two hours, at which time the bed was near equilibrium. For the case shown, sediment transport occurred by bedload transport alone. The initial response of the sediment to the flow is to deposit sediment immediately inside the region of high shear bounding the eddy zone. As time progresses, this deposit grows and moves into the eddy and a deposit forms in the main channel in response to the decelerating flow. These predictions are in good agreement with the observed changes in bed elevation, both in terms of magnitude and timing.

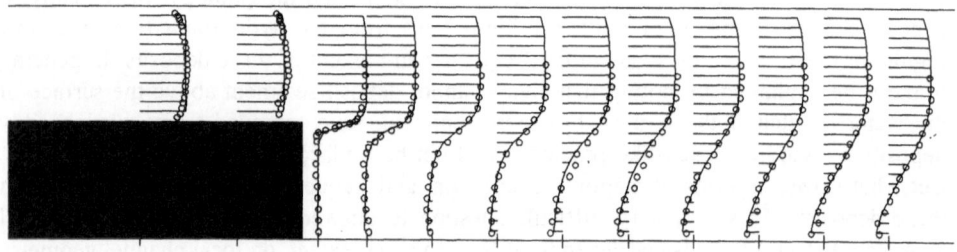

Figure 7. Comparison of computed (vectors and solid lines) and measured (circles) streamwise flow components for the simple flat-bedded laboratory case. The flume is 0.7 m wide, the flow is 0.2 m deep, and the maximum velocity is 0.63 m/s.

In Figure 9, computed bed elevation is shown after two hours of evolution for the cross-section located at the point of maximum elevation of the separation zone deposit both for the case where the full model is employed (a) and for the case when secondary flows are neglected (b); the corresponding measured bed topography is also shown as individual points. Neglecting secondary flows produces approximately the correct form of the deposit, but it seriously underpredicts the growth rate. Thus, the need for a three-dimensional approach is clear.

At present, the model described here is being employed in the Colorado River in Grand Canyon for the assessment of various discharges in maintaining eddy deposits. Figure 10 shows a preliminary comparison between the flow model results and flow measured in the Colorado using the techniques described by McDonald and Nelson (1994). Although the

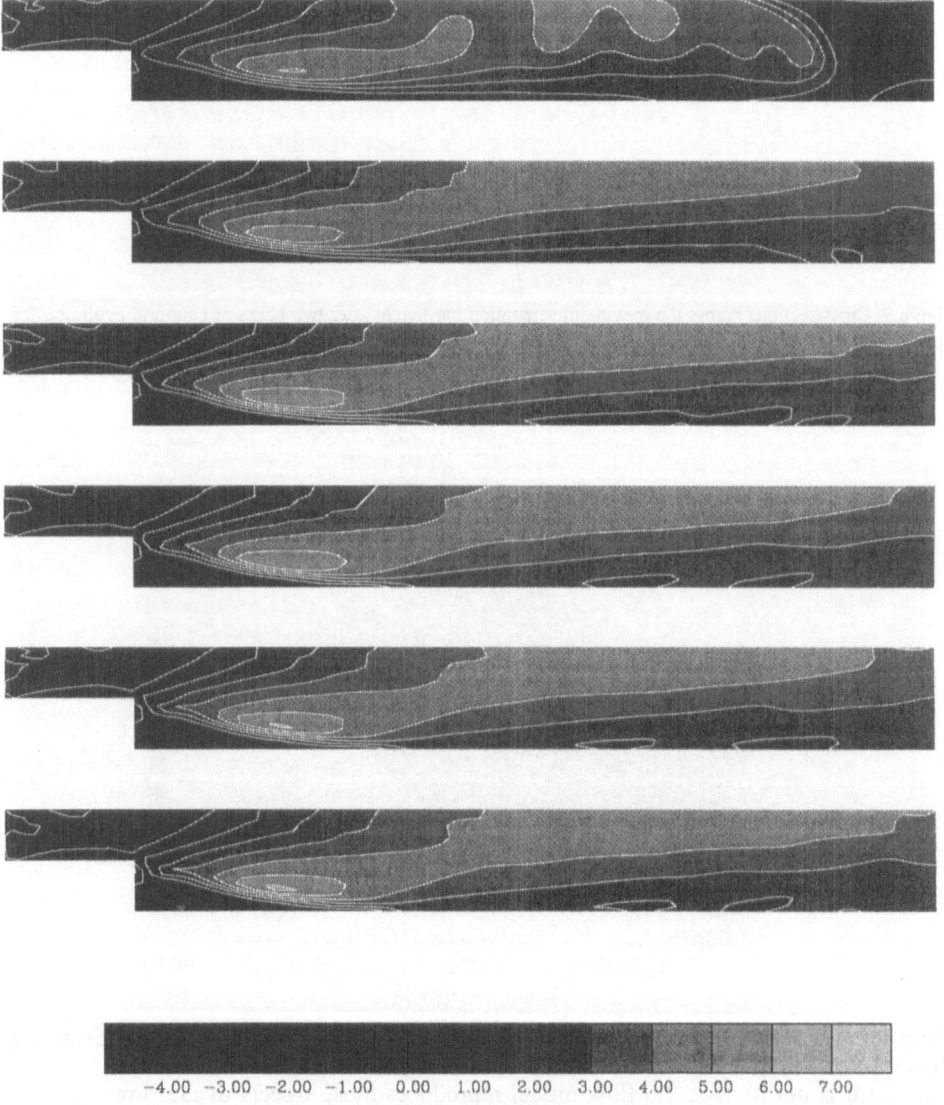

-4.00 -3.00 -2.00 -1.00 0.00 1.00 2.00 3.00 4.00 5.00 6.00 7.00

Figure 8. Temporal evolution of an initially flat bed in the laboratory flume. Starting at the top of the figure, t = 20 min, 40 min, 60 min, 80 min, 100 min and 120 min. Bed sediment is 1 mm sand that moved only as bedload.

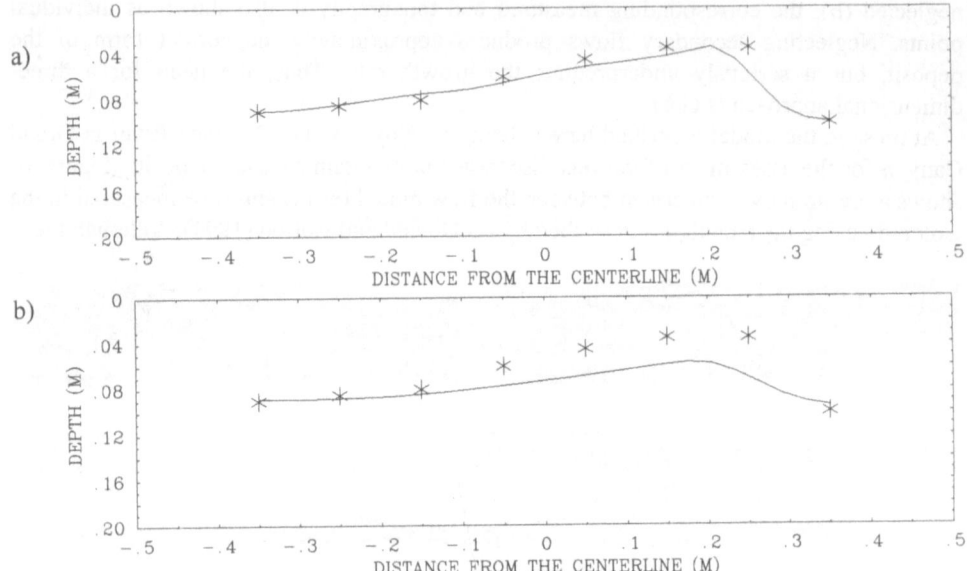

Figure 9. Cross-sections across the maximum separation bar height after two hours of temporal evolution for the full model (a) and the model without secondary flow effects (b).

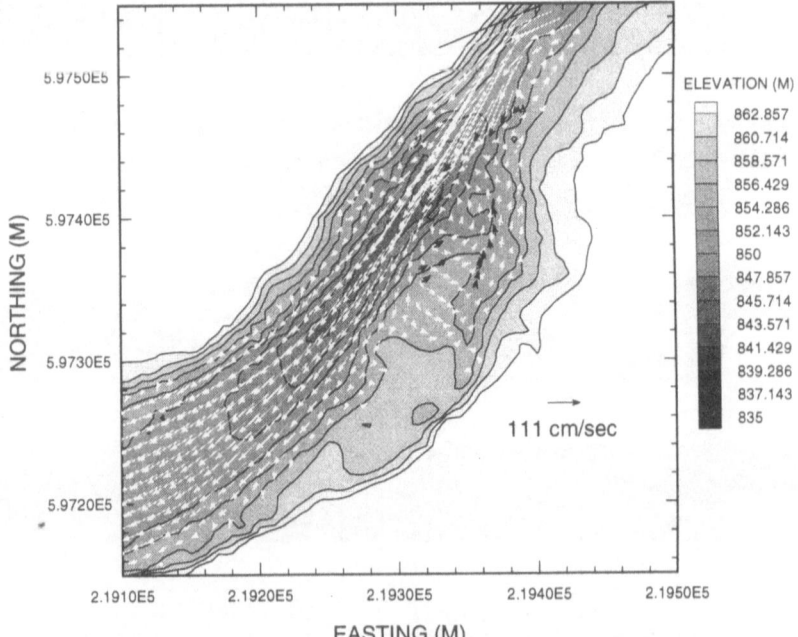

Figure 10. Map of bed contours, computed flow and measured flow in Eminence Break eddy in the Colorado River in Grand Canyon.

agreement is not perfect, the flow model reproduces most aspects of the flow well. At least some of the discrepancy may be associated with the fact that some of the topography used to generate the model results was obtained using aerial photographs that were not coeval with the measurements. The next step in this research is a comparison between model predictions and the results of a 'synthetic' flood, currently scheduled to occur sometime in 1996. Essentially, a full scale version of the laboratory bed evolution experiment will be carried out, with careful monitoring of bathymetric evolution over the course of a high flow event created through unusually high releases from Glen Canyon Dam. With this field testing, the approach can be used along with sediment availability criteria to design flow events that will help to maintain deposits in lateral separation eddies, which are critical to riparian habitat and recreation in the Colorado River in Grand Canyon.

3. Conclusions

The key to rational, equitable management of any environment is a sound, scientifically defensible methodology upon which decisions may be based. Unfortunately, there are commonly issues in rivers related to flow and sediment supply alterations where such a methodology is lacking. However, in situations where good basic data are unavailable, or where data required to resolve an issue cannot be collected due to time or logistical constraints, modern techniques in computational fluid mechanics coupled to sediment transport algorithms can provide an acceptable alternative. Future development of these techniques must concentrate on progressively more accurate descriptions of the turbulence field, and more precise characterizations of sediment transport, especially in complex nonuniform and/or unsteady flows. As more of these approaches are developed and proven, progressively more detailed issues in channel maintenance can be addressed in a manner that balances the need for development and preservation within a scientific framework.

References

Andrews, E.D.: 1994, *Water Resources Research*, **30**, 2241-2250.
Jarrett, R.D.: 1989, in: B.C. Yen (ed.), *Proceedings of the International Conference on Channel Flow and Catchment Runoff*, 599-608.
Limerinos, J.T.: 1970, U. S. Geological Survey Professional Paper 282B.
McDonald, R.R. and Nelson, J.M.: 1994, *Eos Transactions*, **75**, 268.
Nelson, J.M.: 1991, *Eos Transactions*, **72**, 218.
Nelson, J.M., Emmett, W.W., and Smith, J.D.: 1991, in: S. Fan and Y. Kuo (eds), *Proceedings of the Fifth Federal Interagency Sedimentation Conference*, 55-62.
Nelson, J.M. and Smith, J.D.: 1989a, in: S. Ikeda and G. Parker (eds), *River Meandering*. Washington, D.C.: American Geophysical Union, 69-102.
Nelson, J.M. and Smith, J.D.: 1989b, in: S. Ikeda and G. Parker (eds), *River Meandering*. Washington, D.C.: American Geophysical Union, 321-377.
Parker, G., Klingeman, P.C. and McLean, D.G.: 1982, *J. Hydraulics Div. ASCE.*, **108**, 544-571.
Parker, R.S., Nelson, J.M., Elliott, J.G. and Carey, W.P.: 1992, *Eos Transactions*, **73**, 239.
Schmidt, J.C. and Graf, J.B.: 1990, U.S. Geological Survey Professional Paper 1493.
Wiberg, P.L. and Smith, J.D.: 1987, *Water Resources Research*, **23**, 1471-1480.
Wiberg, P.L. and Smith, J.D.: 1991, *Water Resources Research*, **27**, 825-838.

between project conditions and the results of a "synthetic" flood, around a schedule to reach synergism in scale, especially a full scale version of the laboratory bed evolution. Experience will be carried out, with careful monitoring of bathymetric evolution over the course of a nightly event, created through unusually high releases from Glen Canyon Dam. With the wide setting, the approach can be used along with sediment availability studies to design projects that will help to maintain deposits in local separation eddies, which are crucial to channel habitat and recreation in the Colorado River in Grand Canyon.

5. Conclusions

The key to natural equitable management of any environment is a sound scientifically defensible methodology upon which decisions must be based. Engineering these geomorphically based issues related to ROW and sediment supply alterations, since such methodologies are being shown, or in situations where good base data are available. Of these such analytical studies no useful management will it continue to thrive at large project scales. Whenever techniques incorporating fluid mechanics coupled to sediment transport alterations provide answers at large scales, future developments of these analytical numerical approaches on prediction of large bedforms development in the river system. This could solve another characterization of sediment transport, especially at smaller amplitude and lower intensity flows. As more of these approaches are developed and refined, increasingly useful answers related to channel maintenance can be obtained in a manner that balances the need for development and preservation within a scientific framework.

References

[references list — largely illegible]

RESPONSE OF SNOWMELT HYDROLOGY TO CLIMATE CHANGE

K.L. BRUBAKER and A. RANGO

USDA ARS Hydrology Laboratory, BARC-W Building 007 Room 104, Beltsville, Maryland 20705 USA

Abstract. In mountainous regions where the accumulation and melt of seasonal snow cover are important for runoff production, the timing and quantity of water supply could be strongly affected by regional climate change, particularly altered temperature and precipitation regimes. In this paper, the hydrological response to climate change scenarios is examined using a semi-distributed snowmelt runoff model. The model represents an improvement over simple temperature-based models, in that it incorporates the net radiation into the snowpack. Thus it takes into account the basin's topography and slope orientation when computing snowmelt. In general, a warmer climate is expected to shift snowmelt earlier into the winter and spring, decreasing summer runoff. The effects of other potential climate changes (such as precipitation and cloudiness patterns) are explored. The uncertainties in these predictions are discussed.

Keywords: snowmelt, runoff, net radiation, snow cover, climate change, water supply

1. Introduction

The runoff produced by the melting of seasonal snow cover in mountainous basins is an important water resource in many regions of the world. The occurrence of precipitation as snow — rather than rain — and its accumulation during the cold season provide a natural reservoir that releases its stored water during the summer season, when it is needed for both agriculture and hydropower. The timing and quantity of this water supply could be strongly affected by changes in a region's climate, particularly altered temperature and precipitation regimes.

Sustainable water resource planning requires consideration of possible climatic changes that are relevant to system design (Schultz and Hornbogen, 1995). Currently, much attention is being given to a global warming trend due to an increase of greenhouse gases in the Earth's atmosphere. However, the effects of such large-scale change on particular regions are uncertain. In addition, the general circulation models that are used to make predictions of climate change still do not adequately represent important physical processes (such as clouds) and feedback processes that might either counteract or reinforce the warming trend.

Many factors affect the response of snowmelt hydrology to changes in climate; these are:

- changes in snowfall and accumulation
- changes in the timing and amount of energy to melt the snow.

Experiments with mathematical models of snowmelt-runoff production indicate that a warmer climate — all else being equal — would shift runoff earlier into the winter and

Water, Air and Soil Pollution **90**: 335-343, 1996.

spring, at the expense of summer runoff. In addition, a greater fraction of the precipitation would occur as rain rather than snow, bypassing the natural reservoir storage of the snowpack. Thus, the effect of temperature change is to shift the timing of runoff; changes in the region's precipitation regime, however, would affect the runoff volume. A warmer temperature could decrease runoff volume through increased evaporation; however, a small increase in precipitation could offset that decrease (Rango, 1992; McCabe and Hay, 1995).

Two questions must be addressed:

- How will a region's climate change in the future?
- How will snowmelt hydrology respond to that change?

Given the current state of knowledge, the first question remains largely hypothetical. Therefore, this paper addresses the second question. The approach is a series of sensitivity studies, that is, experiments that demonstrate the effect of various aspects of regional climate on the basin snowmelt response. A simple snowmelt-runoff model is used to demonstrate some of the complexities of the problem.

2. Snowmelt Runoff Model

The Snowmelt Runoff Model (SRM) (Martinec et al., 1994) is used worldwide for forecasting melt-season runoff in mountain regions. SRM has been developed for microcomputer application and is ideal for use in data-sparse regions (Kumar et al., 1991), due to its simple data requirements and incorporation of remote sensing to determine snow-covered area. SRM has been used successfully to simulate runoff in over 60 basins ranging from 0.76 to 122,000 km^2 in area.

SRM uses a temperature-index (degree-day) approach to melt snow from a basin's elevation zones. Air temperature has been shown to be the best single meteorological variable for predicting snowmelt (Zuzel and Cox, 1975). The degree-day approach has been criticized as lacking a physical basis in comparison to more complete approaches that consider all aspects of the energy exchange. Nonetheless, degree-day methods remain popular in snowmelt modeling, and perform well when properly applied (Rango and Martinec, 1995).

A new version of SRM incorporating net surface radiation is under development (Kustas et al., 1994); depending on data availability, the user could choose between the simple temperature-index or the slightly more complex temperature and radiation versions. In the degree-day version of SRM, the melt depth (M [cm day^{-1}]) is calculated as:

$$M = aT_d \tag{1}$$

where T_d [°C] is the degree-day index and a [cm °C^{-1} day^{-1}] the degree-day coefficient. In the radiation-based version, Equation (1) is expanded to include a net radiation index (R_d [W m^{-2}]),

$$M = m_Q R_d + a_r T_d \tag{2}$$

In Equation (2), m_Q [(cm day^{-1}) (W m^{-2})$^{-1}$] is a physical constant converting energy to water mass or depth, R_d is the radiation index, and a_r [cm day^{-1} °C^{-1}] is the restricted degree-day coefficient, which is not equal to a in Equation (1) but multiplies the same T_d. One advantage of the radiation-based version is that the restricted degree-day coefficient a_r is not as variable in time as the simple degree-day coefficient a. The net radiation component can take actual radiation measurements or can calculate the net radiation based on meteorological data and topography. In addition to altered temperature and precipitation regimes in a hypothetically changed climate, the radiation-based version can account for changes in humidity and cloudiness, which affect the net radiation received at the snow surface.

For the Degree-Day version of SRM, the basin is subdivided into elevation zones, and a lapse rate is applied to base-station temperatures in order to obtain daily-average temperatures for the zones. In the Radiation version, general orientation (aspect) of the hillslope is a major factor in the amount of solar radiation received. Therefore, in this version, the basin is further divided into aspect/elevation zones.

The melt formulas, Equations (1) and (2), are applied only to the snow-covered fraction of each zone. Thus, a measurement or prediction of snow-covered area is a required input to the melt calculation. For simulations, the required snow-cover data are obtained from remote-sensing images. In forecast mode, SRM automatically evaluates the future course of snow-cover depletion by deriving curves of snow-covered-area versus cumulative melt (modified depletion curves) from records of snow-covered area versus time (conventional depletion curves). This procedure is described in Martinec et al., 1994.

3. Net Radiation

The radiation received at a point at the Earth's surface is composed of both shortwave (solar) and longwave (terrestrial) components, as follows:

$$R_{net} = K_{in} - K_{refl} + L_{atm} - L_{surf} \tag{3}$$

where K_{in} represents the incoming shortwave (solar) radiation that reaches the surface after reflection, backscattering, and absorption by clouds, K_{refl} the portion of that incoming energy that is reflected by the surface, L_{atm} the longwave radiation emitted by the overlying atmosphere (both clear sky and clouds), and L_{surf} the longwave radiation emitted by the surface back to the atmosphere. Equation (3) is a general statement of the radiation balance, and does not explicitly account for the radiational effects of vegetation cover. The radiation index R_d in Equation (2) is set equal to R_{net}, if R_{net} is positive (into the snow), and zero otherwise.

4. Study Site

This study focuses on the Dischma basin in the Swiss Alps (area 43.3 km^2, elevation range 1668-3146 meters above sea level). Data from the year 1977 are used for the sensitivity analysis. Information on snow-covered area is obtained from high-quality aerial photographs of the basin. Detailed meteorological data are available from the

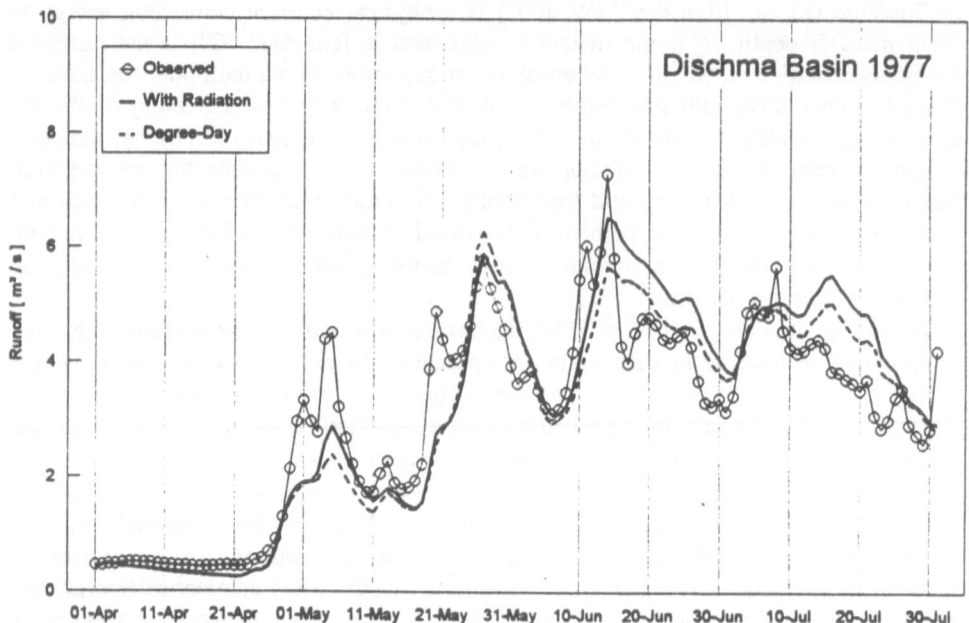

Fig. 1. Observed and model-simulated runoff hydrographs for the Dischma basin, Swiss Alps, in 1977.

Weissfluhjoch research station, allowing fairly accurate estimation of net radiation. The Weissfluhjoch station (elevation 2693 m.a.s.l.) lies near, but not in, the Dischma basin.

5. Snowmelt-Runoff Sensitivity to Climatic Variables

CONTROL (CURRENT CLIMATE) SIMULATIONS

Figure 1 shows model-simulated runoff hydrographs for the Dischma basin, compared to the observed hydrograph for four months of 1977. In the Radiation simulation, the restricted degree-day coefficient was set to a constant value of 0.36 cm $°C^{-1}$ day^{-1}, and in the Degree-Day simulation, the coefficient varied from 0.4 on 1 April to 0.55 on 31 July. Both versions of the model succeed in capturing the general shape, but not the details, of the hydrograph. In the interest of consistency, the changed-climate simulations described below are compared to the current-climate simulations rather than the observed runoff.

WARMER TEMPERATURE

Figure 2(a) shows the response of both model versions to an increase of 3°C in daily-average temperature on every day of the melt season, with precipitation, cloud, and relative humidity remaining the same as in the 1977 observations. Because the relative humidity is unchanged and because saturation specific humidity increases with temperature, the volume of water vapor in the air is increased in the warmer climate scenarios, with a resulting increase in atmospheric longwave emissivity. In these simulations, the snow-covered area at the beginning of the melt season (1 April) is assumed to be the same as under the current climate (this last assumption is relaxed in another simulation, described below). As shown in Figure 2(b), the net change in runoff between the warmer and current climate is nearly the same in the two models. In a

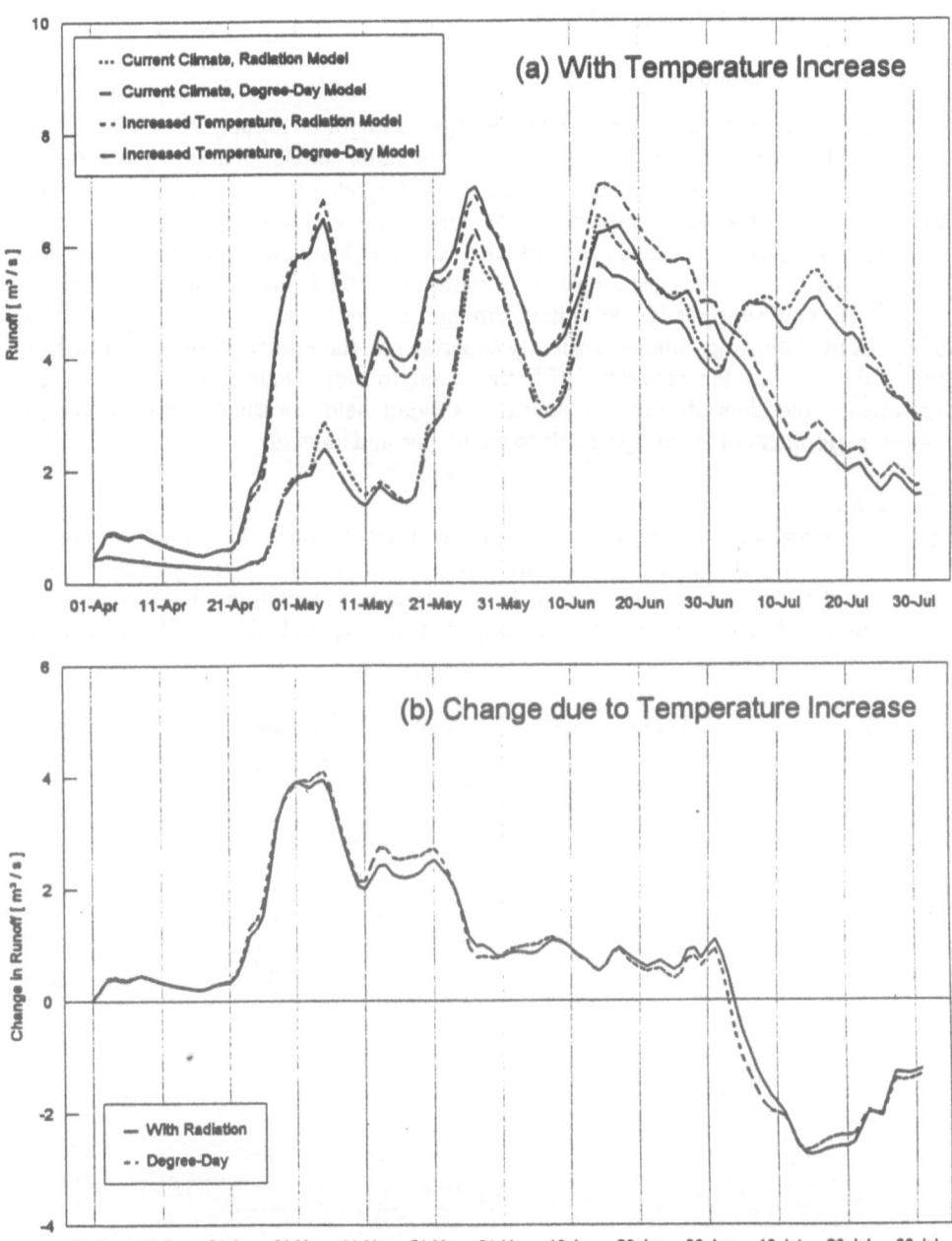

Fig. 2. (a) Simulation hydrographs using two versions of SRM for melt season 1977 in the Dischma basin under current climate conditions and with a 3 °C increase in temperature; (b) the change in runoff due to the increased temperature in both model versions.

warmer climate, the pattern of runoff is shifted, as energy is available for melt earlier in
the season. With the snow cover depleted by this earlier runoff, less snow is available for
melt in the summer months, with a resulting decrease in runoff during those months. For
example, in this experiment, the warmer-climate July runoff represents a 36 percent
decrease with respect to the current climate.

POSSIBLE CHANGES IN CLOUDS UNDER A WARMER CLIMATE

A change in the global and regional climate regimes might be accompanied by a change
in the cloudiness patterns of the region. The Radiation model includes mathematical
expressions for a number of cloud properties, including cloud fraction and cloud optical
thickness. Cloud fraction represents the amount of the sky that is covered by clouds.
Cloud optical thickness depends both on the physical thickness of the clouds and the
frequency distribution of water droplet sizes within the cloud. The distribution of droplet
sizes affects how sunlight is transmitted through and reflected from the clouds (Liou,
1992). In the following climate sensitivity experiments, the Radiation version of SRM is
used to demonstrate the sensitivity of basin runoff to these cloud properties. In these
experiments, the clear-air relative humidity is again held the same as in the current
climate, regardless of the changes made to cloud type and amount.

Cloud amount
Figure 3 shows the simulation hydrographs that result when the cloud fraction is
increased and decreased by 20% of sky-cover, in addition to the 3 °C temperature change,
in the Radiation version of SRM. The model shows some sensitivity to the fraction of sky
covered by cloud. It is somewhat surprising that increased cloud cover leads to more

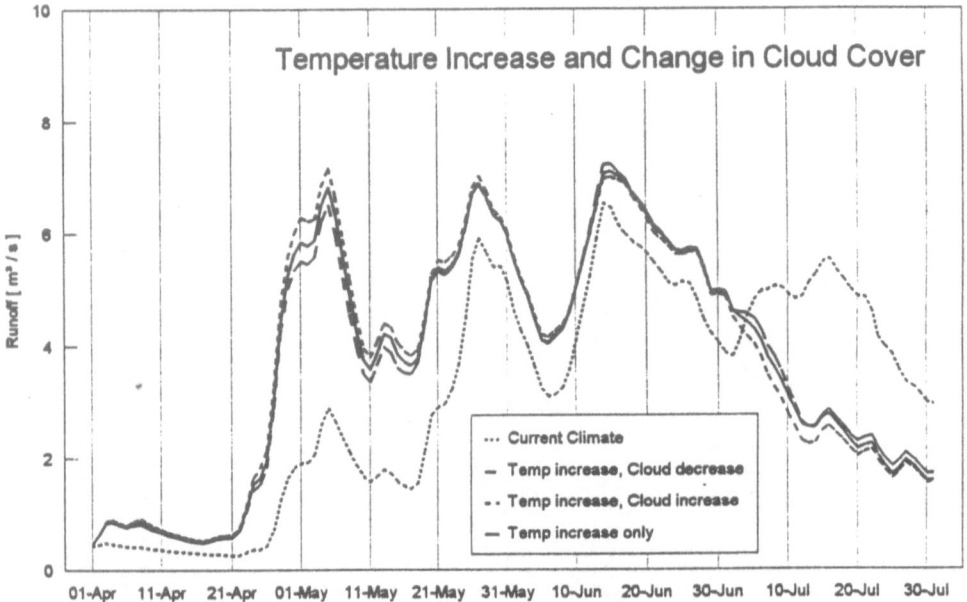

Fig. 3. Model-simulated runoff hydrographs, based on the 1977 melt season in Dischma basin, for the current
climate, and a warmed (by 3 °C) climate, with decreased cloud amount, increased cloud amount, and no change
in cloud amount.

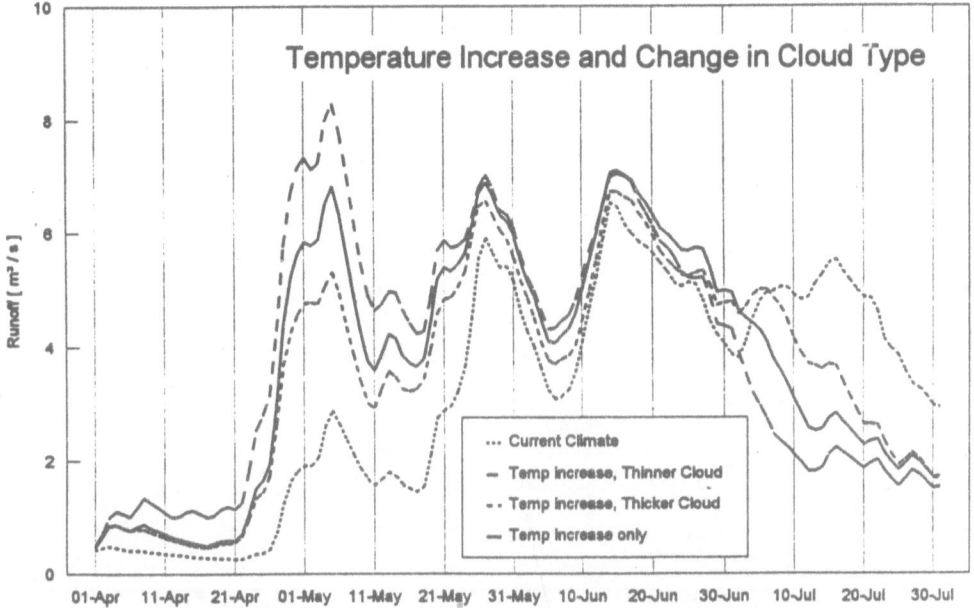

Fig. 4. Simulated climate-change scenarios based on the 1977 melt season in Dischma basin, for a warmer (by 3 °C) climate, with optically thinner, optically thicker, and unchanged clouds.

melt, as reflected by a slightly higher peak in runoff at 4 May. Intuitively, one would expect that as more clouds allow less sunlight, less melt would occur. However, clouds are significant radiators in the longwave, and this unexpected result reflects an increase in the L_{atm} term when cloud cover is increased, offsetting the decrease in the K_{in} term.

Cloud type (thickness)
Cloud type is represented in the SRM Radiation model by a mathematical expression for cloud optical thickness. This term depends on the frequency distribution of water droplet sizes in the cloud (Liou, 1992). Figure 4 shows the simulation results when this expression is changed to represent, respectively, optically thinner clouds (such as low stratus) and optically thicker clouds (such as nimbostratus or fair-weather cumulus), in addition to the 3 °C temperature increase. In these experiments, the cloud-cover fraction is held the same as in the 1977 observations, which is not totally consistent with the assumed change in cloud type. However, the experiment allows us to observe the model's sensitivity to the cloud optical thickness, which affects only the K_{in} term in Equation (4). As expected, optically thinner clouds allow more solar energy to reach the surface, further enhancing the shift in the runoff hydrograph to earlier in the melt season. In the other simulation, optically thicker clouds partially offset the effect of the temperature change.

DECREASED SNOWFALL

Year-round increases in temperature could lead to a greater fraction of winter precipitation falling as rain, rather than snow, and a resulting decrease in the water

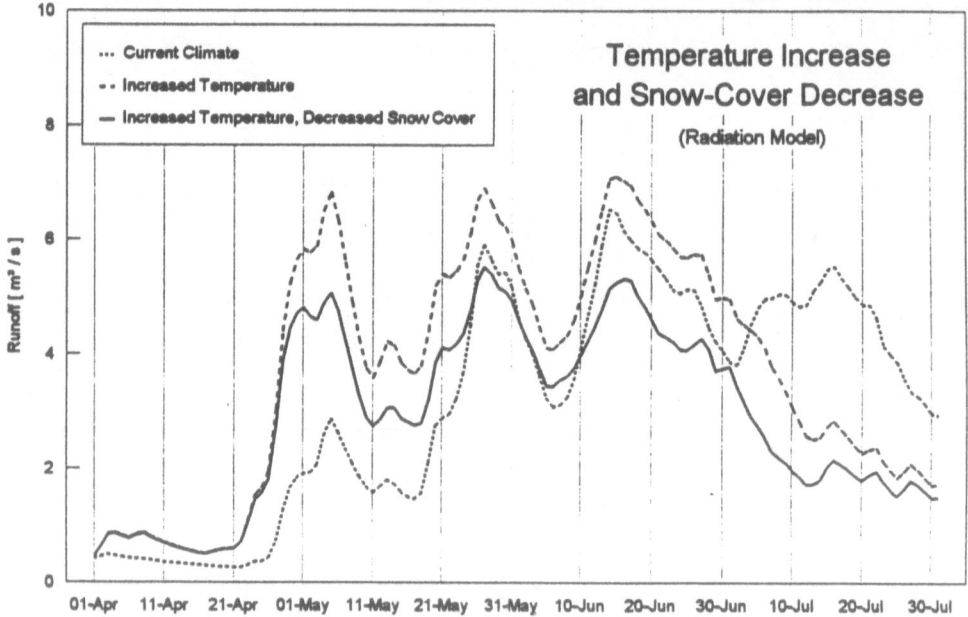

Fig. 5. Simulated runoff hydrographs based on the 1977 melt season for Dischma basin, for current climate and
a warmer (by 3 °C) climate with the same and a decreased snowpack at the beginning of the melt season.

equivalent of the snowpack. SRM is not a snow accumulation model, and the
simulations described in this paper are seasonal rather than year-round. The response of
melt-season runoff to a reduced winter snowpack, in addition to warmer temperatures, is
simulated by shifting the observed snow-covered area (SCA) curves to represent a
decreased snow-water equivalent at the beginning of the melt season. In this experiment,
the curves are simply (and arbitrarily) shifted by one-half month, so that the SCA on 16
April in the current climate becomes the SCA on 1 April in the "changed" climate. The
decline of SCA after that date is adjusted by the modified-depletion-curve technique in
SRM. Rango and Martinec (1994) describe a more rigorous procedure to estimate the
time shift of snow cover, taking into account changes in winter-season precipitation
(snowfall) as well as temperature. The result (Figure 5) is a dramatic decrease in the
seasonal volume of runoff in response to the warmed climate.

6. Summary

In mountainous regions where snowmelt is a major factor in runoff production, a warmer
climate can be expected to shift the runoff hydrograph earlier into the winter and spring
at the expense of summer runoff. However, other climatic changes may accompany
regional warming, including changes in precipitation and cloudiness patterns. The
simulation results presented in this paper are intended as sensitivity studies, and not
predictions of basin response; in addition, because the simplified radiation balance does
not account for vegetation, these results are not necessarily applicable to forested basins.

Simple experiments based on the 1977 melt season in the Dischma basin, Swiss Alps, demonstrate that, due to the important role of clouds in determining the radiation received at the Earth's surface, changes in cloud amount or type might exacerbate or compensate for temperature-driven changes in runoff, depending on the direction and magnitude of these changes. Improved satellite observations of clouds are becoming available (Simpson and Gobat, 1995), which should allow improvements in radiation-based snowmelt simulation. A better understanding of present cloud climatology and cloud physics will improve the scientific community's ability to predict what changes in regional cloud patterns and precipitation are likely to result from large-scale global warming.

Acknowledgments

This research is supported in part by the Electric Power Research Institute, Palo Alto, California.

References

Kumar, V.S., Haefner, H. and Seidel, K.: 1991, *Snow, Hydrology and Forests in High Alpine Areas (Proceedings of the Vienna Symposium), IAHS Publication No. 205*, 101-109.

Kustas, W.P., Rango, A. and Uijlenhoet, R.: 1994, *Water Resources Research*, **30**, 1515-1527.

Liou, K.N.: 1992, *Radiation and Cloud Processes in the Atmosphere.* Oxford University Press, 487 pp.

Martinec, J., Rango, A., and Roberts, R.: 1994, *Snowmelt Runoff Model (SRM) User's Manual*, Geographica Bernensia, P29, Department of Geography, University of Bern, 65 pp.

McCabe, G.J. Jr. and Hay, L.:1995, *Hydrological Sciences – Journal – des Sciences Hydrologiques*, **40**, 303-318.

Rango, A.. *Nordic Hydrology*, 1992, **23**:155-172.

Rango, A. and Martinec, J., 1995, *Water Resources Bulletin*, **31**, 657-669.

Rango, A. and Martinec, J., 1994, *Nordic Hydrology*, **25**, 233-246.

Schultz, G.A. and Hornbogen, M.: 1995, *Modelling and Management of Sustainable Basin-Scale Water Resource Systems, IAHS Publication No. 231*, Wallingford, Oxfordshire, UK, 329-333.

Simpson, J.J. and Gobat, J.I.: 1995, *Remote Sensing of Environment*, **52**, 36-54.

Zuzel, J.F. and Cox, L.M.: 1975, *Water Resources Research*, **11**, 174-176.

INTERNATIONAL CLEAN WATER CONFERENCE
28–30 NOVEMBER, 1995; LA JOLLA, CALFIRONIA

LIST OF PARTICIPANTS

DEAN W. ALGER
Indianapolis Power & Light Company
USA

SYED M. ALI
CA State Water Resources Control Board
USA

GARY ANDES
Electric Power Research Institute
USA

TONY ARMOR
Electric Power Research Institute
USA

CANDY ARQUIETT
Lockheed Martin Energy Systems, Inc.
USA

GREG BEHRENS
Radian Corporation
USA

PHIL BENSON
CH₂M Hill USA

FREDRICK J. BICKERTON Jr.
Duquesne Light Company
USA

GEORGE BOWIE
Tetra Tech, Inc.
USA

C. R. BOZEK
Edison Electric Institute
USA

BOB BROCKSEN
Electric Power Research Institute
USA

W.N. BRODIE
Tennessee Valley Authority
USA

ROBERT E. BROSHEARS
U.S. Geological Survey
USA

KAYE L. BRUBAKER
USDA/ARS Hydrology Laboratory
USA

BRAD BURKE
Southern Company Services, Inc.
USA

JOHN J. CENKNER
Allegheny Power Service Corporation
USA

CARL W. CHEN
Systech Engineering, Inc.
USA

IGOR CHERKO
Omaha Public Power District
USA

WINSTON CHOW
Electric Power Research Institute
USA

PAUL CHU
Electric Power Research Institute
USA

DAEL CLARKSON
Public Service Company of Colorado
USA

LUIGI CONTE
ENEL SpA
Italy

J.H. COPELAND
U.S. Geological Survey
USA

DAVID E. DAUGHERTY
Global Management Services
USA

J.M. DAVE
Jawaharlal Nehru University
India

MARIETTE DE JONG
Kluwer Academic Publishers
The Netherlands

FRED DEPENBROCK
Stone & Webster Management Consultants, Inc.
USA

TERESA DEBONO
Pacific Gas and Electric Company
USA

STEPHEN B. DIXON
Louisville Gas and Electric Company
USA

BRUCE DOBBS
Los Alamos Technical Associates, Inc.
USA

A.G. EKLUND
Radian Corporation
USA

H.E. EVANS
PowerGen
UK

THOMAS J. FEELEY III
U.S Department of Energy
USA

MYRA FRASER
Electric Power Research Institute
USA

DAVID FURUKAWA
Separation Consultants, Inc.
USA

GIAN S. GHUMAN
Savannah State College
USA

GREG GILL
Ontario Hydro
Canada

LEV GINZBURG
Applied Biomathematics
USA

JOHN GOODRICH-MAHONEY
Electric Power Research Institute
USA

MORGAN GORANFLO
Tennessee Valley Authority
USA

WILLIAM GRAYBEAL
Public Service Company of Colorado
USA

MARK HALPERN
Potomac Electric Power Company
USA

YASUTOSHI HARUNA
Electric Power Development Company Ltd.,
Japan

SAM HASHEMI
Potomac Electric Power Company
USA

DAN HEAD
Kentucky Utilities Company
USA

A. HEATH
National Power PLC
UK

RAY HEDRICK
Salt River Project
USA

ARTHUR HEIDRICH JR.
Detroit Edison
USA

DAVID HOFFMAN
CINenergy Corporation
USA

ARTHUR J. HOROWITZ
U.S. Geological Survey
USA

CALVIN HUBER
University of Wisconsin at Milwaukee
USA

HEIDI JAGODZINSKI
Minnesota Power
USA

EDWARD M. KEITH
Northeast Utilities
USA

MARY H. KERR
Sargent & Lundy Engineers
USA

STEVE KIDNEY
Energy Daily & King Communications
USA

BRIANT KIMBALL
U.S. Geological Survey
USA

BILL KOWALSKI
Dairyland Power Cooperative
USA

ELLEN LANUM
Electric Power Research Institute
USA

LORY E. LARSON
Southern California Edison Company
USA

GEORGE H. LEAVESLEY
U.S. Geological Survey
USA

DAVID LEE
Wisconsin Electric Power Company
USA

LEONARD LEVIN
Electric Power Research Institute
USA

CHRISTINE LEW
Tetra Tech, Inc.
USA

ALICIA R. LEWIS
Tennessee Valley Authority
USA

RANDY LEWIS
CINergy Corporation
USA

RON LEWIS
Duke Power Company
USA

TOM LEGRONE
Hughes Training-Link Operations
USA

CHRISTINE LILLIE
Electric Power Research Institute
USA

KATHERINE LINDQUIST
TVA Norris Engineering Laboratory
USA

CARRI LOHSE-HANSON
Minnesota Pollution Control Agency
USA

JULES J. LOOS
Potomac Electric Power Company
USA

MARTIN H. MACH
TRW Space & Technology Division
USA

JOHN F. MACKENZIE
Pacific Gas & Electric Company
USA

R.F. MADDALONE
TRW, Inc.
USA

RICK A. MANCINI
New York State Electric & Gas Corporation
USA

WAYNE J. MANCRONI
Central Hudson Gas & Electric Corporation
USA

JACK MATTICE
Electric Power Research Institute
USA

SHERMAN MAY
Bechtel Corporation
USA

GERRY MAYBACH
Electric Power Research Institute
USA

DAVE MCINTOSH
Electric Power Research Institute
USA

TOM MCCLUSKEY
San Diego Gas & Electric Company
USA

BILLY MCCORMAC
The Journal of Water, Air and Soil Pollution
USA

SANDRA MEIER
New York Power Pool
USA

SUSAN G. METZGER
Lawler, Matusky & Skelly Engineers
USA

WAYNE C. MICHELETTI
Wayne C. Micheletti, Inc.
USA

JOLECIA MITCHELL-MARIGNY
ENTERGY
USA

TOM MOORER
Southern Nuclear Operating Company
USA

SALLY MUIR
San Diego Gas & Electric Company
USA

YUFUS MUSSALLI
Stone & Webster Engineering Corporation
USA

RICHARD O. MYERS
DuPont Company
Denmark

DEBORAH NAGLE
Council on Environmental Quality
USA

J.M. NELSON
U.S. Geological Survey
USA

D.K. NORDSTROM
U.S. Geological Survey
USA

BABU NOTT
Electric Power Research Institute
USA

JOSEPH O'HAGAN
California Energy Commission
USA

DOUGLAS ORR
Radian Corporation
USA

FRANK B. OUDKIRK
Electric Power Research Institute
USA

JOHN PALMER
Southern California Edison Company
USA

DAN PATEL
Southern Company Services, Inc.
USA

STEPHEN PECK
Electric Power Research Institute
USA

MIKA POHJONEN
IVO Imatran Voima Oy
Finland

LINDY PRINTZ
Tennessee Valley Authority
USA

STEPHEN RAGONE
S.E. Ragone & Associates
USA

JAMES K. RICE
James K. Rice Chartered
USA

RICHARD J. ROBERTSON
Northeast Nuclear Energy Company
USA

DAVID W. ROBINSON
Tennessee Valley Authority
USA

DAVID W. RODGERS
Ontario Hydro Technologies
Canada

WILLIAM D. SAKSEN
Orange and Rockland Utilities, Inc.
USA

MILEDY SANTANA
New England Power Company
USA

JOSE SCARANO
ENEL
Italy

KAREN SCHLICHTING
Sierra Pacific Power Company
USA

JUDITH SCOTT
TRW, Inc.
USA

CHRISTIAN SEIGNEUR
ENSR Consulting & Engineering
USA

K.A. SELBY
Puckorius & Associates, Inc.
USA

JACK SHARMA
Ontario Hydro
Canada

HOWARD L. SHELNUTT
Georgia Power Company
USA

RICHARD STERNBERG
National Rural Util. Coop. Finance Corp.
USA

ROGER D. SUNG
Southern California Edison Company
USA

ROBERT SWANEKAMP
Power Magazine
USA

BARBARA TOOLE-O'NEIL
Electric Power Research Institute
USA

JOHN L. TSOU
Electric Power Research Institute
USA

RAY L. TUTTLE
New York State Electric & Gas Corporation
USA

CATHERINE TYRRELL
Santa Monica Bay Restoration Project
USA

KENICHI USHIKOSHI
Shinko Pantec Company Ltd.
Japan

DENISE VALITSKI
Delmarva Power & Light Company
USA

JOHN A. VEIL
Argonne National Laboratory
USA

KENNETH R. WAPMAN
Decision Focus, Inc.
USA

ROBERT G. WETHEROLD
Radian Corporation
USA

JASON M. WILKERSON
LG&E Energy Corporation
USA

HAL M. WILLIAMS
Tennessee Valley Authority
USA

ROBERT J. WILLIAMS
Virginia Power
USA

W.M. WINKLER
Louisville Gas and Electric Company
USA

JOE WISNIEWSKI
Wisniewski & Associates, Inc.
USA

AMY L. WOODIS
Woodis Associates
USA

JAMES R. WRIGHT
Tennessee Valley Authority
USA

JEAN-CLAUDE YOUNAN
South Carolina Electric & Gas Company
USA

KENT D. ZAMMIT
Electric Power Research Institute
USA

MICHAEL J. ZIMMERMAN
Clayton Mittelhauser
USA

AUTHOR INDEX

Arquiett, C. 83
Behrens, G.P. 113
Bowie, G.L. 93
Breitburg, D.L. 93
Brewer, M.S. 303
Brice, T. 173
Brocksen, R.W. 3, 21
Broshears, R.E. 195
Brubaker, K.L. 335
Chen, C. 65
Chow, W. 3, 21
Cole, L. 143
Connor, K. 21
Constantinou, E. 55
Copeland, J.H. 313
Cutter, G.A. 93
Datskou, I. 83, 133
Daugherty, E.D. 3
Demas, C.R. 281
Depenbrock, F. 231
Ekkad, N.V. 295
Evans, D.W. 219
Ferson, S. 71
Garbarino, J.R. 281
Gerke, M. 83
Gherini, S.A 123
Gilmour, C.C. 93
Ginzburg, L.R. 71
Goldstein, R.A. 65, 71
Goldstrohm, D. 173
Goodrich-Mahoney, J.W. 205
Hall, G.E.M. 281
Hauser, G.E. 65
Helm, K.R. 183
Herr, J. 65
Horowitz, A.J. 281
Huber, C.O. 295
Leavesley, G.H. 303
Lemieux, C. 281
Levin, L., 55
Lew, C.S. 123
Lindquist, K. 143

Lohse-Hanson, D. 31
Lum, K.R. 281
Maddalone, R.F. 163, 265
Markstrom, S.L. 303
McGee, M. 143
Mills, W.B. 123
Moch, I. 231
Morris, D. 173
Moses, D.O. 41
Murphy, C. 153
Mussalli, Y.g. 3, 231
Nelson, J.M. 321
Nordstrom, D.K. 257
North, K. 133
Nott, B. 173, 265
O'Neil, B.T. 113
Orr, D.A. 113
Pate, S.R. 173
Peck, S. 11
Porcella, D.B. 93
Puckorius, P.R. 183
Rango, A. 335
Reidel, G.F. 93
Rice, J.K. 265
Rodgers, D.W. 105, 219
Rylant, K.E. 65
Sagona, F.J. 65
Sanders, J.G. 93
Schroder, J. 105
Scott, J.W. 265
Seigneur, C. 55
Selby, K.A. 173, 183
Smay, D. 173
Spaulding, T. 173
Vereecken Sheehan, L. 105, 219
Viger, R.J. 303
Wetherold, R.G. 113
Whiddon, N.T. 269
Wilkinson, K.J. 123
Wisniewski, J. 3
Woodis, A.L. 3
Zimmerman, M. 153

KEY WORD INDEX

acid mine drainage 195
activated carbon 245
aqueous streams 113
ash pond 113

bacterial-based toxicity test 105
best management practices 153
binational 31
bioaccumulation 93
bioassays 71
biogeochemistry, biogeochemical cycling 93

channel maintenance 321
chemical model 257
chloramine 295
COOLADD 173
clean water, clean water act 3, 153
cleaning, cleanup objectives 83, 133
climate change 335
coal 163
Colorado 313
combined cyble power plant 231
conservation 183
constructed wetland treatment systems 205
control strategies 3
convection 313
cooling tower 41
cooling water/treatment manual 173
cost, cost reduction, cost-benefit analysis 83, 133, 173
cost comparison 231
coulometry 295

Daphnia magna 219
dechlorination 295
decision support system 143, 303
decision variables 65
dermal absorption 55
desalination 231
diffuser 41
discharge 113, 123, 153
discharge monitoring 163
dissolved, dissolved oxygen 163, 281
downscaling 313
drinking water 55

EPA 153
ecological risk 3, 71, 123
ecosystem management 11, 71
ecotoxicology 71
effluent toxicity 105
electric power production, electric utilities, cooling system 3, 31, 183, 303
electric transmission 231
environmental sampling 269

filtration, filtration artifacts 281
fish consumption 55
fluid mechanics 321
fresh water 3

general permit 153
geochemical modeling 195
geographic information system 143, 303
geomorphology 321
groundwater, groundwater plume 83, 133

health risk 21, 55, 123
housekeeping 153
hydrologic models 303
hypothetical homesteader risk 83

ingestion 55
instream flows 321
instrumental methods 269
ion exchange 245

kinetics 295

LaMP 31
Lake Superior 31
legislation 11
major elements 281
measurement quality 3
mercury 31
membrane filters 281
mixing zone 41
modeling 3, 313
multiple effect distillation 231
multi-stage flash 231

National Pollutant Discharge Elimination System 153
natural waters 257
net radiation 335
numerical modeling 143

orographic 313
PCBs 31
PISCES database 113
plant siting 231
point source vs. nonpoint source pollution 65
populations 71
population-level endpoints 71
power plant 55, 113
precipitation 313
public risk 83

R& D planning 21
radioactive liquid waste 219
rainbow trout 105
rate 295
reactive solute transport modeling 195
reclaimed water 183
refinery cooling 183
remediation 133
reservoir operations/management 143, 303
residual 295
reverse osmosis 231
risk assessment, risk management 21, 133
rivers 123
runoff 153, 335
sediment transport 321
selenium 93
sensitivity 313
SEQUIL 173
sewage plant effluent 183
sheet-off 153
simulation model 93
snowmelt 335
snow cover 335

sorption 219
source control 153
special designations 31
speciation 257, 269
steam power plant 231
storm water pollution prevention plan 153
sulfite 295
summer 313
surface water 21
sustainable development 11

thermal, thermal compliance 41, 143
toxi-chromotest 105, 219
toxicity experiments, toxicity reduction 93, 105, 219
trace elements 113, 281
trace metals 257
treatment control 153

uranium 163
UV photo-oxidation 105, 219

virtual elimination 31

wastewater treatment 205
water, water analysis 265, 295
water chemistry 205
water discharges 55
water quality 65, 143, 265
water recycling 183
water resource(s)/management/planning 3, 11, 143, 303
water reuse 183
water supply 335
watershed, watershed models 65, 303
worker risk 83, 133

zero discharge 31
§316 (a) variance 41